Lecture Notes in Computational Science and Engineering

75

Editors:

Timothy J. Barth
Michael Griebel
David E. Keyes
Risto M. Nieminen
Dirk Roose
Tamar Schlick

For further volumes:
http://www.springer.com/series/3527

Alexander N. Gorban • Dirk Roose
Editors

Coping with Complexity: Model Reduction and Data Analysis

Springer

Editors

Alexander N. Gorban
University of Leicester
Department of Mathematics
University Road
LE1 7RH Leicester
United Kingdom
ag153@leicester.ac.uk

Dirk Roose
Katholieke Universiteit Leuven
Department of Computer Science
Celestijnenlaan 200a
3001 Leuven
Belgium
dirk.roose@cs.kuleuven.be

ISSN 1439-7358
ISBN 978-3-642-14940-5   e-ISBN 978-3-642-14941-2
DOI 10.1007/978-3-642-14941-2
Springer Heidelberg Dordrecht London New York

Library of Congress Control Number: 2010937907

Mathematics Subject Classification (2010): 34E10, 34E13, 82C05, 93C35, 41A60

© Springer-Verlag Berlin Heidelberg 2011
This work is subject to copyright. All rights are reserved, whether the whole or part of the material is concerned, specifically the rights of translation, reprinting, reuse of illustrations, recitation, broadcasting, reproduction on microfilm or in any other way, and storage in data banks. Duplication of this publication or parts thereof is permitted only under the provisions of the German Copyright Law of September 9, 1965, in its current version, and permission for use must always be obtained from Springer. Violations are liable to prosecution under the German Copyright Law.
The use of general descriptive names, registered names, trademarks, etc. in this publication does not imply, even in the absence of a specific statement, that such names are exempt from the relevant protective laws and regulations and therefore free for general use.

Cover design: deblik, Berlin

Printed on acid-free paper

Springer is part of Springer Science+Business Media (www.springer.com)

Preface

A mathematical model is an intellectual device that works. The models have various purposes and, according to these purposes they differ by the level of simplification. There are many possible ways to construct the mathematical model we need.

We can start from a detailed model – a hypothesis which reflects all our knowledge of the first principles and then go down the stair of simplification.

We can go the opposite way and use a metaphor or an analogy and start from a very simple model, then add more details and develop a better and more complicated intellectual tool for our needs.

We can use plenty of intermediate approaches as well. In 1980, R. Peierls in his seminal paper [1] suggested that one might distinguish seven different types of models and demonstrated various types of confusion that can result if the nature of the model is misunderstood.

He proposed the term *model making*. Such a wording should awake associations with a very applied type of work like "shoe making." The logic of the applied activity of model making seems to be close to engineering. By analogy with chemical or mechanical engineering we can use the term *model engineering* [2]. It is not by chance that many new achievements in mathematical modeling were produced by engineering experts, by interdisciplinary teams or individual researchers who combine mathematical background with an engineering view on the result: a mathematical model is an intellectual device that must work.

Model reduction is one of the main operations in model making. Following Peierls we can state that the main difference between models is "the degree of simplification or exaggeration they involve" [1]. We have to simplify detailed models. It is also necessary to simplify models even before they reach their final form: in reality the process of model simplification and identification should be performed simultaneously.

The technology of model reduction should answer the modern challenges of the struggle with complexity. Many approaches and specific tools are developed during the last decades. In this volume we collect extended versions of selected talks given at the international research workshop: Coping with Complexity: Model Reduction and Data Analysis (Ambleside, Lake District, UK, August 31–September 4, 2009; in conjunction with the A4A6, the 6th Conference on "Algorithms for Approximation," supported by the EPSRC).

The theme of the workshop was deliberately broad in scope and aimed at promoting an informal exchange of new ideas and methodological perspectives in the increasingly important interdisciplinary areas of model reduction, data analysis and approximation in the presence of complexity. Participants had a wide variety of expertise reflecting the interdisciplinary nature of the workshop. The papers collected in this volume may help to circumvent some of the "language barriers" that unnecessarily hinder researchers from different disciplines.

The papers cover various application areas, from chemical engineering, fluid dynamics and quantum chemistry to population dynamics. Nevertheless all papers share a common property: they all present new ideas and methodological innovations and they provide new tools for model engineering.

This volume may appeal to academics and PhD students in applied mathematics and mathematical modeling in physics, chemistry, chemical engineering and other fields of science and engineering.

References

1. Peierls, R. Model-making in physics. Contemporary Physics, **21** (1980) 3–17
2. Model Reduction and Coarse-Graining Approaches for Multiscale Phenomena. Ed. by A.N. Gorban, N. Kazantzis, I.G. Kevrekidis, H.C. Öttinger, C. Theodoropoulos, Springer, Berlin (2006)

May 2010

Alexander N. Gorban
Dirk Roose

Contents

Averaging of Fast-Slow Systems 1
Marshall Slemrod

**The Use of Global Sensitivity Methods for the Analysis,
Evaluation and Improvement of Complex Modelling Systems** 9
Alison S. Tomlin and Tilo Ziehn

**Optimisation and Linear Control of Large Scale Nonlinear
Systems: A Review and a Suite of Model Reduction-Based
Techniques** 37
Constantinos Theodoropoulos

**Universal Algorithms, Mathematics of Semirings and Parallel
Computations** 63
Grigory L. Litvinov, Victor P. Maslov, Anatoly Ya. Rodionov,
and Andrei N. Sobolevski

**Scaling Invariant Interpolation for Singularly Perturbed
Vector Fields (SPVF)** 91
Viatcheslav Bykov, Vladimir Gol'dshtein, and Ulrich Maas

**Think Globally, Move Locally: Coarse Graining of Effective
Free Energy Surfaces** 113
Payel Das, Thomas A. Frewen, Ioannis G. Kevrekidis, and Cecilia
Clementi

**Extracting Functional Dependence from Sparse Data Using
Dimensionality Reduction: Application to Potential Energy
Surface Construction** 133
Sergei Manzhos, Koichi Yamashita, and Tucker Carrington

A Multilevel Algorithm to Compute Steady States of Lattice Boltzmann Models 151
Giovanni Samaey, Christophe Vandekerckhove, and Wim Vanroose

Time Step Expansions and the Invariant Manifold Approach to Lattice Boltzmann Models 169
David J. Packwood, Jeremy Levesley, and Alexander N. Gorban

The Shock Wave Problem Revisited: The Navier–Stokes Equations and Brenner's Two Velocity Hydrodynamics 207
Francisco J. Uribe

Adaptive Simplification of Complex Systems: A Review of the Relaxation-Redistribution Approach 231
Eliodoro Chiavazzo and Ilya Karlin

Geometric Criteria for Model Reduction in Chemical Kinetics via Optimization of Trajectories 241
Dirk Lebiedz, Volkmar Reinhardt, Jochen Siehr, and Jonas Unger

Computing Realizations of Reaction Kinetic Networks with Given Properties 253
Gábor Szederkényi, Katalin M. Hangos, and Dávid Csercsik

A Drift-Filtered Approach to Diffusion Estimation for Multiscale Processes 269
Yves Frederix and Dirk Roose

Model Reduction of a Higher-Order KdV Equation for Shallow Water Waves 287
Tassos Bountis, Ko van der Weele, Giorgos Kanellopoulos, and Kostis Andriopoulos

Coarse Collective Dynamics of Animal Groups 299
Thomas A. Frewen, Iain D. Couzin, Allison Kolpas, Jeff Moehlis, Ronald Coifman, and Ioannis G. Kevrekidis

Self-Simplification in Darwin's Systems 311
Alexander N. Gorban

Author Index 345

Subject Index 347

Contributors

Kostis Andriopoulos Department of Mathematics and Centre for Research and Applications of Nonlinear Systems, University of Patras, 26500 Patras, Greece, kand@aegean.gr

and

Centre for Education in Science, 26504 Platani – Rio, Greece

Tassos Bountis Department of Mathematics and Centre for Research and Applications of Nonlinear Systems, University of Patras, 26500 Patras, Greece, bountis@math.upatras.gr

and

Centre for Education in Science, 26504 Platani – Rio, Greece

Viatcheslav Bykov Karlsruher Institut fuer Technologie, Institut fuer Technische Thermodynamik, Engelbert-Arnold-Strasse 4, Geb. 10.91, 76131 Karlsruhe, Germany, bykov@itt.uni-karlsruhe.de

Tucker Carrington Department of Chemistry, Queens University, Kingston, ON, Canada K7L 3N6, Tucker.Carrington@chem.queensu.ca

Eliodoro Chiavazzo Department of Energetics, Politecnico di Torino, 10129 Torino, Italy, elidoro.chiavazzo@gmail.com

Cecilia Clementi Department of Chemistry, Rice University, Houston, TX 77005, USA, cecilia@rice.edu

Ronald Coifman Department of Mathematics, Yale University, New Haven, CT 06520, USA, coifman@fmah.com

Iain D. Couzin Department of Ecology and Evolutionary Biology, Princeton University, Princeton, NJ 08544, USA, icouzin@princeton.edu

Dávid Csercsik Process Control Research Group, Systems and Control Laboratory, Computer and Automation Research Institute, Hungarian Academy of Sciences 1518, P.O. Box 63, Budapest, Hungary, csercsik@csl.sztaki.hu

Payel Das Department of Chemistry, Rice University, Houston, TX 77005, USA, daspa@us.ibm.com

Yves Frederix Department of Computer Science, K.U.Leuven, Celestijnenlaan 200A, 3001 Leuven, Belgium, yves.frederix@cs.kuleuven.be

Thomas A. Frewen Department of Chemical Engineering, Princeton University, Princeton, NJ 08544, USA, tfrewen@gmail.com

Vladimir Gol'dshtein Department of Mathematics, Ben Gurion University of the Negev, Beer-Sheva 84105, Israel, vladimir@bgu.ac.il

Alexander N. Gorban Department of Mathematics, University of Leicester, Leicester, LE2 2pa UK, ag153@le.ac.uk

Katalin M. Hangos Process Control Research Group, Systems and Control Laboratory, Computer and Automation Research Institute, Hungarian Academy of Sciences 1518, P.O. Box 63, Budapest, Hungary, hangos@sztaki.hu

and

Department of Electrical Engineering and Information Systems, University of Pannonia, 8200, Egyetem u. 10, Veszprém, Hungary

Giorgos Kanellopoulos Department of Mathematics and Centre for Research and Applications of Nonlinear Systems, University of Patras, 26500 Patras, Greece, giorgoskan@lycos.gr

Ilya Karlin Aerothermochemistry and Combustion Systems Lab, ETH Zurich, 8092 Zurich, Switzerland, karlin.ilya@gmail.com

and

School of Engineering Sciences, University of Southampton, SO17 1BJ Southampton, UK

Ioannis G. Kevrekidis Department of Chemical Engineering, Princeton University, Princeton, NJ 08544, USA, yannis@princeton.edu

and

PACM and Mathematics, Princeton University, Princeton, NJ 08544, USA

Allison Kolpas Department of Mathematical Sciences, University of Delaware, Newark, DE 19716, USA, akolpas@math.udel.edu

Dirk Lebiedz Interdisciplinary Center for Scientific Computing (IWR), University of Heidelberg, Im Neuenheimer Feld 368, 69120 Heidelberg, Germany, dirk.lebiedz@biologie.uni-freiburg.de

and

Center for Systems Biology (ZBSA), University of Freiburg, Habsburgerstraße 49, 79104 Freiburg, Germany

Jeremy Levesley Department of Mathematics, University of Leicester, Leicester LE2 2PQ, UK, jli@le.ac.uk

Contributors

Grigory L. Litvinov The J. V. Poncelet Laboratory (UMI 2615 of CNRS), Independent University of Moscow, Moscow, Russia, glitvinov@gmail.com and

D. V. Skobeltsyn Research Institute for Nuclear Physics, Moscow State University, Moscow, Russia

Ulrich Maas Karlsruher Institut fuer Technologie, Institut fuer Technische Thermodynamik, Engelbert-Arnold-Strasse 4, Geb. 10.91, 76131 Karlsruhe, Germany, ulrich.maas@kit.edu

Sergei Manzhos Department of Chemical System Engineering, School of Engineering, University of Tokyo, 7-3-1, Hongo, Bunkyo-ku, Tokyo 113-8656, Japan, sergei@tcl.t.u-tokyo.ac.jp

Victor P. Maslov Faculty of Physics, Moscow State University, Moscow, Russia, v.p.maslov@mail.ru

Jeff Moehlis Department of Mechanical Engineering, University of California, Santa Barbara, CA 93106, USA, moehlis@engineering.ucsb.edu

David J. Packwood Department of Mathematics, University of Leicester, Leicester LE2 2PQ, UK, dp123@le.ac.uk

Volkmar Reinhardt Interdisciplinary Center for Scientific Computing (IWR), University of Heidelberg, Im Neuenheimer Feld 368, 69120 Heidelberg, Germany, Volkmar.Reinhardt@iur.uni-heidelberg.de

Anatoly Ya. Rodionov D. V. Skobeltsyn Research Institute for Nuclear Physics, Moscow State University, Moscow, Russia, ayarodionov@yahoo.com

Dirk Roose Department of Computer Science, K.U.Leuven, Celestijnenlaan 200A, 3001 Leuven, Belgium, Dirk.Roose@cs.kuleuven.be

Giovanni Samaey Scientific Computing, Department of Computer Science, K.U. Leuven, Celestijnenlaan 200A, 3001 Leuven, Belgium, giovanni.samaey@cs. kuleuven.be

Jochen Siehr Interdisciplinary Center for Scientific Computing (IWR), University of Heidelberg, Im Neuenheimer Feld 368, 69120 Heidelberg, Germany, Jochen.Siehr@iwr.uni-heidelberg.de

Marshall Slemrod Department of Mathematics, University of Wisconsin, Madison, WI 53706, USA, slemrod@math.wisc.edu

Andrei N. Sobolevski The J. V. Poncelet Laboratory (UMI 2615 of CNRS), Independent University of Moscow, Moscow, Russia, ansobol@gmail.com and

A. A. Kharkevich Institute for Information Transmission Problems, Moscow, Russia

Gábor Szederkényi Process Control Research Group, Systems and Control Laboratory, Computer and Automation Research Institute, Hungarian Academy of Sciences 1518, P.O. Box 63, Budapest, Hungary, szeder@scl.sztaki.hu

Constantinos Theodoropoulos School of Chemical Engineering and Analytical Science, University of Manchester, Manchester M60 1QD, UK, k.theodoropoulos@manchester.ac.uk

Alison S. Tomlin Energy and Resources Research Institute, School of Process, Environmental and Materials Engineering, University of Leeds, Leeds LS2 9JT, UK, A.S.Tomline@leeds.ac.uk

Jonas Unger Center for Systems Biology (ZBSA), University of Freiburg, Habsburgerstraße 49, 79104 Freiburg, Germany, jonas.unger@zbsa.de

Francisco J. Uribe Departamento de Física, Universidad Autónoma Metropolitana – Iztapalapa, Av San Rafael Atlixco # 186, cp. 14090, México, Distrito Federal, México, paco@xanum.uam.mx

Christophe Vandekerckhove Scientific Computing, Department of Computer Science, K.U. Leuven, Celestijnenlaan 200A, 3001 Leuven, Belgium, christophe.vandekerckhove@gmail.com

Wim Vanroose Department of Mathematics and Computer Science, Universiteit Antwerpen, Middelheimlaan 1, 2020 Antwerpen, Belgium, wim.vanroose@ua.ac.be

Ko van der Weele Department of Mathematics and Centre for Research and Applications of Nonlinear Systems, University of Patras, 26500 Patras, Greece, weele@math.upatras.gr

Koichi Yamashita Department of Chemical System Engineering, School of Engineering, University of Tokyo, 7-3-1, Hongo, Bunkyo-ku, Tokyo 113-8656, Japan, yamasita@chemsys.t.u-tokyo.ac.jp

Tilo Ziehn Department of Earth Sciences, University of Bristol, Bristol BS8 1RJ, UK, tilo.ziehn@bristol.ac.uk

Averaging of Fast-Slow Systems

Marshall Slemrod

Abstract In this note I present an introduction to a new averaging method in which a fast-slow system will generate its own averaging device (a Young measure), this can be used to estimate the fast observables within the system without computing directly at the fast time scale. For small values of the fast time scale ϵ the new algorithm is $O(\frac{1}{\epsilon})$ times faster than a direct computation.

1 Introduction

Most workers in dynamical systems are familiar with the classical averaging methods. In the examples presented in most texts there are two systems, one fast and one slow, coupled through their associated vector fields. The main point being the fast and slow variables are pre-identified. A survey of fast-slow systems may be found in the paper of Givon et al. [1]. In this note I would like to report on a different approach recently pursued by me and my co-authors Artstein, Gear, Kevrekidis, and Titi in our two papers [2, 3]. In those papers following earlier ideas of Artstein et al. [4–9] we presented a new averaging method in which a fast-slow system will generate its own averaging device (a Young measure) which makes a particularly useful tool for both analysis and numerical computations. Of course here I only report on the basic ideas and strategy: proofs, details and numerical results can be found in the cited references.

M. Slemrod
Department of Mathematics, University of Wisconsin, Madison, WI 53706, USA
e-mail: slemrod@math.wisc.edu

A.N. Gorban and D. Roose (eds.), *Coping with Complexity: Model Reduction and Data Analysis*, Lecture Notes in Computational Science and Engineering 75, DOI 10.1007/978-3-642-14941-2_1, © Springer-Verlag Berlin Heidelberg 2011

2 Systems of Ordinary Differential Equations

To begin, considered a system of ordinary differential equations

$$\frac{dU_\epsilon}{dt} = \frac{F(U_\epsilon)}{\epsilon} + G(U_\epsilon) \tag{1}$$
$$U_\epsilon(0) = U(0),$$

where $U_\epsilon \in \mathbb{R}^N$. Hence $\epsilon > 0$ may be thought of as a small parameter. (A reader familiar with the classical Boltzmann equation of statistical mechanics might want to think of F representing the collision term in the Boltzmann equation while G represents transport.) Next assume there is a scalar quantity $V(U)$ which is conserved along fast motion, i.e., $G \equiv 0$, hence

$$\nabla_U V \cdot F \equiv 0 \tag{2}$$

Again using the analogy with the Boltzmann equation V is called a macroscopic measurement or observer: For the Boltzmann equation this could be density, momentum, or energy. But now unlike the Boltzmann equation we make no assumption as to relaxation of the "fast" part of our system

$$\frac{dU^{(0)}}{ds} = F(U^{(0)}), \qquad s = \frac{t}{\epsilon} \tag{3}$$

to an equilibrium as $\epsilon \to 0$. While we don't preclude relaxation we make only the very mild assumption that solutions of (3)

$$U^{(0)}(s) = U^{(0)}\left(\frac{t}{\epsilon}\right) \tag{4}$$

remain in a bounded subset of $\mathbb{R}^N$ for $0 \le s < \infty$. This of course allows for rapid oscillations in (4) as $\epsilon \to 0$ and $0 \le t \le T$.

Now let us continue with the computation of the dynamics of our observer V. We readily see that along trajectories of (1).

$$\frac{dV(U_\epsilon)}{dt} = \frac{1}{\epsilon}\nabla_U V(U_\epsilon) \cdot F(U_\epsilon) + \nabla_U V(U_\epsilon) \cdot G(U_\epsilon)$$
$$= \nabla_U V(U_\epsilon) \cdot G(U_\epsilon) \tag{5}$$

where the orthogonality relation (2) has been used. Integrate (5) to see

$$V(U_\epsilon(t)) - V(U(0))) = \int_0^t \nabla_U V(U_\epsilon) \cdot G(U_\epsilon)d\tau. \tag{6}$$

Averaging of Fast-Slow Systems

Our goal now is to pass to the limit as $\epsilon \to 0$ in (6), to have a rule for the evolution of the observer V.

The next thing to note that since $U^{(0)}$ the solution of the "fast system" (3) is in fact the dominant part of the solution to the fast-slow system (1), to leading order in ϵ we can rewrite (6) as

$$V\left(U^{(0)}\left(\frac{t}{\epsilon}\right)\right) - V(U(0)) = \int_0^t \nabla_\tau V\left(U^{(0)}\left(\frac{\tau}{\epsilon}\right)\right) \cdot G\left(U^{(0)}\left(\frac{\tau}{\epsilon}\right)\right) d\tau$$

$$+ \text{ higher order terms in } \epsilon, \tag{7}$$

where $0 \le t \le T$.

Recall that $U^{(0)}(\frac{t}{\epsilon}) \in K$, a compact subset of $\mathbb{R}^N$, $0 \le t \le T$. Hence we know from weak* compactness of the unit ball in $L^\infty[0, T]$

$$U^{(0)}\left(\frac{t}{\epsilon}\right) \xrightarrow{w*} \overline{U} \tag{8}$$

where $\overline{U} \in L^\infty[0, T]$.

But what about $V\left(U^{(0)}\left(\frac{t}{\epsilon}\right)\right)$?

Certainly for V a continuous function the composite function $V(U^{(0)}(\frac{\cdot}{\epsilon}))$ forms a bounded sequence in $L^\infty[0, T]$ and hence possesses a weak* compact subsequence

$$V\left(U^{(0)}\left(\frac{\cdot}{\epsilon}\right)\right) \xrightarrow{w*} \overline{V}. \tag{9}$$

But a more precise description is afforded by the formula

$$\overline{V}(t) = < V(\lambda), \upsilon_t(\lambda) > \tag{10}$$

$$= \int_K V(\lambda) d\upsilon_t(\lambda)$$

where υ is the Young measure associate with the sequence $U^{(0)}(\frac{\cdot}{\epsilon})$. The Young measure is a probability measure and of course the variable λ in (10) will vary on the support of V which in our case is the compact set $K \subset \mathbb{R}^N$. (A discussion of Young measures may be found in the paper of Tartar [10], the monograph of Dafermos [11], as well as the already cited references.)

Now that we have recalled the representation of weak* limits in terms of the Young measure we can proceed to the advertised limit in (7) to obtain

$$< V(\lambda), \upsilon_t(\lambda) > -V(U(0)) = \int_0^t < \nabla_U V(\lambda) \cdot G(\lambda), \upsilon_\tau(\lambda) > d\tau \tag{11}$$

or

$$\overline{V}(t) - V(U(0)) = \int_0^t < \nabla_U V(\lambda) \cdot G(\lambda), \upsilon_\tau(\lambda) > d\tau \tag{12}$$

Recall that υ will depend on the generating sequence $\{U^{(0)}(\frac{\cdot}{\epsilon})\}$ and hence depends on $U(0)$ so that (11) is not in fact an integrated ordinary differential equation. Nevertheless (11), (12) does produce a rule for obtaining the evolution of $\overline{V}(t)$ based on computable data. This is explained in the next section.

3 The Algorithm

First of all it is clear that the more observers V we have the more information we can recover about U_ϵ. In fact it may be that practically it is only the values of macroscopic observers V that we are really interested in. Again the example of the Boltzmann equation reminds us that it is only the macroscopic quantities of density, momentum, and energy that are of any consequence.

Next at this algorithmic stage it is convenient to rescale (1) with $s = \frac{t}{\epsilon}$ so that (1) becomes

$$\frac{dU_\epsilon}{ds} = F(U_\epsilon) + \epsilon G(U_\epsilon), \tag{13}$$

$$U_\epsilon(0) = U(0).$$

On any s independent interval $0 \le s \le s_0$, the term $\epsilon G(U_\epsilon)$ will be as negligible regular perturbation so the effect of the $\epsilon G(U_\epsilon)$ term will only be noticeable on an interval $0 \le s \le \frac{s_0}{\epsilon}$. Hence any direct computation of the dynamics of (13) would require step sizes of order ϵ (or smaller) to recognize the $\epsilon G(U_\epsilon)$ and the computation run on extremely long intervals. The inefficiency of this procedure suggests a different approach modeled on (11) and (12).

We suggested in [2, 3] that since all we need to advance $\overline{V}$ is a reasonable approximation to the Young measure υ it is sufficient to run the fast equation

$$\frac{dU^{(0)}}{ds} = F(U^{(0)}) \tag{14}$$

as any interval of length independent of ϵ, say $0 \le s \le L$. We expect that exhibits oscillations on this interval we will sample enough of phase space to approximate the Young measure υ. If repeat this process on a nearby interval say $[1, 2]$ formula (12) allows us to compute say both $\overline{V}(.5)$ from the first run and $\overline{V}(1.7)$ from the second run.

Hence we can approximate $\frac{d\overline{V}}{ds}$ as by the first divide difference $\frac{\overline{V}(1.7)-\overline{V}(.5)}{1.2}$. A hence we can use the chord rule

$$\overline{V}(s) = \overline{V}(.5) + (\frac{\overline{V}(1.7) - \overline{V}(.5)}{1.2})(s - .5)$$

to extrapolate value of $\overline{V}$ for large values of s. This is usual called projective integration. Then at some value of s, e.g., $s = 100, s = 101$ we repeat the process. This

Averaging of Fast-Slow Systems

requires a lifting step to find microscopic data U consistent with the values $\overline{V}(100)$, $\overline{V}(101)$ we have just computed from the chord rule. Notice most importantly the parameter ϵ has vanished from the computation.

4 An Example

A very nice example is provided by the system of ordinary differential equations

$$\frac{dU_k}{ds} + \frac{U_k(U_{k+1} - U_{k-1})}{\partial h} = \frac{\epsilon(U_{k+1} - 2U_k + U_{k-1})}{h^2}, \tag{15}$$

$$U_k = U_{k+N} \quad \text{(periodic boundary conditions)}, \tag{16}$$

$$k = 1, \ldots, N \quad \epsilon, k > 0 \quad \epsilon \ll h \quad h = \frac{2\pi}{N}$$

The system given by (15), (16) is the spatial discrete analogue of the KdV-Burgers type equation

$$u_s + u\left(u_x + \frac{h^2 u_{xxx}}{6}\right) = \epsilon u_{xx} \quad 0 \le x \le 2\pi, \tag{17}$$

$$u(0, s) = u(2\pi, s) \tag{18}$$

If we set $t = \epsilon s$ then (15) takes the form (1) with

$$F(U) = \frac{-1}{2h} \begin{pmatrix} U_1(U_2 - U_N) \\ U_2(U_3 - U_N) \\ \vdots \\ U_N(U_1 - U_{N-1}) \end{pmatrix},$$

$$G(U) = \begin{pmatrix} U_2 - 2U_1 + U_N \\ U_3 - 2U_2 + U_1 \\ \vdots \\ U_1 - 2U_N + U_{N-1} \end{pmatrix}.$$

What makes this system so appealing is the ability to produce macroscopic observers. To see this write $A_k^2 = U_k$ so that the fast system

$$\frac{dU}{ds} = F(U)$$

becomes

$$\frac{dA_k}{ds} = A_k(A_{k+1}^2 - A_{k-1}^2) \tag{19}$$

where s now introduces the irrelevant $\frac{1}{2h}$, and $A_k = A_{k+N}$. But we know from the results of Goodman and Lax [12] admits a "Lax pair," i.e., (19) may be written as

$$\frac{dL}{ds} = [B, L]$$

where $[B, L] = BL - LB$, where L and B are N by N matrices. For example in the case $N = 6$, L has the form

$$L = \begin{pmatrix} 0 & A_1 & 0 & 0 & 0 & A_6 \\ A_1 & 0 & A_2 & 0 & 0 & 0 \\ 0 & A_2 & 0 & A_3 & 0 & 0 \\ 0 & 0 & A_3 & 0 & A_4 & 0 \\ 0 & 0 & 0 & A_4 & 0 & A_5 \\ A_6 & 0 & 0 & 0 & A_5 & 0 \end{pmatrix}$$

i.e., a circulant Jacobi matrix. Now it is trivial that

$$\frac{d}{ds} L^p = [B, L^p], \qquad p = 1, 2, \dots \tag{20}$$

Since the trace, $\mathrm{Tr}([B, L^p]) = 0$, we see

$$\mathrm{Tr}\left(\frac{d}{ds} L^p\right) = 0 \tag{21}$$

For p odd this gives no information since $\mathrm{Tr}(L^p) = 0$ but for p even we obtain a sequence of non-trivial invariants of the fast motion, $p = 2, 4, \dots, \frac{N}{2}$. These "observers" combined with the observer

$$A_1^2 A_2^2 \dots A_N^2 = U_1 U_2 \dots U_N$$

yield $\frac{N}{2} + 1$ "observers" for the fast system.

In our paper we use the algorithm of Sect. 3 to compute the values of these fast observers for the full system (15). For small values of ϵ the algorithm was $O(\frac{1}{\epsilon})$ times faster than a direct computation made by resolving the small scale ϵ. I hope this observation suggests the power of our approach and may motivate the interested reader to read our papers [2, 3].

Acknowledgements This research was sponsored by Grant 2004271 from the US-Israel Binational Science Foundation.

References

1. Givon, D., Kupferman, R., Stuart, A.: Extracting macroscopic dynamics: model problems and algorithms. Nonlinearity **17** (2004) R55–R127
2. Artstein, Z., Gear, C.W., Kevrekidis, I.G., Slemrod, M., Titi, E.S.: Analysis and computation of a discrete KdV-Burgers type equation with fast dispersion and slow diffusion. SIAM J. Numerical Analysis (submitted); e-print arXiv:0908.2752v1 [math.NA]
3. Artstein, Z., Kevrekidis, I.G., Slemrod, M., Titi, E.S.: Slow observables of singularly perturbed differential equations. Nonlinearity **20** (2007) 2463–2481
4. Artstein, Z.: Singularly perturbed ordinary differential equations with nonautonomous fast dynamics. Journal of Dynamics and Differential Equations **11** (1999) 297–318
5. Artstein, Z.: On singularly perturbed ordinary differential equations with measure valued limits. Mathematica Bohemica **127** (2002) 139–152
6. Artstein, Z., Linshitz, J., Titi, E.S.: Young measure approach to computing slowly advancing fast oscillations. Multiscale Modeling and Simulation **6** (2007) 1085–1097
7. Artstein, Z., Slemrod, M.: On singularly perturbed retarded functional differential equations. Journal of Differential Equations **171** (2001) 88–109
8. Artstein, Z., Slemrod, M.: The singular perturbation limit of an elastic structure in a rapidly flowing nearly invicid fluid. Quarterly of Applied Mathematics **59** (2001) 543–555
9. Artstein, Z., Vigodner, A.: Singularly pertubed ordinary differential equations with dynamic limits. Proceedings of Royal Society Edinburgh A **126** (1996) 541–569
10. Tartar, L.: Compensated compactness and applications to partial differential equations. In: Knops, R.J. (ed.), Nonlinear Analysis and Mechanics. Heriot-Watt Symposium Vol. IV, pp. 136–212. Pitman, London (1979)
11. Dafermos, C.M.: Hyperbolic conservation laws in continuum physics. Springer, Berlin (2005)
12. Goodman, J., Lax, P.D.: On dispersive difference schemes I. Communications on Pure and Applied Mathematics **41** (1988) 591–613

The Use of Global Sensitivity Methods for the Analysis, Evaluation and Improvement of Complex Modelling Systems

Alison S. Tomlin and Tilo Ziehn

Abstract Models which involve the coupling of complex chemical and physical processes are being increasingly used within engineering design and decision making. Improvements in available compute power have allowed us to represent such processes with increasing levels of model detail. However, our ability to accurately specify the required high dimensional input data often does not keep pace with the development of model structure. The analysis of model uncertainty must therefore form a key part of the evaluation of such models. Furthermore, sensitivity analysis methods, which determine the parameters contributing most to output uncertainty, can inform the process of model improvement. In this paper we show by example that global sensitivity methods, and in particular methods based on quasi-random sampling high dimensional model representation (QRS-HDMR), are capable of contributing to the model evaluation and improvement process by highlighting key parameters and model subcomponents which drive the output uncertainty of complex models. The method of QRS-HDMR will be described and its application within the fields of combustion and reactive pollution dispersion will be demonstrated. The key points addressed in the work are (1) the potential for complexity reduction using QRS-HDMR methods, (2) global vs. local sensitivity indices for exploring the response to parameters in complex non-linear models, (3) the possibilities for parameter tuning or feasible set reduction via comparison of models with experiment whilst incorporating uncertainty/sensitivity analysis, (4) model improvement through parameter importance ranking coupled with further ab initio modelling studies, (5) robustness to model structure. The generation of a meta-model via QRS-HDMR is shown to be a reasonably efficient global sensitivity method for systems where effects are limited to second-order. Where higher order

---

A.S. Tomlin (✉)
Energy and Resources Research Institute, School of Process, Environmental and Materials Engineering, University of Leeds, Leeds LS2 9JT, UK
e-mail: A.S.Tomline@leeds.ac.uk

T. Ziehn
Department of Earth Sciences, University of Bristol, Bristol BS8 1RJ, UK
e-mail: tilo.ziehn@bristol.ac.uk

A.N. Gorban and D. Roose (eds.), *Coping with Complexity: Model Reduction and Data Analysis*, Lecture Notes in Computational Science and Engineering 75, DOI 10.1007/978-3-642-14941-2_2, © Springer-Verlag Berlin Heidelberg 2011

effects exist, simple transformations of model outputs are shown to improve the accuracy of the meta-modelling process.

1 Introduction

Models involving the coupling of complex chemical and physical processes are being increasingly used within engineering design and decision making. Whilst improvements in available compute power have allowed us to represent such processes with increasing levels of model detail, our ability to accurately specify the required high dimensional input data often does not keep pace with the development of model structure. A classic example of this is in chemical kinetic modelling, where the use of automatic mechanism generation has allowed the specification of reaction pathways for more and more complex starting molecules (e.g. fuel oxidation [1]). However, the thermo-chemical parameters for each elementary process often have to be estimated rather than being determined from first principles. As a result, whilst our ability to resolve complex processes numerically, appears to improve over time, our trust in the predictions of such models may not.

The evaluation of such models becomes a key task and our ability to improve their predictions with respect to validation data is a vital part of model development. In order to do this efficiently, analysis of the relative importance of very large numbers of parameters or variables is required. Efforts to improve the model can therefore be focused on parameters of high importance i.e. those which drive the variability in the output predictions. This often turns out to be a small subset of the whole input space [2]. Sensitivity analysis can be employed for this purpose and several methods for its application in parameter importance and model reduction studies have been developed in recent years [3–5]. Local linear methods have been most commonly used since they are usually computationally quite efficient. However, for highly nonlinear models, local methods can sometimes lead to erroneous results, particularly where the model input parameters have large uncertainty ranges [6]. In some cases, local sensitivity coefficients can be quite different for parameter values which are sufficiently far from the nominal ones. Importance analysis or model reduction based on the often uncertain nominal values therefore becomes problematic. This highlights the importance of developing global sensitivity methods that can provide reliable sensitivity indices over wide ranges of model inputs, for high dimensional input spaces.

The global sensitivity analysis method introduced here is that of quasi-random sampling high dimensional model representation (QRS-HDMR [7, 8]). HDMR methods were originally developed to provide a straightforward approach to explore the input-output mapping of a model without requiring large numbers of model runs [9, 10]. They are model replacements or meta-models which can represent input-output relationships for high dimensional models, using low dimensional hierarchical functions. This is possible because usually only low-order correlations between inputs have a significant effect on the outputs. Due to the hierarchical form of the HDMR component functions, sensitivity indices can be determined from them in an automatic way in order to rank the importance of input parameters and to explore the influence of parameter interactions. Extensions to the existing set of HDMR tools were developed in [6, 11] to allow for instance, the exclusion of unimportant HMDR component functions leading to the efficient analysis of large dimensional input spaces. Results from a Matlab based software package, which has been developed for general application to any complex model and which is freely available [8], will be demonstrated. Examples of its application will be provided from the fields of combustion and air pollution modelling in order to illustrate the power of the approach, and to highlight potential differences compared to the use of local sensitivity coefficients.

1.1 Examples of Complex Systems

Combustion models provide a challenge for uncertainty analysis since they usually contain a large number of uncertain parameters, many of which are derived from a limited number of experimental or theoretical studies, or are estimated based on structural additivity relationships by considering similar functional groups [12]. The uncertainty ranges can therefore be quite large for many parameters, and in addition, there is often a lack of consensus within the literature as to accepted nominal values as well as uncertainty ranges. This calls into question the use of local sensitivity information such as is usually available within commercial modelling packages (e.g. CHEMKIN [13]), since these give information in a small region of parameter space close to the suggested nominal values. However, using global methods, parameters of high importance across the input range can be identified and potentially improved by more detailed studies. An example will be shown where global sensitivity methods have been used to successfully highlight a key parameter within a high pressure combustion process, which following being updated via further ab initio studies, led to improvements in model agreement with experimental results.

The accuracy of air pollution models is vital in terms of their ability to predict the possible air quality impacts of pollution management strategies. They are however, another complex case, as they often involve the coupling of many different model components such as emissions, turbulent air flows, dispersion and chemical processing. Their input parameters can be a mixture of model parametrisations, traffic characteristics and measured inputs such as meteorological data and background concentrations of important atmospheric compounds. It is important to assess whether the model structure is robust enough to allow appropriate interpretation of the importance of physical and traffic effects i.e. to establish that the influence of model parametrisations does not dominate over that of physical and emissions processes. Sensitivity analysis can help to answer these questions and the example shown here is of pollution dispersion from a nitrogen oxide source.

2 Global Sensitivity and Uncertainty Analysis Based on QRS-HDMR

2.1 Why Use Uncertainty and Sensitivity Analysis?

In the evaluation of any complex model we may be interested in knowing the confidence that can be placed in its predictions. This would be the equivalent of determining experimental error bars but for model outputs. The sources of such uncertainty can include lack of knowledge of the values of input parameters, errors in the measurement of key physical inputs, or even problems with the model structure itself. The latter may stem from a lack of understanding of the chemical and physical processes, or our inability to represent them at a sufficiently fundamental level such as with turbulence closure models, or parametrisations of the pressure and temperature dependence of kinetic processes. If such a lack of knowledge of model inputs is propagated through the model system then a model output becomes a distribution rather than a single value. Measures such as output variance can then be used to represent output uncertainty [4].

One aim of sensitivity analysis (SA) is to evaluate the relative contribution of the model inputs to the overall output uncertainty and therefore to provide a measure of parameter importance. This could take the form of analysis of variance (ANOVA) decomposition, where a proportion of the overall output variance is attributed to the lack of knowledge of each of the model inputs, pairs of inputs etc. [14]. The most important parameters can then become the subject for further study, either by using *ab inito* modelling or fundamental experiments to improve knowledge of their values, or perhaps for tuning if they dominate the output uncertainty. The degree of confidence with which they can be tuned depends on how much of the output variance can be attributed to their lack of knowledge i.e. by how much they are isolated within the experimental conditions under investigation. SA in many cases facilitates a degree of complexity reduction, since it helps to highlight a few key parameters or sub-models which determine the predicted outputs of highly complex modelling systems. The examples covered here will highlight the fact that many parameters can be fixed at their nominal values with little or no impact on the output variance.

2.2 High Dimensional Model Representation

The high dimensional model representation (HDMR) method is a set of tools explored by [9] in order to express the input-output relationship of a complex model with a large number of input variables. The mapping between the input variables $x_1, \ldots, x_n$ and the output $f(\mathbf{x}) = f(x_1, \ldots, x_n)$ in the domain R^n can be written in the following form:

$$f(\mathbf{x}) = f_0 + \sum_{i=1}^{n} f_i(x_i) + \sum_{1 \leq i < j \leq n} f_{ij}(x_i, x_j) + \ldots + f_{12\ldots n}(x_1, x_2, \ldots, x_n). \quad (1)$$

The Use of Global Sensitivity Methods

Here f_0 denotes the mean effect (zeroth-order), which is a constant. The function $f_i(x_i)$ is a first-order term giving the effect of variable x_i acting independently (although generally non-linearly) upon the output $f(\mathbf{x})$. The function $f_{ij}(x_i, x_j)$ is a second-order term describing the cooperative effects of the variables x_i and x_j upon the output $f(\mathbf{x})$. The higher order terms reflect the cooperative effects of increasing numbers of input variables acting together to influence the output $f(\mathbf{x})$. The HDMR expansion is computationally very efficient if higher order input variable interactions are weak and can therefore be neglected. For many systems a HDMR expansion up to second-order already provides satisfactory results and a good approximation of $f(\mathbf{x})$ [10]. Where not, appropriate transformations of the outputs can sometimes be used (see Sect. 3.3.2) to help build a low-order HDMR model. This at least allows the accurate identification of important parameters.

There are two commonly used HDMR expansions. Cut-HDMR depends on the value of $f(\mathbf{x})$ at a specific reference point $\bar{\mathbf{x}}$ and random sampling RS-HDMR depends on the average value of $\bar{\mathbf{x}}$ over the whole domain, where the average is usually obtained over a suitable random sample. Previous research has shown that often better convergence properties are achieved by using a quasi-random sample rather than a straightforward random sample [8, 15, 16]. QRS-HDMR sampling methods were therefore used in this work.

2.3 QRS-HDMR

In this study we have applied QRS-HDMR, where the zeroth-order term f_0 is approximated by the average value of $\mathbf{x}^{(s)} = (x_1^{(s)}, x_2^{(s)} \ldots, x_n^{(s)})$ for all $s = 1, 2, \ldots, N$

$$f_0 \approx \frac{1}{N} \sum_{s=1}^{N} f(\mathbf{x}^{(s)}) \tag{2}$$

with N the number of sampled model runs. The higher order component functions are approximated by orthonormal polynomials:

$$f_i(x_i) \approx \sum_{r=1}^{k} \alpha_r^i \varphi_r(x_i) \tag{3}$$

$$f_{ij}(x_i, x_j) \approx \sum_{p=1}^{l} \sum_{q=1}^{l'} \beta_{pq}^{ij} \varphi_p(x_i) \varphi_q(x_j) \tag{4}$$

$$\ldots$$

where k, l, l' represent the order of the polynomial expansion, α_r^i and β_{pq}^{ij} are constant coefficients to be determined and $\varphi_r(x_i)$, $\varphi_p(x_i)$ and $\varphi_q(x_j)$ are the basis functions [7]. The approximation of the component functions reduces the sampling effort dramatically so that only one set of quasi-random samples N is necessary in order to determine all QRS-HDMR component functions.

2.4 QRS-HDMR Extensions

The standard RS-HDMR approach was extended by an optimisation method [11], which automatically chooses the best polynomial order for the approximation of each of the component functions. Component functions can also be excluded from the HDMR expansion if they do not make a significant contribution to the modelled output value via the use of a threshold [6, 8]. The aim is to reduce the number of component functions to be approximated by polynomials and therefore to achieve automatic complexity reduction without the use of prior screening methods such as the Morris method [17]. For a second-order HDMR expansion a separate threshold can be defined for the exclusion of the first and second-order component functions.

The exclusion of unimportant component functions has several advantages. Firstly, since the error of the Monte Carlo integration controls the accuracy of the RS-HDMR expansion, it is possible that the inclusion of unnecessary terms can increase the integration error, reducing the accuracy of the HDMR meta-model. Secondly, if the meta-model were to be used for subsequent analysis, the lower number of terms aids its computational efficiency. The exclusion of component functions also provides an immediate level of complexity reduction, before parameter importance ranking has been performed.

The accuracy of the constructed HDMR meta-model can be determined in many different ways. A common approach is to use the relative error (RE) between the response of the real model and the meta-model:

$$\text{RE} = \left| \frac{f(\mathbf{x}^{(s)}) - \hat{f}(\mathbf{x}^{(s)})}{f(\mathbf{x}^{(s)})} \right| \tag{5}$$

where $\hat{f}(\mathbf{x}^{(s)})$ is the approximated output using the RS-HDMR expansion (first or second-order) and $f(\mathbf{x}^{(s)})$ is the output response of the real model. Other methods include the comparison of the probability density function (pdf) and the cumulative distribution function (cdf) respectively for the real model and the meta-model or the calculation of the coefficient of determination (R^2). All of these options are available in the GUI-HDMR software [8].

2.5 Calculation of the Sensitivity Indices

The method of Sobol' [14] is a commonly used method in global SA in order to calculate the partial variances and sensitivity indices. It is conceptually the same as the RS-HDMR approach. However, once the RS-HDMR expansion is calculated,

The partial variances D_i, D_{ij}, ... for sensitivity analysis purposes are easily obtained from [18]:

$$D_i \approx \sum_{r=1}^{k_i} \left(\alpha_r^i\right)^2 \tag{6}$$

$$D_{ij} \approx \sum_{p=1}^{l_i} \sum_{q=1}^{l'_j} \left(\beta_{pq}^{ij}\right)^2 \tag{7}$$

$$\cdots$$

Then by normalising the partial variances by the overall variance D of the model output we finally get the sensitivity indices:

$$S_{i_1,\ldots,i_s} = \frac{D_{i_1,\ldots,i_s}}{D}, \qquad 1 \leq i_1 < \cdots < i_s \leq n \tag{8}$$

The first-order sensitivity index S_i measures the main effect of the input variable x_i on the output, or in other words the fractional contribution of x_i to the variance of $f(\mathbf{x})$. The second-order sensitivity index S_{ij} measures the interaction effect of x_i and x_j on the output and so on. The sensitivity indices can be used for importance ranking for the input parameters for each of the model outputs and therefore as a focus for model improvement.

3 Case Studies

In the previous section the methodology for the calculation of global sensitivity indices based on QRS-HDMR was explained. We will now attempt to highlight some important features of application of the methods to complex and nonlinear modelling systems through the use of example case studies. The features to be explored are (1) the potential for complexity reduction using the extended HDMR methods, (2) the presence of nonlinearity and parameter interactions in complex models and the comparison of global vs. local sensitivity indices, (3) the level of parameter isolation in different cases and possibilities for parameter tuning via comparison with experiment and uncertainty/sensitivity analysis, (4) examples of model improvement through parameter importance ranking coupled with further ab initio modelling studies, (5) the robustness to model structure and parametrisations vs. knowledge of physically measured inputs. The case studies will be drawn from a range of applications in combustion and air pollution modelling.

3.1 The Potential for Complexity Reduction Using the Extended HDMR Methods

3.1.1 Premixed Flame Case Study

In order to explore the application of the extended HDMR methods to systems with large input dimension, the GUI-HDMR software was applied to a model of a one dimensional low pressure premixed methane flame used to investigate the influence of fuel sulphur and nitrogen on the NO concentration ([NO]) within the burnt gas region [6]. A one dimensional steady state reaction advection diffusion model (PRE-MIX [19]) was used with uncertainties in the parametrisation of rate constants and thermodynamic data considered. This led to a study of 176 input parameters (153 reaction rates and 23 enthalpies of formation) with the aim to determine their relative importance in driving the output uncertainty in predicted [NO]. Full details of the model scenario can be found in [2] with the focus here on the fuel-rich scenario with a flame stoichiometry of $\phi = 1.6$ and 0.5% of SO_2 and 1.3% of NH_3 added to the flame. In this case, the mechanism using the nominal parameter values tends to over predict the relative increase in NO mole fractions at the end of the flame on the addition of SO_2, compared to the experiments of [20].

The reaction rates were expressed in standard Arrhenius form using units of cm^3 molecule^{-1} s^{-1} for second-order rate constants. Reactions were treated as reversible, with reverse rates calculated from the appropriate equilibrium constants based on enthalpies of formation (ΔH_f) calculated using NASA polynomials [21]. Because so many of the input parameters were estimated, derived from a low number of measurements, or from single theoretical studies, the input distributions were considered to be uniform between pre-defined minimum and maximum values. The ranges adopted are described in Tomlin [2].

The example served to demonstrate that even for a model with such a high dimensional input space, a QRS-HDMR meta-model could be fitted using a sample size of only $N = 1,024$ and up to tenth-order polynomials. The successful use of such a small sample size was due to the presence of mainly first-order terms in the HDMR expansion. For 176 input parameters the full second-order HDMR expansion consists of 15,577 component functions (1 zeroth-order term + 176 first-order terms + 15,400 second-order terms). It is therefore critical to explore the potential for complexity reduction using the method for excluding unimportant component functions. Using a threshold of 1% for the first and second-order component functions, only five of the 177 first-order component functions and none of the 15,576 second-order component functions were approximated by optimal-order polynomials. The resulting first-order HDMR meta-model model gave 99.05% of the tested samples within the 5% RE range compared to a sample of 2,000 full model runs. The accuracy can also be presented by using a scatter plot for the response of the meta-model and the original model or by comparing their empirical probability density functions as shown in Fig. 1. It is illustrated that the data points in Fig. 1a form a straight line with only a low amount of scatter. The coefficient of determination

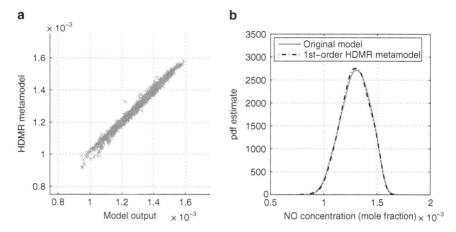

Fig. 1 Accuracy of the RS-HDMR meta-model based on $N = 1{,}024$ model runs for NO concentration (mole fraction) showing (**a**) the scatter plot between the original model response and the first-order RS-HDMR meta-model ($R^2 = 98.50\%$) and (**b**) the probability density function estimate for both the original model and the first-order RS-HDMR meta-model

is $R^2 = 98.50$, which confirms the high accuracy of the first-order HDMR meta-model. The plot of the pdfs in Fig. 1b shows a very good match between the original model and the meta-model. This suggests that despite the high dimensionality of the input space of the model, the predicted [NO] is driven by only a small number of parameters. In this case a high degree of complexity reduction was possible using the extended HDMR tools and without the use of prior screening methods.

3.1.2 Low Temperature Oxidation Study

The low temperature oxidation of fuels and surrogate fuels is an area of growing importance as we attempt to design more efficient engines for use with potential petrol and diesel alternatives. The development of detailed chemical mechanisms for fuel oxidation is therefore receiving much attention and automatic mechanism generation algorithms are becoming increasingly common in an attempt to work towards accurate descriptions of the low temperature kinetics of more and more complex fuels (e.g. [22]). These mechanisms can contain huge numbers of thermo-kinetic parameters and often exhibit highly nonlinear dynamics. The need for complexity reduction in the process of their evaluation is therefore critical, as is the need to highlight parameters which drive their low temperature dynamics since many of the parameters contained within such schemes have been estimated by reference to smaller fuels. There is clearly a need for methods to efficiently highlight important parameters driving the oxidation process as well as tools to improve parameter estimation [23]. Here we present an example of the application of global sensitivity tools to the low temperature oxidation of cyclohexane (a surrogate cyclic alkane) under fuel rich conditions [24].

The kinetic mechanistic foundation of the modelling study was a comprehensive kinetic scheme generated using the automatic mechanism generation algorithm EXGAS [22, 25] comprising 499 species in 1,025 reversible reactions and 1298 irreversible reactions (or 3,348 reactions expressed in irreversible form). However, a model reduction of the scheme was achieved in [26] leading to a scheme comprising 238 irreversible steps whose predictions matched the full scheme over a wide range of conditions. Initial investigations where therefore performed through a global sensitivity analysis of the A-factors for each forward reaction step within the reduced mechanism (i.e. 238 parameters). The simulations were performed in a well mixed vessel, with an initial mixture of cyclohexane + 'air' in 1:2 molar proportions, under isothermal conditions at a temperature of 503 K which is just below the minimum autoignition temperature. The example is therefore representative of a well mixed nonsteady-state box model with highly complex chemistry and complex temporal dynamics. The purpose was to explore why the scheme using the nominal parameter values from [25], appeared to be more reactive than the experiments used for comparison [27], producing a time to maximum reaction rate which was too short and with almost complete oxygen consumption at this time compared to the 50% consumption seen in the experiments. Despite using a reduced mechanism, the input space was still high dimensional at 238 parameters. Since a large number of parameters within the cyclohexane scheme were derived from estimates, input ranges were defined according to a minimum and maximum value, assuming equal probability throughout the ranges. Since the aim was to explore qualitative discrepancies, conservatively large uncertainties of a factor 5 above and below the nominal value for each A-factor were assumed.

Despite the input space being only slightly higher dimensional than the flame model discussed in the previous section, this case study proved to be more difficult in terms of complexity reduction using HDMR. Initial test cases using a small quasi-random sample indicated the presence of higher order interactions between parameters. From experience, accurate fitting of higher order component functions using RS-HDMR methods requires much larger sample sizes than for first-order terms. In this case therefore it proved to be more efficient to employ a global screening method; the Morris One at a Time algorithm [17], to exclude unimportant parameters prior to the application of HDMR. The Morris screening method [17,28] is classified as a global sensitivity experiment since the entire space over which the inputs may vary is covered. It effectively samples a number of random trajectories through a grid covering the input space by varying one parameter at a time, in a random order, by a fixed amount. The method then determines an importance ranking for parameters in terms of their mean effect on output variance. It can also rank the degree of nonlinearity of response to changes in each parameter by ranking the standard deviation of the parameter effect across all trajectories. It does not however provide the functional dependency of the output on individual parameters or parameter pairs and is mainly used as a way of screening out *unimportant* parameters prior to the use of variance based sensitivity methods.

Using the Morris method for the cyclohexane oxidation study, a subset of 33 factors were selected for further analysis. All others were kept at their nominal values during the HDMR analysis. The outputs chosen were the time to maximum reaction rate (τ_{max}) and the percentage oxygen consumption at this time ($O_2(\tau_{max})$). Convergence of the calculated first and second-order sensitivity indices for both outputs was achieved with a sample size of 4,096. Note that this is larger than for the flame case even though we have already screened out some unimportant parameters. This is due to the presence of significant second-order terms. A further degree of complexity reduction was achieved using the HDMR meta-model. Out of the 33 parameters studied, only 15 contributed to the first-order indices for both outputs with interactions between the same group of parameters contributing to the main second-order effects. However, whilst τ_{max} was fairly well described by a second-order meta-model (94% of the total variance), $O_2(\tau_{max})$ was clearly affected by higher order interactions. Only 67% of the total variance of this output was described by the second-order HDMR model.

The inability of the second-order meta-model to capture the higher order effects in the second output is illustrated in Fig. 2 which shows the distribution of $O_2(\tau_{max})$ from 1,000 model runs compared with the predicted HDMR distribution. The figure shows that the original distribution contains a very long tail which suggests a low frequency of values at the extreme of the predicted output values. The second-order model fails to capture this tail effectively. In fact the presence of such a tail can be seen as an indication of higher order effects and in contrast was not seen in the flame model above.

In addition, none of the samples achieved the 50% oxygen consumption seen in the experiments which indicates possible remaining structural problems with the model itself. These could take the form of missing chemical or physical processes and will be discussed further in Sect. 3.3.2.

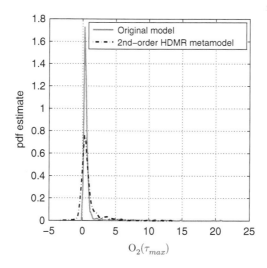

Fig. 2 Accuracy of the RS-HDMR meta-model based on the probability density function (pdf) using 1,000 random points for $O_2(\tau_{max})$ using a second-order HDMR meta-model

3.2 The Presence of Nonlinear Effects

As discussed in the introduction, the most common methods for sensitivity analysis are local ones where the gradient in the neighbourhood of a nominal or reference value for each parameter is used to estimate the response to its variation. Local sensitivity methods are used in the evaluation of models as well as in model reduction strategies i.e. to identify redundant model features that may be removed in order to produce a reduced order model. In many cases however, the value of the input parameter will be highly uncertain and a large range will be feasible. If the response of the chosen output to changes in the parameter is linear across its whole input range, then the use of local sensitivity indices will not be problematic. However, in many case studies there are examples of strongly nonlinear responses to changes in parameters across their input range and we illustrate this point in this section through a number of examples.

3.2.1 Premixed Flame Study

Figure 3 shows the first-order component functions overlaid by the scatter plots for the A-factors for reactions SO + NH = NO + SH and SO + OH = SO_2 + H within the premixed flame study described in Sect. 3.1.1. The component functions indicate the first-order response to changes in the chosen parameter independent from the values of the other parameters. For the reaction SO + NH = NO + SH, the component function shows a linear response across for the A-factor, indicating

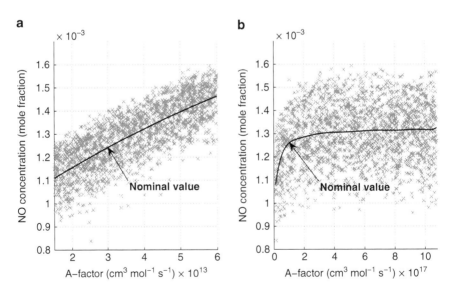

Fig. 3 First-order component functions and scatter plots for (**a**) SO + NH = NO + SH, (**b**) SO + OH = SO_2 + H. The mean f_0 is added to f_i for comparison with scatter plot

that in this case a local sensitivity coefficient at the nominal value would give an accurate picture of the overall response to this parameter across its whole range of uncertainty. The same is not true for the A-factor for reaction $SO + OH = SO_2 + H$ which shows a strong sensitivity at the lower end of its range that begins to saturate at higher values. A local estimate at the nominal value would not in this case give an accurate picture of the response to this parameter across its whole uncertainty range. The example serves to highlight the power of the HDMR meta-model and component functions. The component functions give a strong visual picture of the response to parameter changes across the whole input range. Furthermore, if further work was carried out to improve knowledge of the parameter, thus narrowing its range, the HDMR meta-model could be used to calculate the resulting effect on the overall uncertainty of the model and a new sensitivity coefficient for the parameter.

3.2.2 Low Temperature Oxidation Study

Figure 4 shows the first-order component functions for $O_2(\tau_{max})$ for the A-factors for 2 key competing reaction channels $2C_6H_{11}OO \rightarrow 2C_6H_{11}O + O_2$ and $2C_6H_{11}OO \rightarrow C_5H_{10}CO + C_6H_{11}OH + O_2$. The conditions are as described in Sect. 3.1.2. The scatter plot data is also shown in part (b), where the mean f_0 is added to the first-order component function f_i for comparison. The figure shows a strong nonlinear response to the first reaction channel, with a high sensitivity at the lower end of its range that appears to saturate as the A-factor becomes larger and

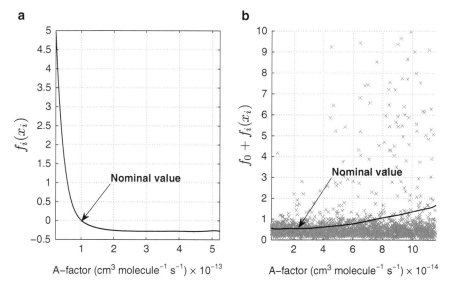

Fig. 4 First-order component function f_i for (**a**) $2C_6H_{11}OO \rightarrow 2C_6H_{11}O + O_2$, and scatter plot for (**b**) $2C_6H_{11}OO \rightarrow C_5H_{10}CO + C_6H_{11}OH + O_2$ and output $O_2(\tau_{max})$. In (**b**) the mean f_0 is added to f_i for comparison with scatter plot

the reaction rate becomes faster. This suggests that as the reaction becomes faster, it potentially moves into a quasi-steady state regime where the overall dynamics of the model is relatively insensitive to its actual rate. The temperature (T) dependant rate constant used in the Nancy scheme is $1.046 \cdot 10^{-13} e^{(365/T)}$ cm^{-3} molecule^{-1} s^{-1}. However, the low temperature (atmospheric chemistry related) measurements of [29] suggested a rate of $0.769 \cdot 10^{-13} e^{(-184/T)}$ cm^{-3} molecule^{-1} s^{-1} which is about a factor of 4 lower than the Nancy parameter value at 503 K. The uncertainty in the rate for this reaction is therefore critical to determining the dynamics of the oxidation process at these low temperatures, and yet kinetic studies of the reaction at relevant temperatures are not available. The sensitivity profile for this parameter also indicates that it may be wrongly defined as a quasi-steady state species if its rate is assumed to be too fast and therefore the use of a local sensitivity measure around the nominal value from the EXGAS scheme could lead to inappropriate model reduction strategies. The positive sensitivity of the competing reaction channel shown in Fig. 4b suggests that the product branching ratio for the two channels is a critical parameter worthy of further study. The global sensitivity analysis therefore highlights the subcomponents of the full chemical scheme which drive the low temperature oxidation at these fuel rich sub critical temperature conditions, namely the peroxy radical chemistry and the fate of the cyclohexyl peroxide. The study also identifies those parameters which require further studies at a fundamental level over a range of temperatures. It is clear that a local linear sensitivity study would perhaps have failed to highlight certain important reactions which have a low sensitivity around the nominal values adopted in the EXGAS scheme.

3.3 Parameter Tuning, Fundamental Studies and Combined Approaches

Optimisation and parameter tuning are commonly used in model development for practical applications (for examples in the field of combustion see [30–32]). They represent a pragmatic approach to model development where parameters are optimised to give the best overall agreement with a chosen set of target experiments. Optimisation is rarely unique and many parameter sets could lead to the same predicted outcomes for chosen targets [33]. Optimisation should not therefore be used as a way of determining parameter values but simply of providing a best fit model, under well defined conditions, given known uncertainties. This approach is reasonable where the conditions for model application are similar to those used during the optimisation process, but can cause difficulties when trying to use the optimised model for new sets of conditions.

Unfortunately models are at their most useful when making predictions for conditions where experiments have not yet been performed and therefore providing new information of use in the design process. A new approach to model improvement and to reducing model uncertainty is therefore required, which minimises the influence of optimisation over specific conditions. In order to achieve this, the range of uncertainty for each of the important parameters in a model should be reduced as far as possible. Such an approach would need to combine information from larger scale experiments (e.g. in combustion: flow tube reactors, rapid compression machines, flames etc.) with experimental or modelling studies at a more fundamental level (e.g. quantum theory calculations).

A perfectly designed experiment for parameter identification would show only sensitivity to this parameter regardless of uncertainties in other model or physical parameters that may be used in the fitting process. The model parameter can then be successfully tuned using the given experiment and the tuned value of the parameter used in wider studies. In practice it is very difficult to achieve this level of parameter isolation and therefore even fundamental experimental studies will lead to a level of uncertainty in the parameter value determined. Such experiments do, however, improve our knowledge of the feasible range of a parameter. Theory calculations and larger scale experiments can also help to achieve this, and therefore, the more studies which contribute to the knowledge of a parameter, the smaller its uncertainty should become as long as the studies are well designed. However, caution should be applied when attempting to quantify parameters from single experiments where the level of parameter isolation is poor as illustrated by the following example.

3.3.1 High Pressure Ignition Delays for Carbon Monoxide (CO) Hydrogen (H$_2$) Mixtures

In some cases larger scale experiments are used to tune parameters without assessment of the effects of overall uncertainties. The following example is included in order to demonstrate where problems can arise in attempting to do this. The case study model was designed to investigate the time to ignition of various hydrogen air mixtures at high pressures. Of particular interest was the potential effect of an increasing extent of substitution of hydrogen by carbon monoxide within the fuel mixture. Careful experiments within a rapid compression machine over a range of pressures from 15 to 50 bar by [34] demonstrated that as the percentage CO increased, then the ignition delay became longer i.e. the mixture became less reactive. However, this behaviour was not demonstrated by a numerical model of the rapid compression machine experiments, despite using a highly detailed kinetic model for CO/H$_2$ combustion [31]. The behaviour of the original model (dotted curve) and the experimental ignition delays (square symbols) are illustrated in Fig. 5. A global sensitivity analysis was therefore performed in order to establish the key kinetic parameters driving the ignition process, with the target output the ignition delay time (t_{ign}).

The analysis revealed the importance of a key CO reaction $HO_2 + CO \rightarrow CO_2 + OH$, which had a strong influence on predicted t_{ign} at these high pressures. The existing data for this reaction was determined using indirect methods (e.g. [36, 37]) and was later adjusted in the work of [38] because of discrepancies between their model and high pressure experimental data. The parameter had not been previously addressed by fundamental theoretical studies and had a high

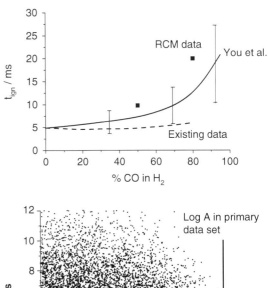

Fig. 5 Experimental vs. modelled time to ignition (t_{ign}) at 30 bar and 1007 K using both the original mechanism [31] and the revised mechanism based on the calculations of [35]

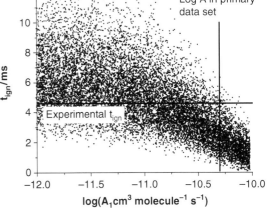

Fig. 6 Monte Carlo based scatter plot of time to ignition vs. the log of the A-factor for reaction $HO_2 + CO \rightarrow CO_2 + OH$ at conditions of 50 bar and 1,040 K

degree of uncertainty. The question therefore arose as to whether the ignition delay experiments could be used to isolate this parameter and therefore to tune its rate and perhaps reduce its uncertainty. A Monte Carlo scatter plot easily demonstrates the degree of isolation of a parameter with respect to a chosen target. Figure 6 shows such a scatter plot for t_{ign} vs. the pre-exponential factor (or A-factor) for the above reaction at conditions of 50 bar and 1,040 K and an initial mixture of $0.8CO + 0.2H_2$. Uncertainties in all of the 76 A-factors in the reaction scheme were adopted in the Monte Carlo simulations. The figure shows that despite being the highest ranked parameter, the A-factor for reaction $HO_2 + CO \rightarrow CO_2 + OH$ is not the only important factor in determining the ignition delay time. The scatter in the figure represents the combined effects of all other uncertainties and shows that the parameter of interest is not well isolated by the ignition delay experiment. In fact the data shows that the experimental t_{ign} is achievable over a wide range of A-factors for the chosen reaction. The comparison does however suggest that the A-factor cannot be greater than that adopted in the original scheme if the experimental ignition delay is to be reached, and is likely to be lower. This limits the range of feasibility for the reaction rate at this temperature and pressure. Further experiments or theoretical calculations are however required in order to provide a better estimate of this parameter over a wide range of conditions.

Following the global sensitivity study, the parameter was re-estimated by [35] using a combination of ab initio electronic structure theory, transition state theory, and master equation modeling. A temperature dependant rate expression was proposed which at 1,040 K was more than a factor of 3 lower than the original, with a suggested uncertainty factor of 2 at high temperatures. Figure 5 shows that the adoption of this new rate significantly improves the ability of the model to represent the influence of CO on the ignition delay times. The overall uncertainty in the scheme (represented by the error bars) is still however quite large, which is not surprising given the scatter in Fig. 6 at the lower end of the range of the A-factor for the key $CO + HO_2$ reaction. Clearly model improvement must be an iterative process with attention now paid to other key reactions such as the pressure dependant reaction $H + O_2 + M \rightarrow HO_2 + M$ which was ranked second highest in the global sensitivity study.

This iterative approach to model improvement can be extremely time-consuming, and in addition, many parameters may be of importance within several model applications to different systems. It would therefore be useful to establish frameworks whereby information from sensitivity studies and new experimental and theoretical calculations could be shared quickly with the aim of presenting the best estimate for a parameter and its possible range. New studies could be quickly incorporated into such a structure in order to reduce overall model uncertainties. Sensitivity information could also be used to suggest new experiments which would optimise the level of parameter isolation for key parameters.

An example of how this may work has been recently proposed in a co-laboratory framework by Frenklach and co-workers [39, 40] for chemical kinetic models. The website PrIMe (Process Informatics ModEl[1]) is intended to act not just as a data depository but also as a source of information on data and model uncertainty which is informed by information from experimental and theoretical work provided by the community. Hence as new data, experiments, theory calculations are uploaded, the uncertainties in estimated rate constants can be collectively reduced i.e. their feasible set is reduced. The feasible set is defined by [40] as expressing the 'collective constraints imposed by theory and experiment' on model parameters. The availability of models with full representation of data uncertainties also allows the better design of future experiments i.e. by performing a hypothetical experiment using a model, it is possible to see the extent of isolation of a particular parameter, or a group of parameters which contribute to the possible measured targets, assuming the model is at least structurally sound.

---

[1] http://www.primekinetics.org

3.3.2 Addressing Structural Problems: The Low Temperature Oxidation Case

What if the model is not structurally sound? Do UA and SA have anything to offer? In some cases, despite including known uncertainties within the rate constants, the model predictions do not overlap with experimental data within the error bars, or do not even reproduce the correct qualitative trends. In this case UA can suggest that there are missing model components or inappropriate parametrisations within the model. In the low temperature cyclohexane oxidation study introduced above, the sensitivity and uncertainty analysis was initiated because of the lack of ability of the model to reproduce the features of quadratic autocatalysis seen in the experiments. However, despite the complexity of the kinetic scheme employed, and the conservative estimates of the uncertainties in the kinetic rate parameters, the tail of the second output distribution was still not close to the experimentally observed 50% consumption of O_2 at the point of maximum reaction rate. This suggested possible structural problems with the model which could include missing chemical or physical model components.

A possible problem within simulations of closed reactors is the ability to accurately represent the termination of radical species at the walls of the reaction vessel which may depend on the type of reactor coating used and the age of the reactor [41]. Wall termination was not included in the first set of simulations performed but has been identified by [41] as providing a possible inhibiting effect on autoignition of fuels at low temperature conditions. It therefore should perhaps be considered as a possible termination process for those radicals that are relatively long lived in the gas phase. Wall reactions were therefore added to the scheme for the limiting case of spatial uniformity for selected species with ranges as estimated in [24] and the HDMR analysis was repeated with now 43 uncertain parameters.

In this case the modelled distribution for $O_2(\tau_{max})$ was much wider than in the first set of simulations as shown in Fig. 7. The tail now does include 50% oxygen consumption and one of the wall termination reactions features in the top five first-order sensitivities. In addition, the inclusion of wall reactions, increases the higher order effects in the model due to parameter interactions. This is perhaps understandable, since the slower the chemical lifetime of a species, the more likely it is to be lost at the vessel wall. Whilst providing interesting insight into the possible interactions between chemical and physical processes, these higher order interactions provide a difficult challenge for the HDMR meta-modelling process. The tail in the distribution represents infrequent values i.e. those resulting from higher order interactions. In this case, the second-order HDMR represents only 51% of the output variance of τ_{max} and 49% of the variance of $O_2(\tau_{max})$. Higher order effects of the important parameters cannot be tested by the second-order meta-model. Simple transformations of the outputs were therefore tested in order to attempt to reduce the influence of the tails in the original distribution. This means that a direct contribution of each parameter to the original output cannot easily be assessed since the inverse transformation of the component functions may not be easy to define. However, if

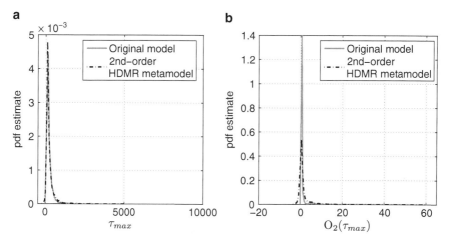

Fig. 7 Pdf distributions of (**a**) τ_{max} and (**b**) $O_2(\tau_{max})$ where the *solid line* is calculated using the full model and the *dotted line* the first and second-order meta-models respectively

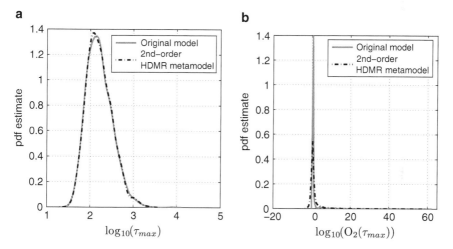

Fig. 8 Pdf distributions for $\log_{10}$ transformation of (**a**) τ_{max} and (**b**) $O_2(\tau_{max})$ where the *solid line* is calculated using the full model and the *dotted line* the first and second-order meta-models respectively

the aim is to identify the group of parameters which determine the majority of the output variance, then such transformations can be useful.

Figure 8 shows the output distribution for the original full model simulations, and the HDMR meta-model for a $\log_{10}$ transformation of τ_{max} and $O_2(\tau_{max})$. The figure shows that the tails of the transformed outputs are shorter for the transformed data and that the HDMR second-order meta-model provides a better fit to the full model data. In this case, the second-order meta-model describes 97% of the variance of

τ_{max} and 71% of the variance of $O_2(\tau_{max})$. This is a marked improvement compared to the use of non-transformed outputs. However, it is still the case that a small number of parameters contribute to the overall variance. For τ_{max} only 13 parameters contribute to the overall variance with 93% now first-order terms including the wall termination for the peroxy and the cyclohexyl peroxide species that were also a key part of the critical kinetic processes. These are the same 13 parameters that contributed to the original time to maximum reaction rate but now the meta-model can describe 97% of $\log_{10}(\tau_{max})$ with mostly first-order components. Thirteen parameters also describe the output variance for $\log_{10}(O_2(\tau_{max}))$, although it is a slightly different set than for output 1. Overall 18 parameters contribute to non-zero component functions within the second-order meta-model and there are no parameters within the second-order terms that do not have non-zero first-order terms. This perhaps suggests that second-order terms are not always necessary if we simply wish to identify important parameters for further study.

The work highlighted a possible missing component of the original model, that of the wall termination of key species. The inclusion of these reactions does allow the qualitative experimental behaviour to be recovered for certain sets of parameter conditions. However, further work would be required in order to reduce the uncertainties in the key inputs to the scheme and to establish better parametrisation of processes occurring at the vessel walls which may involve coupling a more complex physical model (e.g. including diffusion) with the chemical kinetic scheme.

3.4 Model Robustness

The final case study addresses the issue of model fitness for purpose in the area of pollution dispersion and air quality management. For a model to be useful for making strategic decisions about air quality it must be able to properly represent the influence of emissions and atmospheric processes without significant sensitivities to the choice of turbulence model and other internal model parametrisations. This is potentially difficult to achieve since we often must make choices about parametrisations on the basis of available computing power rather than for genuine physical reasons. The problem of turbulence closure is one such example which is common in many fields of reactive flow modelling. Within air pollution models, we often require rapid turnaround times for simulations in order to be able to quickly assess the potential impact of changes to emissions profiles on air quality over wide ranges of meteorological conditions. Modelling approaches such as Large Eddy Simulation (LES), designed to accurately represent short time-scale intermittent processes, can rarely be afforded within operational models. It is therefore useful to assess whether lower levels of turbulence closure, such as using a Reynolds Averaged approach, can give meaningful results on the time and spatial scales of interest within atmospheric dispersion models.

The Use of Global Sensitivity Methods 29

3.4.1 Reactive Plume Scenario

The case study represents a combined modelling system which has been developed for the prediction of atmospheric secondary pollutant formation from emissions of nitrogen oxides from combustion sources. Atmospheric dispersion occurs in a highly turbulent environment where turbulent mixing and chemical transformation can take place on similar time-scales. It is therefore is subject to similar turbulence closure problems found in many engineering applications. Whilst direct numerical simulation (e.g. [42, 43]) and high resolution LES of atmospheric flows (e.g. [44, 45]) have been attempted, in general practical dispersion applications require less computationally expensive turbulence closure models such as Reynolds Averaged models [46]. Questions then arise as to whether such models are fit for purpose i.e. are they capable of accurately describing turbulent chemical interactions when they do not contain representations of all turbulent length and time-scales that are really present in the atmosphere, but rather only an average? Also, given that they must contain parametrisations of such length and time-scales, how robust are model simulations to the method of parametrisation and the parameter values chosen? In order to address these questions the model case presented combines a Reynolds Averaged representation of atmospheric turbulence with a Lagrangian stochastic particle dispersion model with micro-mixing and chemical sub-models [47]. The complex modelling system was used to investigate a reactive plume of nitrogen oxides (NO_x) released into an approximately homogeneous turbulent grid flow doped with ozone (O_3) for comparison against the wind tunnel experiments of [48]. The case was also extended to include possible photolysis reactions which were not present in the original wind tunnel experiments. The full case study is described in [49] but here we address the question of model suitability and robustness.

The main uncertainties within the model occur in parameters defining the turbulence scales, source size, the fraction of NO and NO_2 in the source NO_x and the reaction rate of $NO + O_3 \rightarrow NO_2 + O_2$ and several other temperature dependent and photolysis reactions within the $NO/NO_2/O_3$ chemical scheme. The target outputs of the simulations are the down-wind concentration profiles of the mean non-reactive tracer total NO_x ($\bar{\Gamma}_{NO_x}$), secondary species NO_2 ($\bar{\Gamma}_{NO_2}$) and O_3 ($\bar{\Gamma}_{O_3}$), as well as the root mean square (rms) NO_2 concentrations (γ'_{NO_2}) which provide a measure of concentration fluctuations. The ability of the model to provide accurate predictions of the mean and fluctuating concentrations of the secondary pollutant NO_2 is a useful test of whether the coupled Reynolds Averaged Lagrangian particle approach is a suitable way to represent the main turbulent chemical interactions. The time averaged Reynolds Stress profiles were adopted from the original experiments of [48].

In terms of turbulence model parametrisations, the main parameters are the Lagrangian structure function coefficient c_0, the mixing time-scale coefficient α and the initial source size. The Lagrangian structure function is defined as the ensemble average of the square of the change in Lagrangian velocity and the definition of c_0 is therefore important in determining the effective turbulent diffusion in velocity space. There is some debate within the literature as to whether it's value can be universally defined for all types of turbulent flows with a range of values between 2 and 10 quoted from different studies in [50]. This wide range suggests that a universal value may not exist and that unless rigorous principles can be defined to estimate c_0 for different cases, a certain amount of uncertainty will remain in its parametrisation. It is interesting to establish therefore how sensitive concentration predictions are to the chosen value. Within the model tested, a simple particle mixing model is adopted, that of interaction by exchange with the mean (IEM) concentration [51]. In order to provide generality, the mixing model uses a coefficient α which defines the relationship between the turbulent time-scales (total turbulent kinetic energy and its dissipation rate) and the mixing time-scale at every point in the flow. This simple mixing model is easier to compute than more comprehensive schemes such as the interaction by exchange with the conditional mean (IECM) model, but according to [51] can lead to spurious fluxes in concentration. It is therefore of interest to see if the model predictions are robust to the value chosen for α, and one feature of an acceptable mixing scheme is that mean non-reactive tracer concentrations should not be affected by its parametrisation.

Figure 9a shows that within the uncertainty bounds, the model simulates the plume centre line $\bar{\Gamma}_{NO_x}$ quite well at different distances from the source. The error bars however, are quite large and therefore sensitivity analysis is used to determine the main contributers to the overall output uncertainty. The results demonstrated a

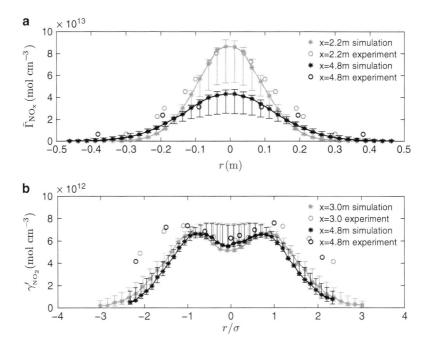

Fig. 9 Radial profiles of (**a**) $\bar{\Gamma}_{NO_x}$ and (**b**) γ'_{NO_2} at different distances x from the point source obtained from simulations and from wind tunnel experiments [48]. Error bars are based on 400 simulations with varying source size, α and c_0

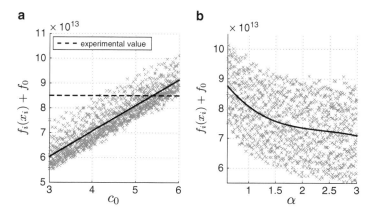

Fig. 10 First-order component functions and scatter plots for (**a**) the structure function coefficient c_0 and (**b**) the mixing time-scale coefficient α with respect to $(\bar{\Gamma}_{NO_x})$ at the plume centre for $x = 2.2$ m. The mean f_0 is added to f_i for comparison with scatter plot

significant influence of c_0 on the simulation of downwind plume centre line passive tracer $(\bar{\Gamma}_{NO_x})$ concentrations, contributing around 80% of the overall variance at several downwind distances. Figure 10a shows the HDMR component function and scatter plot for the response to changes in c_0 illustrating a linear relationship. It appears at first sight that there is a reasonable level of isolation of this model parameter provided by the experiment but there is still a small influence of α as shown in Fig. 10b. This is an undesirable effect since the passive tracer concentrations should not be sensitive to the mixing time-scale constant although it contributes only 20% of the overall variance at the furthest downwind distance. Several issues therefore arise: firstly is the mixing model chosen fit for purpose? Secondly, can the wind tunnel experiment be used to tune the value of c_0 for this type of turbulent flow? The sensitivity to α is not linear and is shown to be higher at the lower end of its range. It may be that the lower limit of 0.6 is unreasonably low. This was explored further in the extended case. In terms of tuning c_0, the sensitivity to α means that the parameter is not fully isolated by the experiment and some scatter in the response is seen. However, it can be argued that the experimentally observed plume centre line concentrations are only recovered using a value of c_0 between 4 and 6 and therefore the experiment may well reduce the feasible range for the parameter is these flow conditions. The value of 5 adopted from [47] through comparison with wind tunnel data from [52] seems to be a reasonable assumption for the Brown and Bilger conditions used here. However, this still does not suggest universality for c_0 since many of the studies referred to by Anfossi [50] suggest values below 5. [53] have also proposed a spatial dependence of c_0 in canopy flows.

The expanded case study attempted to represent more realistic conditions in relation to atmospheric processes, in that photolysis reactions were included and the initial source concentrations of NO_x and background concentrations of O_3 were closer to those that may be found in a polluted urban area for example. The source was also defined as being a combination of NO and NO_2 emissions as would be the case for most combustion sources, with a level of uncertainty in the initial fraction of NO included in the sensitivity study. This variability was intended to mimic possible uncertainties in the primary NO_2 fraction within transport emissions for NO_x [54]. For full details of the rages selected for the parameters see [49]. In this case, the plume centre line mean concentrations of secondary species O_3 and NO_2 were dominated by the influence of physical and chemical parameters rather than the turbulence model parametrisation. For centreline $\bar{\Gamma}_{O_3}$ the background concentration of O_3 was the dominant parameter ($\sim$60% of total variance), with the activation energy for the reaction of O_3 with NO also playing a significant role ($\sim$20%). c_0 and α contributed only 2 and 11% respectively to the total variance, indicating that the model predictions were fairly robust to the parametrisation of turbulent length and time-scales.

The predicted centre line $\bar{\Gamma}_{NO_2}$ and γ'_{NO_2} did overlap with the experimentally observed values within the error bars of the UA, although the plume edge values did not, indicating that the model structure and level of turbulence close were not well suited to predicting intermittent behaviour at the edge of the plume. In particular, γ'_{NO_2} were underestimated by the model at the plume edges, even at the extremes of the error bars (Fig. 9b). In general however, we would be interested in peak NO_2 levels rather than lower concentrations so the model may still be suitable for such purposes. For both $\bar{\Gamma}_{NO_2}$ and γ'_{NO_2}, the main parameter driving the variance was the fraction of NO in the emission source (64% for $\bar{\Gamma}_{NO_2}$, 50% for γ'_{NO_2}). The structure function coefficient had a fairly low influence on the output variance (9% for $\bar{\Gamma}_{NO_2}$, 1.5% for γ'_{NO_2}) and α had a low influence on $\bar{\Gamma}_{NO_2}$ (10%) but quite a strong influence on γ'_{NO_2} (30%). The only chemical parameter with any significance was again the activation energy for the reaction $NO + O_3$. The work suggests that for high NO_x conditions close to sources (e.g. as found in a polluted roadside environment) that the parametrisation of the source would dominate in importance over that of the chemical reaction scheme. The mixing time-scale is also a critical parameter in terms of predicting fluctuations.

Since NO was fixed as 100% of the source in the experiments of [48], the experiments may provide a way to reduce the feasible range for α which for a fixed emission becomes the most important parameter. The predicted γ'_{NO_2} only began to overlap with the experimental values only when α was greater than 1 and a nominal value of 1.5 (twice that suggested in [47]) gave a better agreement for the peak NO_2 predictions without affecting the predictions of NO_x. So, although the experiments of Brown and Bilger do not allow us to fix universal values for the turbulence constants, they allow us to assess whether or not the level of turbulence closure and the selection of parametrisation method leads to a robust model capable of overlapping with the experimentally observed behaviour, and may allow us to limit the feasible range for certain parameters for a specific flow type, namely grid generated isotropic turbulence. In this case, the model allowed the recovery of peak mean concentrations for the passive tracer NO_x and the reactive tracer NO_2. Low mean values at the plume edges were not recovered by the model within the range of uncertainty however, and it is difficult to say if this is due to the structure of the model or experimental error. The range of values for c_0 could be limited to 4–6 and α to 1–3 based on the experimental observations with nominal values of 5 and 1.5 suggested. Similar studies for other sets of conditions (e.g. non-isotropic canopy layer turbulence) would also be useful in order to address the issue of universality for these constants.

4 Overall Conclusions

We have attempted to show by example that global sensitivity methods, and in particular QRS-HDMR, are capable of contributing to the model evaluation and improvement process by highlighting key parameters and model subcomponents which drive the output uncertainty of complex models. We have suggested that complete parameter isolation is difficult to achieve in many experimental setups, but that through the comparison of model outputs with experiment whilst taking into account parameter uncertainty, knowledge about the feasible range of parameters can be gained. The use of importance ranking can also suggest the key parameters which should be further addressed via more fundamental studies. It was shown that theoretical calculations can be used to improve our knowledge of such key parameters, thus reducing overall model uncertainty and improving the agreement between model simulations and experiment. This identification of key parameters is of particular use in reactive flow models where often hundreds of input parameters are included with large uncertainty ranges. However, a degree of complexity reduction is usually possible with often only a small number of input parameters determining the model output variance. Global sensitivity analysis does come at a price in terms of the number of model runs required to achieve accurate results. However, the generation of a meta-model via HDMR has been shown to be reasonably efficient for systems where effects are limited to second-order. Where higher order effects exist, simple transformations of model outputs can be used to improve the meta-modelling process. The set of important parameters can therefore be evaluated accurately without having to resort to extremely large sample sizes. It is hoped that the use of global UA and SA will become a routine part of model evaluation in the future and collaborative web based tools are starting to develop which may facilitate this process.

Acknowledgements The authors would like to thank EPSRC for providing funding for both authors (GR/R76172/01, GR/558904/01) and Professor John Griffiths and Dr. Kevin Hughes for useful data and discussions.

References

1. Battin-Leclerc, F.: Detailed chemical kinetic models for the low-temperature combustion of hydrocarbons with application to gasoline and diesel fuel surrogates. Progress in Energy and Combustion Science **34** (2008) 440–498

2. Tomlin, A.S.: The use of global uncertainty methods for the evaluation of combustion mechanisms. Reliability Engineering and System Safety **91** (2006) 1219–1231
3. Tomlin, A.S., Turányi, T., Pilling, M.J.: Mathematical tools for the construction, investigation and reduction of combustion mechanisms. Pilling, M.J. (ed.) Low Temperature Combustion and Autoignition. Elsevier, Amsterdam (1997) 293–437
4. Saltelli, A.: Sensitivity Analysis. Wiley, NY (2000)
5. Turányi, T.: Sensitivity analysis in chemical kinetics. International Journal of Chemical Kinetics **40** (2008) 685–686
6. Ziehn, T., Tomlin, A.S.: A global sensitivity study of sulphur chemistry in a premixed methane flame model using HDMR. International Journal of Chemical Kinetics **40** (2008) 742–753
7. Li, G., Wang, S.-W., Rabitz, H.: Practical Approaches To Construct RS-HDMR Component Functions. Journal of Physical Chemistry A **106** (2002) 8721–8733
8. Ziehn, T., Tomlin, A.S.: GUI-HDMR – a software tool for global sensitivity analysis of complex models. Environmental Modelling and Software **24** (2009) 775–785
9. Rabitz, H., Aliş, Ö.F., Shorter, J., Shim, K.: Efficient input-output model representations. Computer Physics Communications **117** (1999) 11–20
10. Li, G., Rosenthal, C., Rabitz, H.: High dimensional model representations. Journal of Physical Chemistry A **105** (2001) 7765–7777
11. Ziehn, T., Tomlin, A.S.: Global sensitivity analysis of a 3D street canyon model – Part I: The development of high dimensional model representations. Atmospheric Environment **42** (2008) 1857–1873
12. Benson, S.W.: Thermochemical Kinetics. Methods for the Estimation of Thermochemical Data and Rate Parameters. Wiley, NY (1976)
13. Kee, R.J., Grcar, J.F., Smooke, M.D., Miller, J.A.: A Fortran Program for Modeling Steady Laminar One-Dimensional Premixed Flames. Sandia National Laboratories SAND85-8240 (1985)
14. Sobol', I.M.: Global sensitivity indices for nonlinear mathematical models and their Monte Carlo estimates. Mathematics and Computers in Simulation **55** (2001) 271–280
15. Sobol', I.M.: On the distribution of points in a cube and the approximate evaluation of integrals. USSR Computational Mathematics and Mathematical Physics **7** (1967) 86–112
16. Kucherenko, S.: Application of global sensitivity indices for measuring the effectiveness of quasi-monte carlo methods and paramter estimation. Proceedings of the fifth International Conference on Sensitivity Analysis of Model Output (2007) 35–36
17. Morris, M.D.: Factorial sampling plans for preliminary computational experiments. Technometrics **33** (1991) 161–174
18. Li, G., Wang, S.-W., Rabitz, H., Wang, S., Jaffé, P.: Global uncertainty assessments by high dimensional model representations (HDMR). Chemical Engineering Science **57** (2002) 4445–4460
19. Kee, R., Rupley, F., Miller, J.: Chemkin-II: A Fortran Chemical Kinetics Package for the Analysis of Gas Phase Chemical Kinetics. Sandia National Laboratories SAND89-8009B (1991)
20. Hughes, K.J., Tomlin, A.S., Dupont, V., Pourkashanian, M.: Experimental and modelling study of sulfur and nitrogen doped premixed methane flames at low pressure. Faraday Discussions **119** (2001) 337–352
21. Burcat, A.: Thermochemical Data for Combustion Calculations, Chapter 8. Springer, Berlin (1984)
22. Warth, V., Battin-Leclerc, F., Fournet, R., Glaude, P.A., Come, G.M., Scacchi, G.: Computer based generation of reaction mechanisms for gas-phase oxidation. Computers and Chemistry **24** (2000) 541–560
23. Carstensen, H.H., Dean, A.M.: Rate constant rules for the automated generation of gas-phase reaction mechanisms. Journal of Physical Chemistry A **13** (2009) 367–380
24. Ziehn, T., Hughes, K.J., Griffiths, J.F., Porter, R., Tomlin, A.S.: A global sensitivity study of cyclohexane oxidation under low temperature fuel-rich conditions using HDMR methods. Combustion Theory and Modelling **13** (2009) 589–605

25. Buda, F., Heyberger, B., Fournet, R., Glaude, P.-A., Warth, V., Battin-Leclerc, F.: Modeling of the gas-phase oxidation of cyclohexane. Energy and Fuels **20** (2006) 1450–1459
26. Hughes, K.J., Fairweather, M., Griffiths, J.F., Porter, R., Tomlin, A.S.: The application of the QSSA via reaction lumping for the reduction of complex hydrocarbon oxidation mechanisms. Proceedings of the Combustion Institute **32** (2009) 543–551
27. Snee, T.J., Griffiths, J.F.: Criteria for spontaneous ignition in exothermic, autocatalytic reactions: Chain branching and self-heating in the oxidation of cyclohexane in closed vessels. Combustion and Flame **75** (1989) 381–395
28. Morris, M.D.: Input screening: Finding the important model inputs on a budget. Reliability Engineering and System Safety **91** (2006) 1252–1256
29. Lightfoot, P.D., Cox, R.A., Crowley, R.A., Destriau, M., Hayman, G.D., Jenkin, M.E., Moortgat, G.K., Zabel, F.: Organic peroxy radicals: Kinetics, spectroscopy and tropospheric chemistry. Atmospheric Environment **26** (1992) 1805–1961
30. Frenklach, M., Wang, H., Rabinowitz, M.J.: Optimization and Analysis of large chemical kinetic mechanisms using the solution mapping method – combustion of methane. Progress in Energy and Combustion Science **18** (1992) 47–73
31. Davis, S.G., Joshi, A.V., Wang, H., Egolfopoulos, F.: An optimized kinetic model of H2/CO combustion. Proceedings of the Combustion Institute **30** (2005) 1283–1292
32. Harris, S.D., Elliott, L., Ingham, D.B., Pourkashanian, M., Wilson, C.W.: The optimisation of reaction rate parameters for chemical kinetic modelling of combustion using genetic algorithms. Computer Methods in Applied Mechanics and Engineering **190** (2000) 1065–1090
33. Zsély, I.G., Zador, J., Turanyi, T.: On the similarity of the sensitivity functions of methane combustion models. Combustion Theory and Modelling **9** (2005) 721–738
34. Mittal, G., Sung, C.J., Fairweather, M., Tomlin, A.S., Griffiths, J.F., Hughes, K.J.: Significance of the reaction $HO_2 + CO$ during the combustion of $CO + H_2$ mixtures at high pressures. Proceedings of the combustion Institute **31** (2007) 419–427
35. You, X.Q., Wang, H., Goos, E., Sung, C.J., Klippenstein, S.J.: Reaction kinetics of $CO + HO_2 \rightarrow$ products: Ab initio transition state theory study with master equation modelling. Journal of Physical Chemistry A **111** (2007) 4031–4042
36. Atri, G.M., Baldwin, R.R., Jackson, D., Walker, R.W.: The reaction of OH radicals and HO_2 radicals with carbon monoxide. Combustion and Flame **30** (1977) 1–12
37. Vandooren, J., Oldenhove de Guertechin, L., Van Tiggelen, P.J.: Kinetics in a lean formaldehyde flame. Combustion and Flame **64** (1986) 127–139
38. Mueller, M.A., Yetter, R.A., Dryer, F.L.: Flow reactor studies and kinetic modeling of the $H_2/O_2/NO_x$ and $CO/H_2O/O_2/NO_x$ reactions. International Journal of Chemical Kinetics **31** (1999) 705–724
39. Frenklach, M.: Transforming data into knowledge-process informatics for combustion chemistry. Proceedings of the combustion Institute **31** (2007) 125–140
40. Russi, T., Packard, A., Feeley, R., Frenklach, M.: Sensitivity analysis of uncertainty in model prediction. Journal of Physical Chemistry A **112** (2008) 2579–2588
41. Porter, R., Glaude, P.A., Buda, F., Batin-Leclerc, F.: A tentative modeling study of the effect of wall reactions on oxidation phenomena. Energy and Fuels **22** (2008) 3736–3743
42. Huang, J., Cassiani, M., Albertson, J.D.: Analysis of coherent structures within the atmospheric boundary layer. Boundary-Layer Meteorology **131** (2009) 147–171
43. Coceal, O., Dobre, A., Thomas, T.G., Belcher, S.E.: Structure of turbulent flow over regular arrays of cubical roughness. Journal of Fluid Mechanics **589** (2007) 375–409
44. Xie, Z.T., Coceal, O., Castro, I.P.: Large-eddy simulation of flows over random urban-like obstacles. Boundary-Layer Meteorology **129** (2008) 1–23
45. Letzel, M.O., Krane, M., Raasch, S.: High resolution urban large-eddy simulation studies from street canyon to neighbourhood scale. Atmospheric Environment **42** (2008) 8770–8784
46. Dixon, N.S., Boddy, J.W.D., Smalley, R.J., Tomlin, A.S.: Evaluation of a turbulent flow and dispersion model in a typical street canyon in York, UK. Atmospheric Environment **40** (2006) 958–972
47. Dixon, N.S., Tomlin, A.S.: A Lagrangian stochastic model for predicting concentration fluctuations in urban areas. Atmospheric Environment **41** (2007) 8114–8127

48. Brown, R.J., Bilger, R.W.: An experimental study of a reactive plume in grid turbulence. Journal of Fluid Mechanics **312** (1996) 373–407
49. Ziehn, T., Dixon, N.S., Tomlin, A.S.: The effects of parametric uncertainties in simulations of a reactive plume using a Lagrangian stochstic model. Atmospheric Environment **43** (2009) 5978–5988
50. Anfossi, D., Degrazia, G., Ferrero, E., Gryning, S.E., Morselli, M.G., Trini Castelli, S.: Estimation of the Lagrangian structure function constant c_0 from surface-layer wind data. Boundary-Layer Meteorology **95** (2000) 249–270
51. Sawford, B.L.: Micro-mixing modelling of scalar fluctuations for plumes in homogeneous turbulence flow. Turbulence and Combustion **72** (2004) 133–160
52. Fackrell, J.E., Robins, A.G.: Concentration fluctuations and fluxes in plumes from point sources in a turbulent boundary layer. Journal of Fluid Mechanics **117** (1982) 1–26
53. Poggi, D., Katul, G.G., Cassiani, M.: On the anomalous behavior of the Lagrangian structure function similarity constant inside dense canopies. Atmospheric Environment **42** (2008) 4212–4231
54. Carslaw, D.C.: Evidence of an increasing NO_2/NO_x emissions ratio from road traffic emissions. Atmospheric Environment **39** (2005) 4793–4802

Optimisation and Linear Control of Large Scale Nonlinear Systems: A Review and a Suite of Model Reduction-Based Techniques

Constantinos Theodoropoulos

Abstract The purpose of this paper is twofold: (1) To provide a concise review of methods, recently presented in the literature, which have developed and/or used model reduction technologies for the optimisation and control of large-scale linear and nonlinear systems and (2) to present an overview of the collection of related technologies that have been developed within our group at the University of Manchester concerning the model reduction-based steady-state and dynamic optimisation of large-scale systems, modelled with black-box dynamic and steady state solvers. Furthermore, a new methodology for the linear model predictive control of large-scale non-linear systems will be presented. It relies on adaptive linearisations of the discretised state-space equations using low-order projections of the system's gradients. The tubular reactor has been used as an illustrative example to demonstrate the capabilities of all the above methods due to its high nonlinearity, exhibited through a number of bifurcations at different parameter combinations, and distributed parameter characteristics.

1 Introduction

Model reduction technology is becoming more and more useful nowadays, as becomes obvious from the increasing number of publications, presenting new, or simply using existing model reduction techniques, paradoxically due to (or in spite of) the significant increase in computational power available to the research and to the industrial community alike. While the advent of easily accessible fast computers with large memory capacities facilitates the more detailed modelling of complex processes and phenomena, there is still an even more increasing need for fast computations in order to handle optimisation and control strategies, especially

---

C. Theodoropoulos
School of Chemical Engineering and Analytical Science, University of Manchester,
Manchester M60 1QD, UK
e-mail: k.theodoropoulos@manchester.ac.uk

A.N. Gorban and D. Roose (eds.), *Coping with Complexity: Model Reduction and Data Analysis*, Lecture Notes in Computational Science and Engineering 75, DOI 10.1007/978-3-642-14941-2_3, © Springer-Verlag Berlin Heidelberg 2011

for on-line systems applications. Hence, while simulations with thousands or even millions of unknowns can be routinely handled by modern computational facilities, such detailed simulators cannot be used for optimisation and control, since the corresponding computations would be prohibitively slow to (a) satisfy constraints in process development times (for optimisation applications) and (b) to match the response times of real processes (a common requirement for real-time optimisation and model predictive control applications). Model reduction technologies offer a viable solution bridging the gap between detailed models and fast computations. Some are incorporated as a distinct intermediate step between the (high-dimensional) simulator and the (low-dimensional) optimiser or controller and some are integrated within the corresponding methodology.

In the remainder of this section recent developments and uses of model reduction technologies in optimisation and control strategies will be briefly overviewed and discussed.

1.1 Overview of Recent Developments

The first and most straight-forward concept related to model reduction is the simplification of a detailed model based on physical/chemical principles with the subsequent development of an approximate model, or a sequence of inter-related simplified sub-models, which adequately represent the system's states and/or dynamics. This can be termed as *physical reduction*. In [1] (and references within) issues of model upscaling and downscaling between different levels as well as model aggregation and simplification are discussed. The context here is farm-management optimisation. Efficient upscaling from the process (farm) level to a higher (national) level is achieved by generating a number of simplified models (based on semi-automatic methods) and aggregated models are generated which maintain information on heterogeneity between different local models. The methodology chosen here replaces model variables with constants and the simplified/aggregated models generated are compared with full models, through High Performance Computer (HPC) exhaustive searches. Obviously this reduction requires a substantial off-line computational effort. In [2] an empirical method is presented for a stack of microreactors where the fluid flow and temperature distributions in the channels and walls are averaged and heat exchanges are considered between these fluid and solid phases. Hence, the original high-dimensional partial differential equation (PDE) system is reduced to a small set of ordinary differential equations (ODEs). Another physical reduction approach for the optimal control of large-scale systems is given in [3]. Here the direct multiple shooting method [4], which approximates functions in each time interval with simple linear or piece-wise constant functions is used as an optimiser. Additional reduction is achieved by computing reduced subspaces for the original problem by ignoring (groups of) spatial variables and the corresponding equations (constraints), based on local homogeneities (empirical) information, and projection matrices are constructed. The resulting approximate model is optimised and the solution in mapped back to the full space using projection matrices and interpolation methods as a starting point for optimising the full model.

Alternatively (or even in addition) a multitude of formal mathematical methodologies have been developed for reducing the dimensionality of large-scale systems for optimisation and control applications. Many of these techniques rely on *projection* methods. These, in general, aim to compute, either a priori or adaptively, 'slow', low-dimensional manifolds of the systems' dominant dynamic behaviour and then to project the original systems onto these reduced subspaces, producing reduced systems. The issue for optimisation and control is to adequately compute such manifolds for the parametric ranges required to obtain representative search spaces. Furthermore, special classes of model reduction approaches have been developed specifically for control applications, however, many of these techniques are geared towards linear systems. All these methods constitute the class of *mathematical reduction* methods.

An important model reduction method for control applications is *balanced truncation*, which uses observability and controllability criteria to assess the relative importance of the states of linear time invariant systems. Based on the corresponding measures, projection matrices can be constructed, which are used to truncate the state space of the original system. Systems' sparsity has been exploited to increase the efficiency of the implementation of this method for very large systems as well as parallelisation techniques. (see e.g. [5] and references within). In [6] a good overview of the method and of related advances in the literature is given. In the same paper the balanced truncation method is employed for the control of semi-discretised Stokes equations. Preservation of the stability for the reduced system is shown and the upper bound of the approximation is also computed. The balanced truncation method has also been recently used [7] for linear systems exhibiting stochastic jumps combined with non-convex low-order optimisation to appropriately modify the resulting reduced system to meet certain constraints.

Optimisation technology has been used as a model reduction tool for model predictive control (MPC) applications. In [8] a constrained optimisation algorithm is presented, which is used to compute a low-dimensional basis of the system and the corresponding system states, which minimise the error between the resulting reduced and the original full model. This is a computationally demanding off-line reduction method, which can be further relaxed by making assumptions on the structure of the reduced basis. The reduced model is then coupled with a multi-parametric MPC framework [9], where explicit solutions of the linear MPC problem at a number of different conditions and parameters are solved off-line and are tabulated, to minimise the on-line computations, thus significantly reducing the on-line computational load. In this sense, multi-parametric control can be also viewed as a model reduction method for control applications.

For the optimisation of systems with an infinite number of constraints exhibiting uncertainty, a 'scenario approach' based on random sampling which reduces the system optimisation to a smaller finite number of constraints has been developed [10]. A fixed number of constraints, corresponding to an uncertainty parameter, is randomly chosen and the corresponding reduced optimisation problem is solved. It is also shown that for convex problems all constraints (even the ones not accounted for in the reduced problem) are satisfied. Robust and optimal control problems are also thus handled. In [11] model reduction for optimisation is performed through surface response methods using a radial basis functions network. Appropriate sample points for this response surface are obtained from the original detailed model, while the (multi-objective) optimiser in this framework, which acts upon the reduced model (the response surface) is the multi-objective particle swarm method, a meta-heuristics-based genetic algorithm. Sample points are iteratively added through the procedure to refine the obtained response surface.

Model reduction in the context of control has yielded a number of methodologies to address large-scale problems, including H_∞, and H_2 [12] methods, which in general aim to compute robust controllers, by solving an optimisation problem over the corresponding space, often using linear matrix inequalities (LMI).

Model reduction in the framework of robust linear control has been also formulated as a Hankel norm problem [13], which gives a good approximation in the H_∞ norm (see [14] and references within). In this paper standard linear matrix inequalities (LMI) are used to set up the problem and model reduction is achieved using standard LMI solvers. Moreover, H_∞ model reduction has been studied for a number of systems in the context of controller design, where the resulting reduced model has an a priori prescribed error bound (see [15] and references within). In [15] switched stochastic systems are reduced using average dwell time and Liapunov functions to ensure the stability of the error and an optimisation problem is solved to perform the actual projection in order to obtain the reduced system. In [16] a branch and bound algorithm has been used to minimise the error between the full and the reduced model for continuous and discretised in time models. Hankel singular values and canonical forms have been also used to speed-up convergence. The problem of H_∞ norm minimisation for linear systems with uncertainty (both continuous and discrete) was addressed in [17] using a two-phase iterative procedure, including branch-and-bound optimisation with tight bounds and checks for possible modification of the uncertainty polytope thus constructed. In [18] reduction of the controller for linear discrete-time systems is tackled using an iterative process including convex optimisation techniques and extra sufficient conditions to ensure stability of the resulting closed-loop system. H_∞ methods have been extended to the model reduction for control of singular systems with the use of covariance matrices [19]. The goal was to maintain the impulsive behaviour of the reduced order singular systems and connections with *normal* linear systems were drawn. Furthermore, in [20] the covariance approximation problem has been extended to include a variable parameter matrix and genetic algorithm-based optimisation has been used to reduce the error between the full and the reduced model.

Covariance controller design has been used to construct finite-dimensional approximations for the feedback control of a multi-scale nonlinear system, a sputtering process, modelled by macroscopic (PDE-based) and mesoscopic kinetic Monte Carlo (kMC) simulators [21]. Here, state covariances of the stochastic model are used to fit the parameters of an equivalent stochastic PDE.

For systems with fast and slow time-scales, singular perturbation analysis has been used [22] to yield quasi-steady state approximations for the fast time-scales resulting in reduced DAE systems. These can be further decomposed to intermediate and slow time scales. The approach has been applied to a recompression distillation system.

The concept of *passivity* in systems theory and in particular in control has been used for years (see [23]). Recently it has also been exploited for the stabilisation of nonlinear reaction-transport systems discretising the system in a number of sub-systems and using coarse-graining to account for spatial variations [24]. Passivity is important in control systems for noise attenuation and also for robustness purposes [25]. Model reduction methods for linear systems preserving passivity using rational interpolation and Krylov projection methods were developed in [26], while in [27] the passivity-preserving model reduction schemes developed implemented interpolation at spectral zeros of the system's transfer function, through the computation of invariant subspaces. Moreover, in [28] analytic interpolation with degree constraint was exploited for both discrete and continuous systems and a combination of the methods in [27] and [28] was proposed.

Computing the most important modes of the system, by identifying patterns in the system's responses is at the heart of the proper orthogonal decomposition (POD) method [29]. POD relies on identifying off-line the principal components of the system and hence it is relevant to principal component analysis (PCA). In fact, the two terms POD and PCA are often used intermittently. In POD, appropriate *representative* samples of data need to be collected and the corresponding covariance matrices are constructed. Eigenvalue analysis of these matrices provides the number of modes, which represent the system with desired accuracy, i.e. capture the system's energy for the parametric space investigated. *Global* empirical global basis functions are then computed using the method of snapshots [30] or directly from Singular Value Decomposition of the matrix. A Galerkin projection of the original systems onto the computed eigenfunctions will produce the reduced system as a linear combination of time coefficients and of the basis functions. POD is a very popular model reduction technique and has been used extensively for optimisation and control applications.

Since the sampling process is largely empirical and representative samples are critical for optimisation and control (where wide parametric ranges are considered in general) several methods have been reported to guide the parametric process. An optimisation process has been proposed in [31] to find the points of maximum difference between the full and reduced models and to update the sample appropriately. This *greedy* optimisation algorithm is reduced in itself since it uses error estimators instead of the full model for the constraints. In [32] the case of linear full models was examined and the optimisation problem constructed was an unconstrained one. In [33] an adaptive greedy algorithm was introduced. In [34] the POD global basis functions are computed through a goal-oriented optimisation problem. For linear relation between states and outputs an equivalent unconstrained optimisation problem is solved using adjoint methods. An adaptive iterative method for updating the basis functions during the feedback control of distributed processes with a recursive addition of snapshots was addressed in [35]. In [36] time coefficients were first computed solving small unconstrained optimisation problems, and they were then interpolated and used in a process optimisation procedure. The method was applied to the parameter optimisation of underwater bubble explosions. Multi-objective optimisation has been performed for nonlinear structure dynamics [37] using a two-level reduction: The first level used POD to reduce the full models and the second level used adaptive surface response methods on the reduced models to produce polynomial equations that were used in the optimisation procedure. In [38] POD as well as Laplacian spectral decomposition (LSD) have been used to construct reduced models, which were used to design state observers for tubular reactors, while in [39] POD-based reduced models were used in the optimal control of convection-diffusion-reaction processes. PCA has been used in [40] to reduce high-dimensional kMC simulators in conjunction with self-organised mappings, which group similar surface patterns to construct discrete state spaces. The reduced models were employed in the optimisation of a film deposition process. PCA has also been recently implemented in a large-scale nonlinear dynamic optimisation software, which uses collocation and finite elements in conjunction with non-linear programming techniques [41].

Krylov methods used to efficiently solve large-scale sparse linear systems provide efficient model reduction platforms. [42] discusses the combined use of balanced truncation SVD-based methods and iterative Krylov solvers, which are used to produce a rational interpolation of certain moments of the reduced model, to obtain reduced models of linear dynamic systems. Krylov methods have also been proposed [43] to solve Lyapunov matrix equations for large-scale linear time-invariant systems.

1.2 A Novel Framework for Optimisation and Control of Nonlinear Systems

We have recently developed a suite of optimisation and control methodologies based on the *equation-free* approach (see [44] where an overview of this method and of relevant applications for multi-scale systems are extensively discussed). The underlying methodology is based on the Recursive Projection Method (RPM) [45], developed to accelerate and stabilise fixed-point procedures. The main idea is to recursively compute a first order Krylov approximation of the invariant subspace of the system directly from its successive iterates. Based on the natural separation of time scales that many systems of engineering and scientific interest exhibit, the dimension of the dominant modes (which are captured by the computed invariant subspace) is low compared to the original high-dimensional system, resulting in the computation of a low-order *dominant* subspace. A schematic of the eigenspectrum of a dynamic system depicting a clear separation of scales between the dominant (red) and the rest of the eigenvalues (blue) is depicted in Fig. 1.

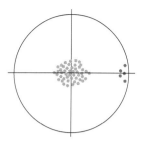

Fig. 1 Eigenspectrum of system's dynamics depicting a gap between the dominant the rest of the eigenmodes

Projecting the system onto this invariant subspace low-order projections of the full-scale Jacobians are produced and are then used to perform Newton iterations. These are subsequently combined with the fixed-point iterations into a coupled Newton–Picard process to significantly speed-up and stabilise the convergence to the fixed point even if this is unstable. This idea of computing, essentially on-line, a low-dimensional approximation of the system's subspace has been further developed in our group, in a series of methodologies presented in the following sections.

2 Steady State Optimisation Using Black-Box Dynamic Solvers

The model reduction methodology presented in this section [46] is beneficial for the gradient-based optimisation of large-scale nonlinear systems simulated with black-box dynamic solvers and/or explicit time integrators (e.g. Runge–Kutta methods of any order etc.). In these cases the (large-scale) Jacobian and consecutively Hessian matrices needed in the optimisation algorithms are either not accessible by the user (for black-box simulators) or, in the case of explicit solvers, do not actually exist. It should be mentioned here, that stochastic methods, which can easily interface with input/output simulators, usually require excessive computational time due to the large number of function calls. Obtaining Gradient and Hessian matrices through direct numerical perturbations, on the other hand, would be an enormous task for large systems. Finally, dynamic solvers cannot always be integrated to the corresponding steady state (as in the case of instabilities and multiple steady states) hence function evaluations for steady state optimisation might not be possible.

2.1 The Recursive Projection Reduced Hessian Algorithm

Our optimisation methodology, the Recursive Projection Reduced Hessian (RPRH) algorithm [46], is based on the reduced Hessian algorithm [47, 48], which provides an efficient method to optimise systems with few degrees of freedom, using low-dimensional projections of the system's Hessians onto the subspace of the decision variables. However, this method still requires the construction and inversion of large system Jacobians. Taking into account that the black-box dynamic simulator can be represented by a fixed-point procedure, since we are interested in the steady states of the integrator, we can write it as follows:

$$\mathbf{u} = \mathbf{G}(\mathbf{u}, \mathbf{z}) \tag{1}$$

u being the states and z the degrees of freedom of the problem. The optimisation problem considered then is the following:

$$\min_{\mathbf{u}, \mathbf{z}} f(\mathbf{u}, \mathbf{z}) \quad \text{s.t.} \quad \mathbf{u} - \mathbf{G}(\mathbf{u}, \mathbf{z}) = 0. \tag{2}$$

In this work [46] we have first coupled the dynamic solver with the RPM to accelerate convergence to the corresponding steady states (even to unstable ones) through Newton–Picard iterations, using H, the low-dimensional restriction of the system's Jacobian, G_u onto the invariant subspace $\mathbf{P}$.

$$H = PG_u P \tag{3}$$

Here P is the orthogonal projector of $\mathbf{P}$. In practice a low-dimensional basis, $\hat{Z}$, for the subspace P is constructed either through the RPM first order Krylov approximation [45] or through proper subspace iterations, while H is computed through a few numerical directional perturbations. A coordinate basis, $Z^\star$ of the subspace of the decision variables can then be constructed, using only the low-dimensional H without the need to construct and to invert large Jacobians:

$$Z^\star = \begin{bmatrix} \hat{Z} (I - H)^{-1} \hat{Z}^T G_z \\ I \end{bmatrix} \tag{4}$$

From (4) we can get:

$$Z^\star = \begin{bmatrix} \hat{Z} & 0 \\ 0 & I \end{bmatrix} \begin{bmatrix} (I - H)^{-1} \hat{Z}^T G_z \\ I \end{bmatrix} = Z_{ext} Z_r \tag{5}$$

Here Z_{ext} is the basis, $\hat{Z}$ of the subspace $\mathbf{P}$ extended to include the space of degrees of freedom and Z_r the 'reduced' coordinate basis computed only with the use of the low-dimensional Jacobian H. The RPRH procedure is then given by the pseudo-code in Fig. 2. From step 5 in Fig. 2 and (5) it can be seen that the reduced Hessian is computed from:

$$\hat{B}_R = Z^{\star T} B Z^\star = Z_r^T Z_{ext}^T B Z_{ext} Z_r \tag{6}$$

It can be easily seen that this denotes a double-projection step firstly onto the *dominant* invariant subspace and secondly onto the subspace of the decision variables.

Optimisation and Linear Control of Large Scale Nonlinear Systems   45

1.	Choose initial guesses for x_0 and B_0
2.	Compute $x, H, \hat{Z}$ using the dynamic simulator within RPM
3.	Evaluate $f, \nabla f$
4.	Compute the basis $Z* = \begin{bmatrix} \hat{Z}(I-H)^{-1}\hat{Z}^T G_z \\ I \end{bmatrix}$
5.	Calculate the reduced Hessian: $\hat{B}_R = Z*^T B Z*$
6.	Solve the unconstrained QP subproblem: $\min\left(Z*^T \nabla f\right) p_{z*} + \frac{1}{2} p_{z*}^T \hat{B}_R p_{z*}$ s.t. $x^L - x \le Z* p_{z*} \le x^U - x$
7.	Compute an estimate for the Lagrange multipliers: $(I-H)\varphi = -Z*^T Y^T \nabla f$
8.	Update the solution: $x = x + Z* p_{z*}$
9.	Check for convergence. If convergence is not achieved go to (2).

Fig. 2 Pseudo-code depicting the RPRH procedure

This double projection side-steps the need for any of the large-scale first and second derivative matrices and significantly reduces the computational cost of the optimisation procedure. Furthermore, from step 7 in Fig. 2 it can be seen that a low-dimensional projection of the Lagrange multipliers is needed, which can be directly computed from H and $\hat{Z}$ incurring further computational savings.

2.2 Case Study

The tubular reactor where an exothermic reaction $A \rightarrow B$ occurs [49] was used as an illustrative case study to demonstrate the capabilities of RPRH since it has a rich parametric behaviour including a number of bifurcations (turning points, Hopf points), instabilities and multiple steady states making the optimisation problem challenging. The dynamic PDEs for the mass and energy balances are given below:

$$\frac{\partial x_1}{\partial t} = \frac{1}{Pe_1} \frac{\partial^2 x_1}{\partial y^2} - \frac{\partial x_1}{\partial y} + Da\left(1 - x_1\right) \exp\left(\frac{x_2}{1 + \frac{x_2}{\gamma}}\right) \quad (7a)$$

$$\frac{\partial x_2}{\partial t} = \frac{1}{Le Pe_2} \frac{\partial^2 x_2}{\partial y^2} - \frac{1}{Le} \frac{\partial x_2}{\partial y} - \frac{\beta x_2}{Le} + CDa\left(1 - x_1\right) \exp\left(\frac{x_2}{1 + \frac{x_2}{\gamma}}\right) + \frac{\beta x_{2_w}}{Le} \quad (7b)$$

with boundary conditions

$$\frac{\partial \mathbf{x_1}}{\partial y} - Pe_1 \mathbf{x_1} = 0, \quad \frac{\partial \mathbf{x_2}}{\partial y} - Pe_2 \mathbf{x_2} = 0 \quad \text{at} \quad y = 0 \tag{8a}$$

$$\frac{\partial \mathbf{x_1}}{\partial y} = 0, \quad \frac{\partial \mathbf{x_2}}{\partial y} = 0 \quad \text{at} \quad y = 1 \tag{8b}$$

Here, Da is the Damköhler number, Le the Lewis number, Pe_1 and Pe_2 are the Peclet numbers for mass and heat transport, β a dimensionless heat transfer coefficient, B_1 is the dimensionless adiabatic temperature rise, x_{2_w} the dimensionless adiabatic wall temperature, and y the dimensionless longitudinal coordinate. The parameter values were: $Le = 1.0$, $Pe_1 = 5.0$, $Pe_2 = 5.0$, $\gamma = 20.0$, $\beta = 1.50$, $C = 12.0$ and $x_{2_w} = 0.0$. The system was discretised in 250 computational nodes with finite differences, resulting in a 500 states ODE system. The ODEs were then integrated using explicit 4th order Runge–Kutta so that Jacobian and Hessian matrices were not directly available. A number of computational results were produced considering cases with one degree of freedom, (Da), three degrees of freedom (three cooling zones which can be independently controlled through $x_{2_{w_j}}$, $j = 1 \ldots 3$ – see (19)) and six degrees of freedom $(Da, Pe_1 = Pe_2, Le$ and $x_{2_{w_j}}$, $j = 1 \ldots 3)$. The optimal points computed in all cases had excellent agreement with standard reduced Hessian methods at a fraction of the computational cost. Furthermore, the behaviour of the algorithm for different subspace sizes, m, and different reporting times, T, was investigated. It was found that shorter reporting times and larger subspaces ($m = 10$ as opposed to $m = 6$) increased the speed of the solution since they required less iterations despite the slightly increased cost of computing larger subspaces. The issue of computing feasible points, i.e. solving the steady state system at each iteration vs. performing a single Newton step at each iteration, therefore computing feasible points only near the optimum was investigated. It was found that the latter approach needs more iterations to converge, however the overall computational times are shorter than in the former case. Furthermore, the issue of using forward vs. central differences for the numerical computation of the reduced Jacobians was also studied. While the performance of the algorithm was similar in both cases, central differences should exhibit a more stable behaviour albeit at extra computational cost (see [46] for extensive results). It was therefore concluded that RPRH could very efficiently handle black-box integrators and effectively turn them to steady state optimisers.

3 Dynamic Optimisation with Black-Box Integrators

3.1 Overview of the Methodology and Algorithmic Details

The model reduction-based optimisation methodology presented in the previous section has been extended to handle the gradient-based dynamic optimisation of large-scale systems simulated with input/output integrators [50]. Our optimisation problem then becomes:

$$\min_{\mathbf{u}(t),\mathbf{z}} \int_{t_0}^{t_f} \Phi\left(t, \mathbf{u}\left(t\right), \mathbf{z}\right) dt \tag{9a}$$

$$\text{s.t. } \frac{\partial \mathbf{u}}{\partial t} = \mathbf{F}\left(t, \mathbf{u}\left(t\right), \mathbf{z}\right) \tag{9b}$$

$$\mathbf{u}\left(t_0\right) = \mathbf{u}_0 \tag{9c}$$

The meaning of u and z is the same as in the previous section. A methodology was developed based on the multiple-shooting approach [4] which has been shown to be more robust than the single-shooting method, which can get hampered by numerical instabilities especially in the case of highly non-linear problems. Multiple shooting requires the partitioning of the time horizon in a number of intervals, the system is integrated over each interval and continuous profiles in time over the whole horizon are enforced with appropriate constraints linking the states at the end of each (internal) interval to those at the beginning of the next interval. The gradients of these constrains then form a block-structure Jacobian. The corresponding NLP problem is, hence, very large and increases as the number of time intervals chosen increases. Obviously the most computationally costly step here is the computation of the very large-scale system gradients and of the corresponding Hessian matrices. While sensitivity analysis and automatic differentiation can be used to speed-up this bottleneck there is still need for improvement. In this work we combined the multiple-shooting technique with model reduction technology using Newton–Picard iterations [51] to compute projections of the large block-Jacobians onto the (low-dimensional) dominant subspaces of each time interval. Then reduced Hessian methods [48] were employed to perform a second projection onto the subspace of the (relatively few) control parameters of the problem. This double projection results in a significantly reduced unconstrained quadratic subproblem, which is solved at each iteration, with significant computational gains. Discretising the dynamic optimisation problem given in (9) in a number of N time intervals $[t_i, t_{i+1}]$ for the multiple shooting formulation we need to integrate the dynamic system in (9b) to get:

$$\mathbf{u}_{i+1} = \mathbf{G}\left(t_i, \mathbf{u}_i, t_{i+1}, \mathbf{z}\right) \tag{10}$$

where $\mathbf{u}_{i+1}$ is the solution of the integration at t_{i+1} and $\mathbf{G}\left(t_i, u_i, t_{i+1}, z\right)$ is a non-expansive map. To ensure continuity of the time profiles over the whole time horizon the following constrains are imposed:

$$\mathbf{r}_{i+1}\left(t_i, \mathbf{u}_i, t_{i+1}, \mathbf{u}_{t+1}, \mathbf{z}\right) = \mathbf{u}_{i+1} - \mathbf{G}\left(t_i, \mathbf{u}_i, t_{i+1}, \mathbf{z}\right) = 0. \tag{11}$$

Hence, the dynamic optimisation problem is reformulated as:

$$\min_{u_1,\ldots,u_N,z} \sum_{i=0}^{N-1} \Phi\left(t_i, u_i, t_{i+1}, u_{i+1}, z\right) \tag{12a}$$

$$\text{s.t. } \mathbf{r}_{i+1}\left(t_i, \mathbf{u}_i, \mathbf{u}_{i+1}, t_{i+1}, \mathbf{z}\right) = 0 \ \forall i = 0,\ldots,N-1 \tag{12b}$$

$$\mathbf{u}(t_0) = \mathbf{u}_0 \tag{12c}$$

where (12a) is the discretised integral of the objective function in (9a). The Jacobian of the constraints (12b) with respect to the state variables $\mathbf{u}$ has the following structure:

$$J_u = \begin{bmatrix} I & 0 & \cdots & & 0 \\ -\frac{\partial \mathbf{G}(t_1, u_1, t_2, \mathbf{z})}{\partial u_1} & I & \cdots & & 0 \\ \vdots & \vdots & \ddots & & \vdots \\ \vdots & 0 & & I & 0 \\ 0 & & \cdots & -\frac{\partial \mathbf{G}(t_{N-1}, u_{N-1}, t_N, \mathbf{z})}{\partial u_{N-1}} & I \end{bmatrix} \tag{13}$$

The gradient vector with respect to the degrees of freedom is given by:

$$J_z = \begin{bmatrix} \frac{\partial \mathbf{G}(t_0, u_0, t_1, z)}{\partial \mathbf{z}} \\ \vdots \\ \frac{\partial \mathbf{G}(t_{N-1}, u_{N-1}, t_N, z)}{\partial \mathbf{z}} \end{bmatrix} \tag{14}$$

Each block of the Jacobian J_u is an $n \times n$ matrix and its total size is $Nn \times Nn$, while the size of J_z is Nn. The Newton–Picard procedure developed in [51] is then employed to (a) compute the low-order dominant subspaces $\mathbf{P_i}$ for each time interval and to (b) compute solutions for the continuity constraints, performing Newton iterations on the dominant subspaces and Picard iterations on their orthogonal complements $\mathbf{Q_i}$. In this procedure we have assumed that the dominant subspaces $\mathbf{P}_i$ in all time intervals are of size m and subspace iterations [52] have been used for the computation of the corresponding dominant eigenmodes, V_i for each time interval. Thus, the restrictions of the Jacobian blocks $\frac{\partial \mathbf{G}(t_i, u_i, t_{i+1}, z)}{\partial u_i}$ on the subspaces $\mathbf{P_i}$, $H_i = P_i \frac{\partial \mathbf{G}(t_i, u_i, t_{i+1}, z)}{\partial u_i} P_i$ can be then computed with a few directional numerical perturbations. Central derivatives have been used for this purpose. This way the reduced block Jacobian $\overline{W}_u$ of size $Nm \times Nm$ can be calculated with computational efficiency.

$$\overline{W}_v = \begin{bmatrix} I & 0 & 0 & \cdots & 0 \\ -H_1 & I & 0 & \cdots & 0 \\ \vdots & \vdots & \ddots & \cdots & \vdots \\ \vdots & 0 & -H_{N-2} & I & 0 \\ 0 & \cdots & 0 & -H_{N-1} & I \end{bmatrix} \tag{15}$$

Furthermore, the projection of the vector of gradients with respect to the degrees of freedom onto the dominant subspace of each time interval is given by:

Optimisation and Linear Control of Large Scale Nonlinear Systems

$$\overline{W}_z = \begin{bmatrix} V_1^T & \cdots & 0 \\ \vdots & \ddots & \vdots \\ 0 & \cdots & V_N^T \end{bmatrix} \begin{bmatrix} \frac{\partial G(t_0,u_0,t_1,z)}{\partial z} \\ \vdots \\ \frac{\partial G(t_{N-1},u_{N-1},t_N,z)}{\partial z} \end{bmatrix}. \tag{16}$$

The low-dimensional coordinate basis for the reduced Hessian step is then given by:

$$\overline{Z} = \begin{bmatrix} -\overline{W}_v^{-1}\overline{W}_z \\ I \end{bmatrix}, Y = \begin{bmatrix} I \\ 0 \end{bmatrix}. \tag{17}$$

where $\overline{Z} \in \mathbb{R}^{N(m) \times dof}$ and $Y \in \mathbb{R}^{(N(n)+dof) \times n}$. A reduced unconstrained QP sub-problem is then solved, the search direction, $\overline{d}_z$ is computed and the solution is updated. The projection of the Lagrange multipliers onto the dominant subspaces needed to update the reduced Hessian is given by:

$$\phi = -\overline{W}_v^{-1} \begin{bmatrix} V_1^T & \cdots & 0 \\ \vdots & \ddots & \vdots \\ 0 & \cdots & V_N^T \end{bmatrix} Y^T \nabla \Phi \tag{18}$$

where $\phi \in \mathbb{R}^{N(m)}$. A pseudo-code of our model-reduction-based dynamic optimisation process is given in Fig. 3.

---

1. Choose initial guesses for the dependent and independent variables: $\begin{bmatrix} u^0 \\ z^0 \end{bmatrix}$

2. Use the Newton-Picard algorithm with the dynamic solver for the calculation of $\overline{W}_v$ and $\overline{W}_z$

3. Compute the basis Z: $Z = \begin{bmatrix} -\overline{W}_v^{-1}\overline{W}_z \\ I \end{bmatrix}$ and the Reduced Hessian

4. Solve the QP subproblem:

$$\min_{d_{\overline{z}}} \left(\overline{Z}^T \nabla f \right)^T + \frac{1}{2} d_{\overline{z}}^T \left(\overline{Z}^T B \overline{Z} \right) d_{\overline{z}} \quad \text{s.t.} \quad \begin{bmatrix} \left(u^L - u^k\right) \\ \left(z^L - z^k\right) \end{bmatrix} \leq \overline{Z} d_{\overline{z}} \leq \begin{bmatrix} \left(u^U - u^k\right) \\ \left(z^U - z^k\right) \end{bmatrix}$$

5. Calculate the m Lagrange multipliers, φ, from: $\overline{W}_v \varphi = - \begin{bmatrix} V_1^T & \cdots & 0 \\ \vdots & \ddots & \vdots \\ 0 & \cdots & V_N^T \end{bmatrix} Y^T \nabla f$

6. Update the solution vector: $\begin{bmatrix} u^{k+1} \\ z^{k+1} \end{bmatrix} = \begin{bmatrix} u^k \\ z^k \end{bmatrix} + \overline{Z} d_{\overline{z}}$

7. Check for convergence. If convergence is not achieved go to step (2)

Fig. 3 Pseudo-code depicting the dynamic optimisation algorithm

3.2 Case Study

The tubular reactor problem given by (7) and initial conditions $\mathbf{x}_1(0) = \mathbf{x}_2(0) = 0$ was also used to illustrate the dynamic optimisation methodology. Here also three cooling zones were considered, given by (19).

$$x_{2_w}(y) = \sum_{j=1}^{3} \left[H \left(y - y_{j-1} \right) - H \left(y \right) - y_j \right] x_{2_{w_j}}(t) \tag{19}$$

where $H(\cdot)$ is the Heaviside function, $y_0 = 0$, $y_1 = 1/3$, $y_2 = 2/3$, $y_3 = 1$ and $x_{2_{w_j}}(t)$, $j = 1,\ldots,3$ is the dimensionless temperature at each cooling zone. A parametrisation of the control functions using combinations of two types of curves with a total of six parameters as in [53] was used.

$$x_{2_{w_j}}(t) = x_{2_{w_j}}(t_f) - \left[x_{2_{w_j}}(t_f) - x_{2_{w_j}}(t_0) \right] \left[1 - \frac{t}{t_f} \right]^{A_1} \tag{20}$$

$$x_{2_{w_j}}(t) = x_{2_{w_j}}(t_0) - \left[x_{2_{w_j}}(t_0) - x_{2_{w_j}}(t_f) \right] \left[\frac{t}{t_f} \right]^{A_2} \tag{21}$$

The optimisation problem was defined as finding the optimal jacket temperature dynamics that would maximise exit conversion at a given time $t_f = 1.5$. The same parameters as in the previous section were also used here as well as the same FD-based integrator. A number of optimisation studies were performed for this system using 5 and 10 time intervals and different dimensions for the dominant subspaces ($m = 6$ and $m = 8$) (see [50]). Here we present a slightly different set of results, in essence a different local optimum. The number of time intervals chosen were $N = 5$ and the size of the subspaces in all intervals was $m = 6$. The bounds for the state variables and for the parameters were: $0 \leq \mathbf{x}_1 \leq 1, 0 \leq \mathbf{x}_2 \leq 20, 0 \leq x_{2w_i}(t_0) \leq 6$, $0 \leq x_{2w_i}(t_f) \leq 3, 0 \leq x_{2w_i}(t_{int}) \leq 3, 0 \leq t_{int} \leq 3, 1 \leq A_1 \leq 2, 1 \leq A_2 \leq 2$. The procedure converged in 12 iterations and the optimal exit value of x_1 at t_f found was $x_1(t_f) = 0.99697871$. In Table 1 the optimal parameters for the control function corresponding to the three cooling jackets are shown and the corresponding cooling profile dynamics for each jacket are depicted in Fig. 4.

Table 1 Optimal parameters for the control function x_{2w_j} for each cooling jacket

Parameter	Jacket $j = 1$	Jacket $j = 2$	Jacket $j = 3$
$x_{2_{w_j}}(t_0)$	6.0	0.0	6.0
$x_{2_{w_j}}(t_f)$	3.0	3.0	0.0
t_{int}	1.5	0.498	1.3712
$x_{2_{w_j}}(t_{int})$	3.0	3.0	3.0
A_1	1.0	1.999	1.206
A_2	2.0	2.0	2.0

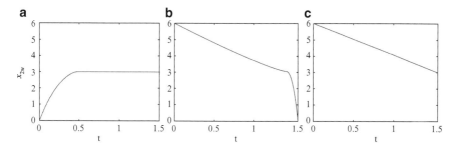

Fig. 4 Control profiles for (**a**) Cooling jacket 1, (**b**) Cooling jacket 2 and (**c**) Cooling jacket 3, for $N = 5$ and $m = 6$

The results presented here and in [50] have been compared with conventional optimisation results using dynamic simulators, with excellent agreement. It should also be noted that in [50] the application of this optimisation methodology to a microscopic kinetic Monte Carlo dynamic simulator was applied verifying the applicability of this technology to multi-scale simulators.

4 Steady State Optimisation with Black-Box Steady State Solvers

A natural extension of the optimisation methodologies presented in the two previous sections arises from the need to perform steady state optimisation for large-scale systems, however, using steady state (input/output) simulators. This is an obvious requirement in process design and also in many engineering applications, where sophisticated simulators exist, but they are not, in general, amenable for optimisation and control applications. The methodology we have developed for this purpose [54] is geared towards solvers using iterative linear algebra (Krylov methods), however even direct solvers can benefit from this technique as will be demonstrated below. The two-step projection approach and the combination of deterministic NLP reduced Hessian-based optimisation with the on-line computation of dominant system subspaces is at the heart of this technique also. Furthermore, the direct incorporation of non-linear inequality constraints has been considered in the algorithm as well as certain algorithmic variations that reduce computational cost.

4.1 Methodology and Algorithmic Details

The optimisation problem considered in this case is given by:

$$\min_{\mathbf{u},\mathbf{z}} f(\mathbf{u},\mathbf{z}) \quad \text{s.t.} \quad \mathbf{G}(\mathbf{u},\mathbf{z}) = 0 \quad \text{and} \quad u^L \leq u \leq u^U, \quad z^L \leq z \leq z^U. \quad (22)$$

Here **u** and **z** have the same meaning as in the previous sections and $\mathbf{G} \in \mathbb{R}^{N \times dof} \rightarrow \mathbb{R}$ and is twice differentiable. The model reduction methodology consists of the following steps: The input/output solver is used to compute a steady state solution. A basis $\hat{Z}$ for the invariant subspace $\mathbf{P}$ of this steady state is then computed through subspace iterations using only the residual vector of the steady state solution, without need to provide the system's large-scale Jacobian. The reduced Jacobian is then computed through a small number of fast numerical perturbations:

$$H = \hat{Z}^T G_u \hat{Z} \tag{23}$$

The coordinate basis needed to employ the reduced Hessian optimisation is given by:

$$Z = \begin{bmatrix} -\hat{Z} H^{-1} \hat{Z}^T \nabla_z \mathbf{G}^T \\ I \end{bmatrix}, Y = \begin{bmatrix} I \\ 0 \end{bmatrix}. \tag{24}$$

Here we can also easily show that Z is derived from a double projection firstly onto the basis of the system's invariant subspace at the steady state and secondly onto the subspace of the system's decision variables. An unconstrained optimisation problem is then solved to obtain the new search direction with respect to the degrees of freedom, $\mathbf{p}_z$, (it can be shown that the search direction with respect to the state variables, $\mathbf{p}_y$ vanishes under certain conditions) and a projection of the Lagrange multipliers, $\phi \in \mathbb{R}^m$ onto the invariant subspace is computed by:

$$H\phi = -\hat{Z} Y^T \nabla f \tag{25}$$

The steps of this optimisation algorithm are given in Fig. 5.

To compute an estimate of the Lagrange multipliers for the next iteration, one needs to update the basis Z. Computing, however, the dominant subspace is the most expensive step of the algorithm. To reduce the basis computations we make

1.	Choose initial guesses for x_0 and B_0
2.	Compute the steady state using the black-box simulator and evaluate $x, f, \nabla f, \hat{Z}, H$
3.	Compute the basis Z : $Z = \begin{bmatrix} -\hat{Z}\mathbf{H}^{-1}\hat{Z}^T\nabla_z\mathbf{G}^T \\ I \end{bmatrix}^T$
4.	Calculate the reduced Hessian, $\hat{B}_R$, using Z
5.	Solve the QP subproblem: $\min_{\mathbf{p}_z} \left(Z^T\nabla f_z\right)^T \mathbf{p}_z + 1/2\mathbf{p}_z^T B_R\mathbf{p}_z$ s.t. $(\mathbf{x}^L - \mathbf{x}) \le Z\mathbf{p}_z \le (\mathbf{x}^U - \mathbf{x})$
6.	Update $\hat{Z}, H$ and calculate an estimate of the Lagrange multipliers: $H\phi = -\hat{Z}^T Y^T \nabla f$
7.	Update the solution: $\mathbf{x} = \mathbf{x} + Z\mathbf{p}_z$
8.	Check for convergence. If convergence is not achieved go to (2).

Fig. 5 Pseudo-code depicting the steady state model reduction-based optimisation method

Optimisation and Linear Control of Large Scale Nonlinear Systems

the assumption that the basis calculated after the QP step is a good approximation of the basis at the corresponding feasible point. This way Z is updated only once per step and the calculation of the Lagrange multipliers (step 6) can be moved to the next iteration, after step 2. This assumption incurs loss of accuracy. Hence, in practice we suggest the use of the modified version for the first few iterations while the original version, depicted in Fig. 5 should be used in the vicinity the optimum, where the cost for updating Z is small.

We have also here considered the steady state optimisation problem including inequality constraints. Then (22) now becomes:

$$\min_{\mathbf{u}, \mathbf{z}} f(\mathbf{u}, \mathbf{z}) \quad \text{s.t.} \quad \mathbf{G}(\mathbf{u}, \mathbf{z}) = 0, \quad \mathbf{W}(\mathbf{u}, \mathbf{z}) \leq 0, \tag{26}$$

$$u^L \leq u \leq u^U, \quad z^L \leq z \leq z^U.$$

Here, we adopted an approach based on constraint aggregation (KS) functions [55]. The KS function aggregates all inequality constraints and replaces them with a single one. The two equivalent forms of the KS function are:

$$KS(W_j) = \frac{1}{\rho} ln[\sum exp(\rho W_j)] \quad \text{and} \tag{27a}$$

$$KS(W_j) = M + \frac{1}{\rho} ln[\sum exp(\rho(W_j - M))] \tag{27b}$$

The summation in (27) is over all the inequality constraints ρ is a parameter and $M = \max(W_j)$. The KS function can be incorporated in the original objective function, f, in order to eliminate all inequality constraints. In our formulation we use a projection of the KS function onto the dominant subspace. This incurs little extra cost to the overall optimisation, as it does not affect the calculation of the basis which is the most expensive step. It is worth mentioning that the algorithm convergence is better when the variables used in the KS function are scaled. That also simplifies the selection of parameters ρ and M.

4.2 Case Study

The tubular reactor problem is also given here as an example to provide continuity with the previous sections. The left-hand side of the (7) was set to 0, the problem was also here discretised with Finite Differences in 250 computational nodes and the resulting algebraic equations were solved using a direct Newton–Raphson solver. The problem with the three cooling zones which can be varied independently through x_{2w_i}, $i = 1, \ldots 3$ was also solved here with the objective to maximise the exit concentration x_1. The parameters considered were, $Da = 0.1$, $Le = 1.0$, $Pe_1 = Pe_2 = 7.0$, $\gamma = 10.0$, $\beta = 1.50$, $C = 12.0$, while $\mathbf{x}_1$ was bounded between 0 and 1 and $\mathbf{x}_2$ between 0 and 8. 7 iterations were required for our scheme to converge and the final value of the objective function was $f = 0.975$. The convergence data for each iteration including the values of the three decision variables, those of the objective function as well as the error $||Z^\star \mathbf{p}_z||$ are given in Table 2. The corresponding convergence curve is shown in Fig. 6, where a fast quadratic-like convergence is observed after the third iteration. The modification to the process mentioned above led to a 15% speed-up of the algorithm compared to its original version, while both versions were significantly faster (more than 2.5 times) than the original reduced Hessian method. The corresponding computed optimal profiles for x_1, x_2 and x_{2w_i} are shown in Fig. 7.

A second case was also studied based on the same tubular reactor model. Here a set of non-linear inequalities were included

$$(x_2|_k + x_2|_{k+1})^2 \leq 160, \quad \text{all} \quad k \in \{125 \ldots 130\} \tag{28}$$

where k denotes the number of the computational node. The optimisation method for this case converged after 12 iterations. The optimum f was 0.9932 and all the inequality constraints are satisfied at this point. The corresponding optimal profiles for this case as well as those of the equivalent problem without the nonlinear inequality constraints are shown in Fig. 8.

Table 2 Convergence data for each iteration for the model reduction steady state optimisation procedure

Iteration	$x_{2w,1}$	$x_{2w,2}$	$x_{2w,3}$	f	$||Z^{*T}\nabla F||$	$||Z^* p_z||$
1	2.000	2.000	2.000	0.885	8.133E-02	1.543E+00
2	2.067	2.052	2.024	0.963	2.304E-02	2.410E+01
3	4.000	2.099	0.000	0.984	1.244E-02	1.257E+01
4	4.000	2.799	4.000	0.975	1.270E-02	2.626E-01
5	4.000	2.951	4.000	0.975	1.271E-02	1.660E-02
6	4.000	2.947	4.000	0.975	1.271E-02	2.580E-04
7	4.000	2.947	4.000	0.975	1.274E-02	1.210E-05

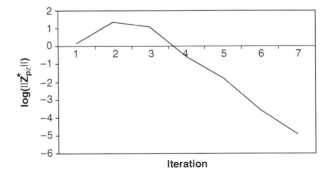

Fig. 6 Convergence curve for the model reduction steady state optimisation procedure

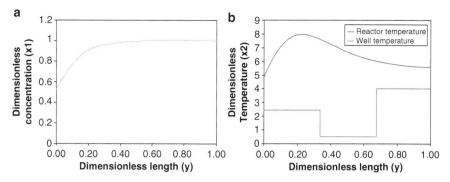

Fig. 7 Profiles of dimensionless (**a**) concentration, x_1, (**b**) temperature x_2 and x_{2w_i} at the optimum point

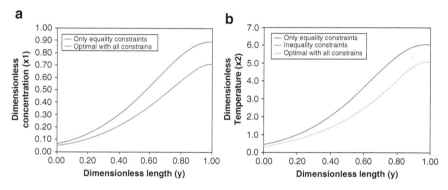

Fig. 8 Profiles of dimensionless (**a**) concentration, x_1, (**b**) temperature x_2 and x_{2w_i} at the optimum point with and without nonlinear inequality constraints

As it can be seen from the above results the steady state optimisation methodology can significantly reduce computational times for the optimisation of black-box steady state simulators. We are currently developing a penalty-function based approach to handle nonlinear inequality constraints. Furthermore, we have demonstrated this method for Krylov-based iterative linear algebra solvers. These results are being submitted for publication in scientific journals.

5 Model Reduction and Linear Control of Non-Linear Systems

The model reduction framework underlying the technologies presented in the previous sections is employed here to construct linear MPC algorithms for non-linear large-scale systems. There are many MPC implementations, all of which compute control actions as the solutions of finite horizon open-loop optimal control problems. The standard MPC formulation requires a linear system model and linear constraints. However, most systems of engineering and scientific interest are nonlinear. Nonlinear MPC is limited to rather small systems [56] since its computational cost is high for on-line applications. Linearisation at the set point may produce a poor approximation for the time horizon considered. Self-tuning control algorithms can account for the nonlinearity of the system [57]. Even if the model is linear and accurate, computational cost can still be large for large-scale systems. A framework for LQR controller design exploiting the dominant subspace of the system by performing a linearisation at the stationary point has been presented [58]. However, a stationery point may not exist or the current state may be very far from it. Hence, the linearized model would be an insufficient approximation of the full one.

5.1 Methodology and Algorithmic Details

Our MPC methodology is based on successive adaptive linearisation steps, hence it approximates the system better than a single linear model. Furthermore compared to data-driven model reduction methods, such as POD it eliminates the off-line basis computations as well as the need to empirically sample the parameter space. The general form of a non-linear system is as follows:

$$\dot{\tilde{x}} = f(\tilde{x}(t), \tilde{u}(t), t), \quad \tilde{x}(t_0) = \tilde{x}_0 \tag{29}$$
$$\tilde{y} = g(\tilde{x}(t), \tilde{u}(t), t)$$

Here, $\tilde{x}(t)$ are state, $\tilde{u}(t)$ input and $\tilde{y}(t)$ output variables, which can be considered as perturbations of the corresponding reference variables:

$$\tilde{x}(t) = x_{ref} + x(t), \quad \tilde{u}(t) = u_{ref} + u(t) \tag{30}$$

Linearising the system of (30) around the reference point yields the following linear system:

$$\dot{x}(t) = (I - J)x(t) + Y u(t) \tag{31}$$
$$y(t) = \Psi x(t) + \Theta u(t)$$

where $J = (\partial f/\partial \tilde{x})_{x_{ref}, y_{ref}}$, $Y = (\partial f/\partial \tilde{u})_{x_{ref}, y_{ref}}$, $\Psi = (\partial g/\partial \tilde{x})_{x_{ref}, y_{ref}}$, $\Theta = (\partial g/\partial \tilde{u})_{x_{ref}, y_{ref}}$. From the continuous system in (32) the discrete-time one can be computed [59]:

$$\dot{x}(k + 1) = K x(k) + L u(k) \tag{32}$$
$$y(k + 1) = M x(k) + N u(k)$$

Optimisation and Linear Control of Large Scale Nonlinear Systems

Here successive linearisations around the current state are considered, i.e. the linearisation matrices K, L, M, N change in time. This is obviously too computationally expensive for a large-scale system and even more so for input/output systems. A model reduction step is then performed here where a basis, $\hat{Z}$ for the (typically) low-dimensional invariant subspace of the current state is computed through efficient subspace iterations. Projecting the linearised system in (32) we obtain the following low-dimensional system in terms of the reduced state variables $\xi = \hat{Z}x$:

$$\dot{\xi}(t) = (I - H)\xi(t) + \hat{Z}Yu(t) \tag{33a}$$
$$y(t) = \Psi\hat{Z}\xi(t) + \Theta u(t) \tag{33b}$$

Here H is the reduced Jacobian of the system as it has been defined in (23) which can also be written in an equivalent discrete state-space form:

$$\dot{\xi}(k + 1) = A\xi(k) + Bu(k) \tag{34a}$$
$$y(k + 1) = C\xi(k) + Du(k) \tag{34b}$$

The reduced linear discrete system in (34) has now a standard form for the application of linear MPC over a (receding) time horizon. Once the current outputs are measured, the system is reduced and linearised is state-space form as in (34) and the control move computed through the solution of a reduced unconstrained optimisation problem, which actually will calculate a sequence of control decisions of which the first one will only be implemented. This procedure is repeated over the number of partitions of the time horizon.

5.2 Case Study

We have used this methodology to stabilise the sustained oscillations (Hopf bifurcation) of a tubular reactor with recycle. The modeling equations (7) (discretised in 250 finite differences) with boundary conditions from (8a) were also used here while the recycle, r, conditions are given by the boundary conditions at the entrance of the reactor, $y = 0$:

$$\frac{\partial x_1}{\partial y} = -Pe_1[(1 - r)x_1|_{t=0} + rx_1|_{y=1} - rx_1|_{y=0}] \tag{35a}$$

$$\frac{\partial x_2}{\partial y} = -Pe_2[(1 - r)x_2|_{t=0} + rx_2|_{y=1} - rx_2|_{y=0}] \tag{35b}$$

For $r = 0$ the system is stable, while it exhibits sustain oscillations (Hopf bifurcation) for $r = 0.5$. The control objective is to stabilise the reactor at $r = 0.5$ using five cooling zones so that it behaves like the system at $r = 0$. The system param-

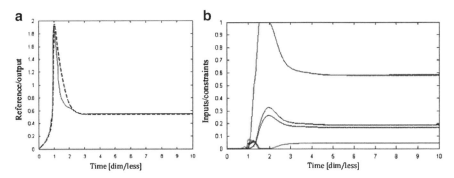

Fig. 9 Profiles of (**a**) control output (*solid line*) vs. reference output (*dashed line*) (**b**) temperatures of the five cooling zones

eters used were: $Da = 0.1$, $Le = 1.0$, $Pe_1 = Pe_2 = 7.0$, $\gamma = 10.0$, $\beta = 1.50$, $C = 12.0$, while the control horizon and the prediction horizon were set equal to 7.

The reduced subspaces computed were of size $m = 4$. Noise disturbances were also considered for the output. In Fig. 9a the reference system profile (dashed line) is shown against the closed loop system output (dimensionless temperature at the exit) profile for $r = 0.5$ and the efficient stabilisation of the system is obvious. In Fig. 9b The control profiles for the five cooling zones in time are shown. Here only low dimensional system gradients were computed while the system was adaptively linearised resulting in very good accuracy of the controller with minimal computational cost. It should be noted that a number of concise results can be found in [60], while a longer version is submitted as a journal paper.

6 Conclusions

A series of model reduction-based methodologies for the gradient-based optimisation and optimal control of nonlinear large-scale systems simulated with different types of input/output simulators has been presented. Both steady state and dynamic optimisation have been considered. In the former case both dynamic integrators and steady state simulators have been used as the simulators of choice. The model reduction technology is essentially based on adaptively identifying, through subspace iterations and/or Arnoldi methods, the low-dimensional dominant system subspaces and subsequently computing, with computational efficiency, low-dimensional projections of Jacobian and Hessian matrices onto these subspaces. These methods were then coupled with existing reduced Hessian algorithms to produce efficient optimisation methodologies, which significantly reduce the computational requirements for handling large-scale black-box systems. We have also shown that these methods can handle both equality and nonlinear inequality constraints. The control algorithm is also based on reduced adaptive linearisations of the system around the states of interest using the same dominant subspace computation methodology. The tubular reactor has been used throughout as the nonlinear system of choice to maintain continuity between the different sections and to demonstrate the features of the techniques presented in a uniform way. We believe that these technologies can help to accomplish a number of systems tasks in an efficient and *automated* way, since they quite easily treat simulators as black-boxes in all cases.

Acknowledgements I would like to acknowledge here the work of my former and current students Eduardo Luna-Ortiz and Ioannis Bonis, as well as the work of my post-doctoral associate Weiguo Xie. The financial support of the EU projects CONNECT (COOP-2006-31638) and CAFE (KBBE-212754) is also gratefully acknowledged.

References

1. Giibons, J.M., Wood, A.T.A., Craigon, J., Ramsden, S.J., Crout, N.M.J.: Semi-automatic reduction and upscaling of large models: A farm management example. Ecological Modelling **221** (2010) 950–958
2. Hardt, S., Baier, T.: Mean-field model for heat transfer in multichannel microreactors. AIChE J **53** (2007) 1006–1016
3. Sager, S., Brandt-Pollmann, U., Diehl, M., Lebiedz, D, Bock, H.G.: Exploiting system homogeneities in large scale optimal control problems for speedup of multiple shooting based SQP methods. Computers and Chemical Engineering **31** (2007) 1181–1186
4. Bock, H., Plitt, K.: A multiple shooting algorithm for direct solution of optimal control problems. Proceedings of ninth IFAC World Congress, Pergamon Press, Budapest (1984) 243–247
5. Benner, P.: Solving large-scale control problems. IEEE Control Systems Magazine **24** (2004) 44–59
6. Stykel, T.: Balanced truncation model reduction for semidiscretized Stokes equation. Linear Algebra and its Applications **415** (2006) 262–289
7. Kotsalis, G., Megretski, A, Dahleh, M.A.: Balanced truncation for a class of stochastic jump linear systems and model reduction for hidden Markov models. IEEE Transactions on Automatic Control **53** (2008) 2543–2557
8. Hovland, S., Gravdahl, J.T., Willcox, K.E.: Explicit model predictive control for large-scale systems via model reduction. Journal of Guidance, Control, and Dynamics **31** (2008) 918]–926
9. Pistikopoulos, E.N., Dua, V., Bozinis, N.A., Bemporad, A., Morari, M.: On-line optimization via off-line parametric optimization tools. Computers and Chemical Engineering **26**, (2001) 175–185
10. Campi, M.C., Garatti, S., Prandini, M.: The scenario approach for systems and control design. Annual Reviews in Control **33** (2009) 149–157
11. Kawarabayashi, M., Tsuchiya, J., Yasuda, K.: Integrated optimization by multi-objective particle swarm optimization. IEEJ Transactions on Electrical and Electronic Engineering **5** (2010) 79–81
12. Huang, X.X., Yan, W.Y., Teo, K.L.: H2 near-optimal model reduction. IEEE Transactions on Automation Control **AC-26** (2001) 1279–1284
13. Glover, K.: All optimal Hankel-norm approximations of linear multivariable systems and their L1-error bounds. International Journal of Control **39** (1984) 1115–1193
14. Sandberg, H., Lanzon, A., Anderson, B.D.O.: Model approximation using magnitude and phase criteria: Implications for model reduction and system identification. International Journal of Robust Nonlinear Control **17** (2007) 435–461
15. Wu, L., Ho W.C.D., Lam, J.: H∞ model reduction for continuous-time switched stochastic hybrid systems. International Journal of Systems Science **40** (2009) 1241–1251

16. Assunção, E., Marchesi, H.F., Teixeira, M.C.M., Peres, P.L.D.: Global optimization for the H∞-norm model reduction problem. International Journal of Systems Science **38** (2007) 125–138

17. Gonçalves, E.N., Palhares, R.M., Takahashi, R.H.C., Chasin, A.N.V.: Robust model reduction of uncertain systems maintaining uncertainty structure. International Journal of Control **82** (2009) 2158 –2168

18. Hofer, A.: Two controller reduction methods based on convex optimization. Mathematical and Computer Modelling of Dynamical Systems **14** (2008) 451–468

19. Jing, W., Liu, W., Zhang, Q.L.: Model reduction for singular systems via covariance approximation. Optimal Control Applications and Methods **25** (2004) 263–278

20. Xu, D., Zhang, Y., Zhang, Q.: A remark on model reduction for singular systems via covariance approximation. Optimal Control Applications and Methods **27** (2006) 293–298

21. Hua, G., Loub, Y., Christofides, P.D.: Model parameter estimation and feedback control of surface roughness in a sputtering process. Chemical Engineering Science **63** (2008) 1800–1816

22. Jogwar, S.S., Daoutidis, P.: Dynamics and control of vapor recompression distillation. Journal of Process Control **19** (2009) 1737–1750

23. van der Schaft, A.: L2-gain and passivity techniques in nonlinear control. Lecture Notes in Control and Information Science 218. Springer, London (1996)

24. Ruszkowski, M., Garcia-Osorio, V., Ydstie, B.E.: Passivity based control of transport reaction systems. AIChE Journal **51** (2005) 3147–3166

25. Gao, H., Chen, T., Chai, T.: Pasivity and passification for networked control systems. SIAM Journal on Control and Optimization **46** (2007) 1299–1322

26. Antoulas, A.C.: A new result on passivity preserving model reduction. Systems and Control Letters **54** (2005) 361–374

27. Sorensen, D.C.: Passivity preserving model reduction via interpolation of spectral zeros. Systems and Control Letters **54** (2005) 347–360

28. Fanizza, G., Karlsson, J., Lindquist, A., Nagamune, R.: Passivity-preserving model reduction by analytic interpolation. Linear Algebra and its Applications **425** (2007) 608–633

29. Holmes, P., Lumley, J.L., Berkooz, G.: Turbulence, Coherent Structures, Dynamical Systems and Symmetry. Cambridge Monographs on Mechanics (1998)

30. Sirovich, L.: Turbulence and the dynamics of coherent structures, parts i–iii. Quarterly of Applied Mathematics **45** (1987) 561–590

31. Veroy, K, Patera, A.: Certified real-time solution of the parametrized steady incompressible Navier–Stokes equations: rigorous reduced-basis a posteriori error bounds. International Journal for Numerical Methods in Fluids **47** (2005) 773–788

32. Bashir, O., Willcox, K., Ghattas, O., van Bloemen Waanders, B., Hill, J.: Hessian-based model reduction for large-scale systems with initial-condition inputs. International Journal for Numerical Methods in Engineering **73** (2008) 844–868

33. Bui-Thanh, T., Willcox, K., Ghattas, O.: Model reduction for large-scale systems with high-dimensional parametric input space. SIAM Journal on Scientific Computing (2008) **30**, 3270–3288

34. Bui-Thanh, T., Willcox, K., Ghattas, O., van Bloemen Waanders, B.: Goal-oriented, model-constrained optimization for reduction of large-scale systems. Journal of Computational Physics **224** (2007) 880–896

35. Varshney, A., Pitchaiah, S., Armaou, A.: Feedback control of dissipative pde systems using adaptive model reduction. AIChE Journal **55** (2009) 906–918

36. My-Ha, D., Lim, K.M., Khoo, B.C., Willcox, K.: Real-time optimization using proper orthogonal decomposition: Free surface shape prediction due to underwater bubble dynamics. Computers and Fluids **36** (2007) 499–512

37. Bouazizi, M.-L., Ghanmia, S., Bouhaddi, N.: Multi-objective optimization in dynamics of the structures with nonlinear behavior: Contributions of the metamodels. Finite Elements in Analysis and Design **45** (2009) 612–623

38. Garcia, M.R., Vilas, C., Banga, J.R., Alonso, A.A.: Exponential observers for distributed tubular (bio)reactors. AIChE Journal **54** 2943–2956

39. Li, M., Christofides, P.D.: Optimal control of diffusion-convection-reaction processes using reduced-order models. Computers and Chemical Engineering **32** (2008) 2123–2135
40. Oguz, C., Gallivan, M.A.: Optimization of a thin film deposition process using a dynamic model extracted from molecular simulations. Automatica **44** (2008) 1958–1969
41. Lang, Y.-D., Biegler, L.T.: A software environment for simultaneous dynamic optimization. Computers and chemical engineering **31** (2007) 931–942
42. Gugercin, S.: An iterative SVD-Krylov based method for model reduction of large-scale dynamical systems. Linear Algebra and its Applications **428** (2008) 1964–1986
43. Heyouni, M., Jbilou, K.: Matrix Krylov subspace methods for large scale model reduction problems Applied Mathematics and Computation **181** (2006) 1215–1228
44. Kevrekidis, I.G, Gear, C.W., Hyman, J.M., Kevrekidis, P.G, Runborg, O., Theodoropoulos, C.: Equation-free coarse-grained multiscale computation: Enabling microscopic simulators to perform system-level tasks. Communications in Mathematical Sciences **1** (2003) 715–762
45. Shroff, G., Keller, H.B.: Stabilization of unstable procedures: The recursive projection method. SIAM Journal on Numerical Analysis **30** (1993) 1099–1120
46. Luna-Ortiz, E., Theodoropoulos, C.: An input/output model reduction-based optimization scheme for large-scale systems. Multiscale Modeling and Simulation **4** (2005) 691–708
47. Nocedal, J., Overton, M.L.: Projected Hessian updating algorithms for nonlinearly constrained optimization. SIAM Journal on Numerical Analysis **22** (1985) 821–850
48. Schmid, C., Biegler, L.T.: Acceleration of reduced Hessian methods for large-scale nonlinear programming. Comput. Chem. Eng. **17** (1993) 451–463
49. Jensen, K.F., Ray, W.H.: The bifurcation behavior of tubular reactors. Computers and Chemical Engineering **37**, (1982) 199–222
50. Theodoropoulos, C., Luna-Ortiz, E.: A reduced input/output dynamic optimization method for macroscopic and microscopic systems. In: Gorban, A., Kazantzis, N., Kevrekidis, I., Ottinger, H.C., Theodoropoulos, C. (eds.) Model Reduction and Coarse-Graining Approaches for Multiscale Phenomena. Springer, Berlin (2006) 535–560
51. Lust, K., Roose, D., Spence, A., Champneys, A.R.: An adaptive Newton–Picard algorithm with subspace iteration for computing periodic solutions. SIAM Journal on Scientific Computing **19** (1998) 1188–1209
52. Bai, Z., Stewart, G.W.: Algorithm 776: SRRIT: A Fortran subroutine to calculate the dominant invariant subspace of a nonsymmetric matrix. ACM Transactions on Mathematical Software **23** (1997) 494–513
53. Choong, K.L., Smith, R.: Optimization of batch cooling crystallization. Chemical Engineering Science **59** (2004) 313–327
54. Bonis, I., Theodoropoulos, C.: A model reduction-based optimisation framework for large-scale simulators using iterative solvers. Computer Aided Chemical Engineering **25** (2008) 545–550
55. Raspandi, C.G., Bandoni, J.A., Biegler, L.T.: New strategies for flexibility analysis and design under uncertainty. Computers and Chemical Engineering **24** (2000) 2193–2209
56. Henson, M.A.: Nonlinear model predictive control: current status and future directions. Computers and Chemical Engineering **23** (1998) 187–202
57. Zhu, Q.M., Warwick, K., Douce, J.L.: Adaptive general predictive controller for nonlinear systems. IEE Proceedings D Control Theory and Applications **138** (1991) 33–40
58. Armaou, A, Theodoropoulos, C., Kevrekidis, I.G: Equation-free gaptooth-based controller design for distributed/complex multiscale processes. Computers and Chemical Engineering **29** (2005) 691–708
59. Franklin, G.F., Powell, J.D., Workman, M.L.: Digital Control of Dynamic Systems, 2nd edn. Addison-Wesley, MA (1990)
60. Bonis, I., Theodoropoulos, C.: A reduced linear model predictive control algorithm for nonlinear distributed parameter systems. Computer Aided Chemical Engineering **28** (2010) 553–558

Universal Algorithms, Mathematics of Semirings and Parallel Computations

Grigory L. Litvinov, Victor P. Maslov, Anatoly Ya. Rodionov, and Andrei N. Sobolevski

Abstract This is a survey paper on applications of mathematics of semirings to numerical analysis and computing. Concepts of universal algorithm and generic program are discussed. Relations between these concepts and mathematics of semirings are examined. A very brief introduction to mathematics of semirings (including idempotent and tropical mathematics) is presented. Concrete applications to optimization problems, idempotent linear algebra and interval analysis are indicated. It is known that some nonlinear problems (and especially optimization problems) become linear over appropriate semirings with idempotent addition (the so-called idempotent superposition principle). This linearity over semirings is convenient for parallel computations.

G.L. Litvinov (✉)
The J. V. Poncelet Laboratory (UMI 2615 CNRS), Independent University of Moscow,
Moscow, Russia
e-mail: glitvinov@gmail.com
and
D. V. Skobeltsyn Research Institute for Nuclear Physics, Moscow State University, Moscow,
Russia

V.P. Maslov
Faculty of Physics, Moscow State University, Moscow, Russia
e-mail: v.p.maslov@mail.ru

A.Ya. Rodionov
D. V. Skobeltsyn Research Institute for Nuclear Physics, Moscow State University, Moscow,
Russia
e-mail: ayarodionov@yahoo.com

A.N. Sobolevski
The J. V. Poncelet Laboratory (UMI 2615 of CNRS), Independent University of Moscow,
Moscow, Russia
e-mail: ansobol@gmail.com
and
A. A. Kharkevich Institute for Information Transmission Problems, Moscow, Russia

A.N. Gorban and D. Roose (eds.), *Coping with Complexity: Model Reduction and Data Analysis*, Lecture Notes in Computational Science and Engineering 75, DOI 10.1007/978-3-642-14941-2_4, © Springer-Verlag Berlin Heidelberg 2011

1 Introduction

It is well known that programmers are the laziest persons in the world. They constantly try to write less and get more. The so-called art of programming is just the way to achieve this goal. They started from programming in computer codes, found this practice boring and invented assembler, then macro assembler, then programming languages to reduce the amount of work (but not the salary). The next step was to separate algorithms and data. The new principle "Algorithm + data structure = program" was a great step in the same direction. Now some people could work on algorithms and express those in a more or less data structure independent manner; others programmers implement algorithms for different target data structures using different languages.

This scheme worked, and worked great, but had an important drawback: for the same algorithm one has to write a new program every time a new data type is required. Not only is this a boring process, it's also consuming time and money and worse, it is error-prone. So the next step was to find how to write programs in a way independent of data structures. Object oriented languages, such as C++, Java and many others (see, e.g., [47, 62]) opened precisely such a way. For C++, templates and STL give even more opportunities. This means that for a given algorithm one can vary a data structure while keeping the program unchanged. The principle was changed to "Algorithm + data structure family = program."

But how does this approach work? What constitutes a "data structure family"? The answer is this: operations. Any algorithm manipulates data by means of a set of basic operations. For example for sorting the comparison operation is required; for scalar products and matrix multiplications, the operations of addition and multiplication are required. So to use a new data structure with the same "generic" program one has to implement the required operations for this data.

The hope was that some standard libraries of generic programs would cover almost all programmers' needs and programming would be very much like Lego constructing. Well, to some extent it worked, but not as much as it was expected. Why? For different reasons. Some of them are economical. Why invest time and money in solving generic needs when a fast patch exists? Indeed, who would buy the next release if the current one is perfect? Another reason: there is not a great variety of possible data structures for most popular algorithms. In linear algebra one can use floating numbers, double precision numbers, infinite precision floating numbers, rational numbers, rational numbers with fix precision, Hensel codes, complex numbers, integers. Not much.

Mathematics of semirings gives a new approach to the generic programming. It parameterized algorithms. The new principle is "Parameterized algorithm + data structure family = program." What are these parameters? They are operations (e.g., addition and multiplication). Sounds great, but does it really work? Yes. And we will show how and why. For example, we will show how the same old algorithm of linear algebra transformed by a simple redefinition of operations, can be applied to different problems in different areas. For example, the same program for solving systems of linear equations can be applied to the shortest path problem and other optimization problems (and interval versions of the problems) by means of simple redefinitions of the addition and multiplication operations. The algorithm was changed (by changing parameters, i.e. operations), the data structure was changed, the nature of the problem was changed, but the program was not changed!

There are deep mathematical reasons (related to mathematics of semirings, idempotent and tropical mathematics) why this approach works. We will briefly discuss them in this paper.

The concept of a generic program was introduced by many authors; for example, in [36] such programs were called "program schemes." In this paper, we discuss *universal algorithms* implemented in the form of generic programs and their specific features. This paper is closely related to papers [38–40, 42, 43], in which the concept of a universal algorithm was defined and software and hardware implementation of such algorithms was discussed in connection with problems of idempotent mathematics [31, 32, 37–46, 49–53, 55, 71, 72]. In the present paper the emphasis is placed on software and hardware implementations of universal algorithms, computation with arbitrary accuracy, universal algorithms of linear algebra over semirings, and their implementations.

We also present a very brief introduction to mathematics of semirings and especially to the mathematics of *idempotent semirings* (i.e., semirings with idempotent addition). Mathematics over idempotent semirings is called *idempotent mathematics*. The so-called idempotent correspondence principle and idempotent superposition principle (see [38–43, 49–53]) are discussed.

There exists a correspondence between interesting, useful, and important constructions and results concerning the field of real (or complex) numbers and similar constructions dealing with various idempotent semirings. This correspondence can be formulated in the spirit of the well-known N. Bohr's *correspondence principle* in quantum mechanics; in fact, the two principles are intimately connected (see [38–40]). In a sense, the traditional mathematics over numerical fields can be treated as a "quantum" theory, whereas the idempotent mathematics can be treated as a "classical" shadow (or counterpart) of the traditional one. It is important that the idempotent correspondence principle is valid for algorithms, computer programs and hardware units.

In quantum mechanics the *superposition principle* means that the Schrödinger equation (which is basic for the theory) is linear. Similarly in idempotent mathematics the (idempotent) superposition principle (formulated by Maslov) means that some important and basic problems and equations that are nonlinear in the usual sense (e.g., the Hamilton–Jacobi equation, which is basic for classical mechanics and appears in many optimization problems, or the Bellman equation and its versions and generalizations) can be treated as linear over appropriate idempotent semirings, see [38–41, 48–51].

Note that numerical algorithms for infinite-dimensional linear problems over idempotent semirings (e.g., idempotent integration, integral operators and transformations, the Hamilton–Jacobi and generalized Bellman equations) deal with the corresponding finite-dimensional approximations. Thus idempotent linear algebra is the basis of the idempotent numerical analysis and, in particular, the *discrete optimization theory*.

Carré [10, 11] (see also [7, 23–25]) used the idempotent linear algebra to show that different optimization problems for finite graphs can be formulated in a unified manner and reduced to solving Bellman equations, i.e., systems of linear algebraic equations over idempotent semirings. He also generalized principal algorithms of computational linear algebra to the idempotent case and showed that some of these coincide with algorithms independently developed for solution of optimization problems; for example, Bellman's method of solving the shortest path problem corresponds to a version of Jacobi's method for solving a system of linear equations, whereas Ford's algorithm corresponds to a version of Gauss–Seidel's method. We briefly discuss Bellman equations and the corresponding optimization problems on graphs.

We stress that these well-known results can be interpreted as a manifestation of the idempotent superposition principle.

We also briefly discuss interval analysis over idempotent and positive semirings. Idempotent internal analysis appears in [45, 46, 49]. Later many authors dealt with this subject, see, e.g., [12, 20, 27, 57, 58, 77].

It is important to observe that intervals over an idempotent semiring form a new idempotent semiring. Hence universal algorithms can be applied to elements of this new semiring and generate interval versions of the initial algorithms.

Note finally that idempotent mathematics is remarkably simpler than its traditional analog.

2   Universal Algorithms

Computational algorithms are constructed on the basis of certain primitive operations. These operations manipulate data that describe "numbers." These "numbers" are elements of a "numerical domain," i.e., a mathematical object such as the field of real numbers, the ring of integers, or an idempotent semiring of numbers (idempotent semirings and their role in idempotent mathematics are discussed in [10, 11, 22–26, 31, 32, 37–43] and below in this paper).

In practice elements of the numerical domains are replaced by their computer representations, i.e., by elements of certain finite models of these domains. Examples of models that can be conveniently used for computer representation of real numbers are provided by various modifications of floating point arithmetics, approximate arithmetics of rational numbers [44], and interval arithmetics. The difference between mathematical objects ("ideal" numbers) and their finite models (computer representations) results in computational (e.g., rounding) errors.

An algorithm is called *universal* if it is independent of a particular numerical domain and/or its computer representation. A typical example of a universal algorithm is the computation of the scalar product (x, y) of two vectors $x = (x_1, \ldots, x_n)$ and $y = (y_1, \ldots, y_n)$ by the formula $(x, y) = x_1 y_1 + \cdots + x_n y_n$.

Universal Algorithms, Mathematics of Semirings and Parallel Computations 67

This algorithm (formula) is independent of a particular domain and its computer implementation, since the formula is well-defined for any semiring. It is clear that one algorithm can be more universal than another. For example, the simplest Newton–Cotes formula, the rectangular rule, provides the most universal algorithm for numerical integration; indeed, this formula is valid even for idempotent integration (over any idempotent semiring, see below and [4, 32, 38–43, 48–51]). Other quadrature formulas (e.g., combined trapezoid rule or the Simpson formula) are independent of computer arithmetics and can be used (e.g., in the iterative form) for computations with arbitrary accuracy. In contrast, algorithms based on Gauss–Jacobi formulas are designed for fixed accuracy computations: they include constants (coefficients and nodes of these formulas) defined with fixed accuracy. Certainly, algorithms of this type can be made more universal by including procedures for computing the constants; however, this results in an unjustified complication of the algorithms.

Computer algebra algorithms used in such systems as Mathematica, Maple, REDUCE, and others are highly universal. Most of the standard algorithms used in linear algebra can be rewritten in such a way that they will be valid over any field and complete idempotent semiring (including semirings of intervals; see below and [45, 46, 49], where an interval version of the idempotent linear algebra and the corresponding universal algorithms are discussed).

As a rule, iterative algorithms (beginning with the successive approximation method) for solving differential equations (e.g., methods of Euler, Euler–Cauchy, Runge–Kutta, Adams, a number of important versions of the difference approximation method, and the like), methods for calculating elementary and some special functions based on the expansion in Taylor's series and continuous fractions (Padé approximations) and others are independent of the computer representation of numbers.

3 Universal Algorithms and Accuracy of Computations

Calculations on computers usually are based on a floating-point arithmetic with a mantissa of a fixed length; i.e., computations are performed with fixed accuracy. Broadly speaking, with this approach only the relative rounding error is fixed, which can lead to a drastic loss of accuracy and invalid results (e.g., when summing series and subtracting close numbers). On the other hand, this approach provides rather high speed of computations. Many important numerical algorithms are designed to use floating-point arithmetic (with fixed accuracy) and ensure the maximum computation speed. However, these algorithms are not universal. The above mentioned Gauss–Jacobi quadrature formulas, computation of elementary and special functions on the basis of the best polynomial or rational approximations or Padé–Chebyshev approximations, and some others belong to this type. Such algorithms use nontrivial constants specified with fixed accuracy.

Recently, problems of accuracy, reliability, and authenticity of computations (including the effect of rounding errors) have gained much attention; in part, this fact is related to the ever-increasing performance of computer hardware. When errors in initial data and rounding errors strongly affect the computation results, such as in ill-posed problems, analysis of stability of solutions, etc., it is often useful to perform computations with improved and variable accuracy. In particular, the rational arithmetic, in which the rounding error is specified by the user [44], can be used for this purpose. This arithmetic is a useful complement to the interval analysis [54]. The corresponding computational algorithms must be universal (in the sense that they must be independent of the computer representation of numbers).

4 Mathematics of Semirings

A broad class of universal algorithms is related to the concept of a semiring. We recall here the definition of a semiring (see, e.g., [21, 22]).

4.1 Basic Definitions

Consider a semiring, i.e., a set S endowed with two associative operations: *addition* $\oplus$ and *multiplication* $\odot$ such that addition is commutative, multiplication distributes over addition from either side, $\mathbf{0}$ (resp., $\mathbf{1}$) is the neutral element of addition (resp., multiplication), $\mathbf{0} \odot x = x \odot \mathbf{0} = \mathbf{0}$ for all $x \in S$, and $\mathbf{0} \neq \mathbf{1}$. Let the semiring S be partially ordered by a relation $\preccurlyeq$ such that $\mathbf{0}$ is the least element and the inequality $x \preccurlyeq y$ implies that $x \oplus z \preccurlyeq y \oplus z$, $x \odot z \preccurlyeq y \odot z$, and $z \odot x \preccurlyeq z \odot y$ for all $x, y, z \in S$; in this case the semiring S is called *positive* (see, e.g., [22]).

A semiring S is called a *semifield* if every nonzero element is invertible.

A semiring S is called *idempotent* if $x \oplus x = x$ for all $x \in S$, see, e.g., [4, 5, 7, 9, 10, 21–24, 31, 37–43]. In this case the addition $\oplus$ defines a *canonical partial order* $\preccurlyeq$ on the semiring S by the rule: $x \preccurlyeq y$ iff $x \oplus y = y$. It is easy to prove that any idempotent semiring is positive with respect to this order. Note also that $x \oplus y = \sup\{x, y\}$ with respect to the canonical order. In the sequel, we shall assume that all idempotent semirings are ordered by the canonical partial order relation.

We shall say that a positive (e.g., idempotent) semiring S is *complete* if it is complete as an ordered set. This means that for every subset $T \subset S$ there exist elements $\sup T \in S$ and $\inf T \in S$.

The most well-known and important examples of positive semirings are "numerical" semirings consisting of (a subset of) real numbers and ordered by the usual linear order $\leqslant$ on $\mathbf{R}$: the semiring $\mathbf{R}_+$ with the usual operations $\oplus = +, \odot = \cdot$ and neutral elements $\mathbf{0} = 0, \mathbf{1} = 1$, the semiring $\mathbf{R}_{\max} = \mathbf{R} \cup \{-\infty\}$ with the operations $\oplus = \max, \odot = +$ and neutral elements $\mathbf{0} = -\infty, \mathbf{1} = 0$, the semiring $\widehat{\mathbf{R}}_{\max} = \mathbf{R}_{\max} \cup \{\infty\}$, where $x \preccurlyeq \infty, x \oplus \infty = \infty$ for all $x, x \odot \infty = \infty \odot x = \infty$

Universal Algorithms, Mathematics of Semirings and Parallel Computations 69

if $x \neq \mathbf{0}$, and $\mathbf{0} \odot \infty = \infty \odot \mathbf{0}$, and the semiring $S_{\max,\min}^{[a,b]} = [a,b]$, where $-\infty \leqslant a < b \leqslant +\infty$, with the operations $\oplus = \max$, $\odot = \min$ and neutral elements $\mathbf{0} = a$, $\mathbf{1} = b$. The semirings $\mathbf{R}_{\max}$, $\widehat{\mathbf{R}}_{\max}$, and $S_{\max,\min}^{[a,b]} = [a,b]$ are idempotent. The semirings $\widehat{\mathbf{R}}_{\max}$, $S_{\max,\min}^{[a,b]}$, $\widehat{\mathbf{R}}_{+} = \mathbf{R}_{+} \bigcup \{\infty\}$ are complete. Remind that every partially ordered set can be imbedded to its completion (a minimal complete set containing the initial one).

The semiring $\mathbf{R}_{\min} = \mathbf{R} \bigcup \{\infty\}$ with operations $\oplus = \min$ and $\odot = +$ and neutral elements $\mathbf{0} = \infty$, $\mathbf{1} = 0$ is isomorphic to $\mathbf{R}_{\max}$.

The semiring $\mathbf{R}_{\max}$ is also called the *max-plus algebra*. The semifields $\mathbf{R}_{\max}$ and $\mathbf{R}_{\min}$ are called *tropical algebras*. The term "tropical" initially appeared in [68] for a discrete version of the max-plus algebra as a suggestion of Ch. Choffrut, see also [26, 55, 61, 72].

Many mathematical constructions, notions, and results over the fields of real and complex numbers have nontrivial analogs over idempotent semirings. Idempotent semirings have become recently the object of investigation of new branches of mathematics, *idempotent mathematics* and *tropical geometry*, see, e.g., [5, 13, 15–19, 25–28, 31, 32, 35, 37–46, 48–53, 55, 71, 72].

4.2 Closure Operations

Let a positive semiring S be endowed with a partial unitary *closure operation* $*$ such that $x \preccurlyeq y$ implies $x^* \preccurlyeq y^*$ and $x^* = \mathbf{1} \oplus (x^* \odot x) = \mathbf{1} \oplus (x \odot x^*)$ on its domain of definition. In particular, $\mathbf{0}^* = \mathbf{1}$ by definition. These axioms imply that $x^* = \mathbf{1} \oplus x \oplus x^2 \oplus \cdots \oplus (x^* \odot x^n)$ if $n \geqslant 1$. Thus x^* can be considered as a "regularized sum" of the series $x^* = \mathbf{1} \oplus x \oplus x^2 \oplus \ldots$; in an idempotent semiring, by definition, $x^* = \sup\{\mathbf{1}, x, x^2, \ldots\}$ if this supremum exists. So if S is complete, then the closure operation is well-defined for every element $x \in S$.

In numerical semirings the operation $*$ is defined as follows: $x^* = (1 - x)^{-1}$ if $x \prec 1$ in $\mathbf{R}_{+}$, or $\widehat{\mathbf{R}}_{+}$ and $x^* = \infty$ if $x \succcurlyeq 1$ in $\widehat{\mathbf{R}}_{+}$; $x^* = \mathbf{1}$ if $x \preccurlyeq \mathbf{1}$ in $\mathbf{R}_{\max}$ and $\widehat{\mathbf{R}}_{\max}$, $x^* = \infty$ if $x \succ \mathbf{1}$ in $\widehat{\mathbf{R}}_{\max}$, $x^* = \mathbf{1}$ for all x in $S_{\max,\min}^{[a,b]}$. In all other cases x^* is undefined. Note that the closure operation is very easy to implement.

4.3 Matrices Over Semirings

Denote by $\mathrm{Mat}_{mn}(S)$ a set of all matrices $A = (a_{ij})$ with m rows and n columns whose coefficients belong to a semiring S. The sum $A \oplus B$ of matrices $A, B \in \mathrm{Mat}_{mn}(S)$ and the product AB of matrices $A \in \mathrm{Mat}_{lm}(S)$ and $B \in \mathrm{Mat}_{mn}(S)$ are defined according to the usual rules of linear algebra: $A \oplus B = (a_{ij} \oplus b_{ij}) \in \mathrm{Mat}_{mn}(S)$ and

$$AB = \left(\bigoplus_{k=1}^{m} a_{ij} \odot b_{kj} \right) \in \mathrm{Mat}_{ln}(S),$$

where $A \in \mathrm{Mat}_{lm}(S)$ and $B \in \mathrm{Mat}_{mn}(S)$. Note that we write AB instead of $A \odot B$.

If the semiring S is positive, then the set $\mathrm{Mat}_{mn}(S)$ is ordered by the relation $A = (a_{ij}) \preccurlyeq B = (b_{ij})$ iff $a_{ij} \preccurlyeq b_{ij}$ in S for all $1 \leqslant i \leqslant m, 1 \leqslant j \leqslant n$.

The matrix multiplication is consistent with the order $\preccurlyeq$ in the following sense: if $A, A' \in \mathrm{Mat}_{lm}(S)$, $B, B' \in \mathrm{Mat}_{mn}(S)$ and $A \preccurlyeq A'$, $B \preccurlyeq B'$, then $AB \preccurlyeq A'B'$ in $\mathrm{Mat}_{ln}(S)$. The set $\mathrm{Mat}_{nn}(S)$ of square $(n \times n)$ matrices over a [positive, idempotent] semiring S forms a [positive, idempotent] semiring with a zero element $O = (o_{ij})$, where $o_{ij} = \mathbf{0}, 1 \leqslant i, j \leqslant n$, and a unit element $E = (\delta_{ij})$, where $\delta_{ij} = \mathbf{1}$ if $i = j$ and $\delta_{ij} = \mathbf{0}$ otherwise.

The set Mat_{nn} is an example of a noncommutative semiring if $n > 1$.

The closure operation in matrix semirings over a positive semiring S can be defined inductively (another way to do that see in [22] and below): $A^* = (a_{11})^* = (a_{11}^*)$ in $\mathrm{Mat}_{11}(S)$ and for any integer $n > 1$ and any matrix

$$A = \begin{pmatrix} A_{11} & A_{12} \\ A_{21} & A_{22} \end{pmatrix},$$

where $A_{11} \in \mathrm{Mat}_{kk}(S)$, $A_{12} \in \mathrm{Mat}_{k\,n-k}(S)$, $A_{21} \in \mathrm{Mat}_{n-k\,k}(S)$, $A_{22} \in \mathrm{Mat}_{n-k\,n-k}(S)$, $1 \leqslant k \leqslant n$, by definition,

$$A^* = \begin{pmatrix} A_{11}^* \oplus A_{11}^* A_{12} D^* A_{21} A_{11}^* & A_{11}^* A_{12} D^* \\ D^* A_{21} A_{11}^* & D^* \end{pmatrix}, \tag{1}$$

where $D = A_{22} \oplus A_{21} A_{11}^* A_{12}$. It can be proved that this definition of A^* implies that the equality $A^* = A^* A \oplus E$ is satisfied and thus A^* is a "regularized sum" of the series $E \oplus A \oplus A^2 \oplus \dots$.

Note that this recurrence relation coincides with the formulas of escalator method of matrix inversion in the traditional linear algebra over the field of real or complex numbers, up to the algebraic operations used. Hence this algorithm of matrix closure requires a polynomial number of operations in n.

4.4 Discrete Stationary Bellman Equations

Let S be a positive semiring. The *discrete stationary Bellman equation* has the form

$$X = AX \oplus B, \tag{2}$$

where $A \in \mathrm{Mat}_{nn}(S)$, $X, B \in \mathrm{Mat}_{ns}(S)$, and the matrix X is unknown. Let A^* be the closure of the matrix A. It follows from the identity $A^* = A^* A \oplus E$ that the matrix $A^* B$ satisfies this equation; moreover, it can be proved that for positive

Universal Algorithms, Mathematics of Semirings and Parallel Computations 71

semirings this solution is the least in the set of solutions to (2) with respect to the partial order in $\mathrm{Mat}_{ns}(S)$.

4.5 Weighted Directed Graphs and Matrices Over Semirings

Suppose that S is a semiring with zero $\mathbf{0}$ and unity $\mathbf{1}$. It is well-known that any square matrix $A = (a_{ij}) \in \mathrm{Mat}_{nn}(S)$ specifies a *weighted directed graph*. This geometrical construction includes three kinds of objects: the set X of n elements $x_1, \ldots, x_n$ called *nodes*, the set Γ of all ordered pairs (x_i, x_j) such that $a_{ij} \neq \mathbf{0}$ called *arcs*, and the mapping $A \colon \Gamma \to S$ such that $A(x_i, x_j) = a_{ij}$. The elements a_{ij} of the semiring S are called *weights* of the arcs.

Conversely, any given weighted directed graph with n nodes specifies a unique matrix $A \in \mathrm{Mat}_{nn}(S)$.

This definition allows for some pairs of nodes to be disconnected if the corresponding element of the matrix A is $\mathbf{0}$ and for some channels to be "loops" with coincident ends if the matrix A has nonzero diagonal elements. This concept is convenient for analysis of parallel and distributed computations and design of computing media and networks (see, e.g., [4, 53, 73]).

Recall that a sequence of nodes of the form

$$p = (y_0, y_1, \ldots, y_k)$$

with $k \geq 0$ and $(y_i, y_{i+1}) \in \Gamma$, $i = 0, \ldots, k-1$, is called a *path* of length k connecting y_0 with y_k. Denote the set of all such paths by $P_k(y_0, y_k)$. The weight $A(p)$ of a path $p \in P_k(y_0, y_k)$ is defined to be the product of weights of arcs connecting consecutive nodes of the path:

$$A(p) = A(y_0, y_1) \odot \cdots \odot A(y_{k-1}, y_k).$$

By definition, for a "path" $p \in P_0(x_i, x_j)$ of length $k = 0$ the weight is $\mathbf{1}$ if $i = j$ and $\mathbf{0}$ otherwise.

For each matrix $A \in \mathrm{Mat}_{nn}(S)$ define $A^0 = E = (\delta_{ij})$ (where $\delta_{ij} = \mathbf{1}$ if $i = j$ and $\delta_{ij} = \mathbf{0}$ otherwise) and $A^k = AA^{k-1}$, $k \geq 1$. Let $a_{ij}^{(k)}$ be the (i, j)th element of the matrix A^k. It is easily checked that

$$a_{ij}^{(k)} = \bigoplus_{\substack{i_0 = i, i_k = j \\ 1 \leq i_1, \ldots, i_{k-1} \leq n}} a_{i_0 i_1} \odot \cdots \odot a_{i_{k-1} i_k}.$$

Thus $a_{ij}^{(k)}$ is the supremum of the set of weights corresponding to all paths of length k connecting the node $x_{i_0} = x_i$ with $x_{i_k} = x_j$.

Denote the elements of the matrix A^* by $a_{ij}^{(*)}$, $i, j = 1, \ldots, n$; then

$$a_{ij}^{(*)} = \bigoplus_{0 \leqslant k < \infty} \bigoplus_{p \in P_k(x_i, x_j)} A(p).$$

The closure matrix A^* solves the well-known *algebraic path problem*, which is formulated as follows: for each pair (x_i, x_j) calculate the supremum of weights of all paths (of arbitrary length) connecting node x_i with node x_j. The closure operation in matrix semirings has been studied extensively (see, e.g., [1, 2, 5–7, 10, 11, 15–17, 22–26, 31, 32, 46] and references therein).

Example 1. (The shortest path problem) Let $S = \mathbf{R}_{\min}$, so the weights are real numbers. In this case

$$A(p) = A(y_0, y_1) + A(y_1, y_2) + \cdots + A(y_{k-1}, y_k).$$

If the element a_{ij} specifies the length of the arc (x_i, x_j) in some metric, then $a_{ij}^{(*)}$ is the length of the shortest path connecting x_i with x_j.

Example 2. (The maximal path width problem) Let $S = \mathbf{R} \cup \{\mathbf{0}, \mathbf{1}\}$ with $\oplus = \max$, $\odot = \min$. Then

$$a_{ij}^{(*)} = \max_{\substack{p \in \bigcup_{k \geqslant 1} P_k(x_i, x_j)}} A(p), \quad A(p) = \min(A(y_0, y_1), \ldots, A(y_{k-1}, y_k)).$$

If the element a_{ij} specifies the "width" of the arc (x_i, x_j), then the width of a path p is defined as the minimal width of its constituting arcs and the element $a_{ij}^{(*)}$ gives the supremum of possible widths of all paths connecting x_i with x_j.

Example 3. (A simple dynamic programming problem) Let $S = \mathbf{R}_{\max}$ and suppose a_{ij} gives the *profit* corresponding to the transition from x_i to x_j. Define the vector $B = (b_i) \in \text{Mat}_{n1}(\mathbf{R}_{\max})$ whose element b_i gives the *terminal profit* corresponding to exiting from the graph through the node x_i. Of course, negative profits (or, rather, losses) are allowed. Let m be the total profit corresponding to a path $p \in P_k(x_i, x_j)$, i.e.,

$$m = A(p) + b_j.$$

Then it is easy to check that the supremum of profits that can be achieved on paths of length k beginning at the node x_i is equal to $(A^k B)_i$ and the supremum of profits achievable without a restriction on the length of a path equals $(A^* B)_i$.

Example 4. (The matrix inversion problem) Note that in the formulas of this section we are using distributivity of the multiplication $\odot$ with respect to the addition $\oplus$ but do not use the idempotency axiom. Thus the algebraic path problem can be posed for a nonidempotent semiring S as well (see, e.g., [66]). For instance, if $S = \mathbf{R}$, then

Universal Algorithms, Mathematics of Semirings and Parallel Computations

$$A^* = E + A + A^2 + \cdots = (E - A)^{-1}.$$

If $\|A\| > 1$ but the matrix $E - A$ is invertible, then this expression defines a regularized sum of the divergent matrix power series $\sum_{i \geq 0} A^i$.

There are many other important examples of problems (in different areas) related to algorithms of linear algebra over semirings (transitive closures of relations, accessible sets, critical paths, paths of greatest capacities, the most reliable paths, interval and other problems), see [1, 2, 4, 5, 10–13, 15–18, 20, 22–27, 31, 32, 45, 46, 53, 57, 58, 63–68, 74–77].

We emphasize that this connection between the matrix closure operation and solution to the Bellman equation gives rise to a number of different algorithms for numerical calculation of the closure matrix. All these algorithms are adaptations of the well-known algorithms of the traditional computational linear algebra, such as the Gauss–Jordan elimination, various iterative and escalator schemes, etc. This is a special case of the idempotent superposition principle (see below).

In fact, the theory of the discrete stationary Bellman equation can be developed using the identity $A^* = AA^* \oplus E$ as an additional axiom without any substantive interpretation (the so-called *closed semirings*, see, e.g., [7, 22, 36, 66]).

5 Universal Algorithms of Linear Algebra Over Semirings

The most important linear algebra problem is to solve the system of linear equations

$$AX = B, \tag{3}$$

where A is a matrix with elements from the basic field and X and B are vectors (or matrices) with elements from the same field. It is required to find X if A and B are given. If A in (3) is not the identity matrix I, THEN system (3) can be written in form (2), i.e.,

$$X = AX + B. \tag{2'}$$

It is well known that the form (2) or (2′) is convenient for using the successive approximation method. Applying this method with the initial approximation $X_0 = 0$, we obtain the solution

$$X = A^* B, \tag{4}$$

where

$$A^* = I + A + A^2 + \cdots + A^n + \cdots \tag{5}$$

On the other hand, it is clear that

$$A^* = (I - A)^{-1}, \tag{6}$$

if the matrix $I - A$ is invertible. The inverse matrix $(I - A)^{-1}$ can be considered as a regularized sum of the formal series (5).

The above considerations can be extended to a broad class of semirings.

The closure operation for matrix semirings $\mathrm{Mat}_n(S)$ can be defined and computed in terms of the closure operation for S (see section 3 above); some methods are described in [1, 2, 7, 10, 11, 22–25, 32, 35, 42, 46, 65–67]. One such method is described below (*LDM*-factorization).

If S is a field, then, by definition, $x^* = (1 - x)^{-1}$ for any $x \neq 1$. If S is an idempotent semiring, then, by definition,

$$x^* = \mathbf{1} \oplus x \oplus x^2 \oplus \cdots = \sup\{\mathbf{1}, x, x^2, \ldots\}, \tag{7}$$

if this supremum exists. Recall that it exists if S is complete, see Sect. 2.

Consider a nontrivial universal algorithm applicable to matrices over semirings with the closure operation defined.

Example 5: Semiring LDM-Factorization

Factorization of a matrix into the product $A = LDM$, where L and M are lower and upper triangular matrices with a unit diagonal, respectively, and D is a diagonal matrix, is used for solving matrix equations $AX = B$. We construct a similar decomposition for the Bellman equation $X = AX \oplus B$.

For the case $AX = B$, the decomposition $A = LDM$ induces the following decomposition of the initial equation:

$$LZ = B, \qquad DY = Z, \qquad MX = Y. \tag{8}$$

Hence, we have

$$A^{-1} = M^{-1} D^{-1} L^{-1}, \tag{9}$$

if A is invertible. In essence, it is sufficient to find the matrices L, D and M, since the linear system (8) is easily solved by a combination of the forward substitution for Z, the trivial inversion of a diagonal matrix for Y, and the back substitution for X.

Using (8) as a pattern, we can write

$$Z = LZ \oplus B, \qquad Y = DY \oplus Z, \qquad X = MX \oplus Y. \tag{10}$$

Then

$$A^* = M^* D^* L^*. \tag{11}$$

A triple (L, D, M) consisting of a lower triangular, diagonal, and upper triangular matrices is called an *LDM-factorization* of a matrix A if relations (10) and (11) are satisfied. We note that in this case, the principal diagonals of L and M are zero.

The modification of the notion of *LDM*-factorization used in matrix analysis for the equation $AX = B$ is constructed in analogy with the construct suggested by Carré in [10, 11] for LU-factorization.

We stress that the algorithm described below can be applied to matrix computations over any semiring under the condition that the unitary operation $a \mapsto a^*$ is applicable every time it is encountered in the computational process. Indeed, when

Universal Algorithms, Mathematics of Semirings and Parallel Computations 75

constructing the algorithm, we use only the basic semiring operations of addition $\oplus$ and multiplication $\odot$ and the properties of associativity, commutativity of addition, and distributivity of multiplication over addition.

If A is a symmetric matrix over a semiring with a commutative multiplication, the amount of computations can be halved, since M and L are mapped into each other under transposition.

We begin with the case of a triangular matrix $A = L$ (or $A = M$). Then, finding X is reduced to the forward (or back) substitution.

Forward substitution

We are given:

- $L = \|l_j^i\|_{i,j=1}^n$, where $l_j^i = \mathbf{0}$ for $i \leq j$ (a lower triangular matrix with a zero diagonal);
- $B = \|b^i\|_{i=1}^n$.

It is required to find the solution $X = \|x^i\|_{i=1}^n$ to the equation $X = LX \oplus B$. The program fragment solving this problem is as follows.

```
for i = 1 to n do
{    xⁱ := bⁱ;
     for j = 1 to i − 1 do
          xⁱ := xⁱ ⊕ (lʲᵢ ⊙ xʲ); }
```

Back substitution

We are given

- $M = \|m_j^i\|_{i,j=1}^n$, where $m_j^i = \mathbf{0}$ for $i \geq j$ (an upper triangular matrix with a zero diagonal);
- $B = \|b^i\|_{i=1}^n$.

It is required to find the solution $X = \|x^i\|_{i=1}^n$ to the equation $X = MX \oplus B$. The program fragment solving this problem is as follows.

```
for i = n to 1 step −1 do
{    xⁱ := bⁱ;
     for j = n to i + 1 step −1 do
          xⁱ := xⁱ ⊕ (mʲᵢ ⊙ xⁱ); }
```

Both algorithms require $(n^2 - n)/2$ operations $\oplus$ and $\odot$.

Closure of a diagonal matrix

We are given

- $D = \mathrm{diag}(d_1, \ldots, d_n)$;
- $B = \|b^i\|_{i=1}^n$.

It is required to find the solution $X = \|x^i\|_{i=1}^n$ to the equation $X = DX \oplus B$. The program fragment solving this problem is as follows.

for $i = 1$ to n do
$\quad x^i := (d_i)^* \odot b^i$;

This algorithm requires n operations $*$ and n multiplications $\odot$.

General case

We are given

- $L = \|l_j^i\|_{i,j=1}^n$, where $l_j^i = \mathbf{0}$ if $i \le j$;
- $D = \mathrm{diag}(d_1, \ldots, d_n)$;
- $M = \|m_j^i\|_{i,j=1}^n$, where $m_j^i = \mathbf{0}$ if $i \ge j$;
- $B = \|b^i\|_{i=1}^n$.

It is required to find the solution $X = \|x^i\|_{i=1}^n$ to the equation $X = AX \oplus B$, where L, D, and M form the *LDM*-factorization of A. The program fragment solving this problem is as follows.

FORWARD SUBSTITUTION
for $i = 1$ to n do
$\{\ x^i := b^i$;
$\quad$ for $j = 1$ to $i - 1$ do
$\quad x^i := x^i \oplus (l_j^i \odot x^j)$; $\}$
CLOSURE OF A DIAGONAL MATRIX
for $i = 1$ to n do
$\quad x^i := (d_i)^* \odot b^i$;
BACK SUBSTITUTION
for $i = n$ to 1 step -1 do
$\{$ for $j = n$ to $i + 1$ step -1 do
$\quad x^i := x^i \oplus (m_j^i \odot x^j)$; $\}$

Note that x^i is not initialized in the course of the back substitution. The algorithm requires $n^2 - n$ operations $\oplus$, n^2 operations $\odot$, and n operations $*$.

LDM-factorization

We are given

- $A = \|a_j^i\|_{i,j=1}^n$.

It is required to find the *LDM*-factorization of A: $L = \|l_j^i\|_{i,j=1}^n$, $D = \mathrm{diag}(d_1, \ldots, d_n)$, and $M = \|m_j^i\|_{i,j=1}^n$, where $l_j^i = \mathbf{0}$ if $i \le j$, and $m_j^i = \mathbf{0}$ if $i \ge j$.

Universal Algorithms, Mathematics of Semirings and Parallel Computations

The program uses the following internal variables:

- $C = \|c_j^i\|_{i,j=1}^n$;
- $V = \|v^i\|_{i=1}^n$;
- d.

INITIALISATION
for $i = 1$ to n do
    for $j = 1$ to n do
        $c_j^i = a_j^i$;
MAIN LOOP
for $j = 1$ to n do
{   for $i = 1$ to j do
        $v^i := a_j^i$;
    for $k = 1$ to $j - 1$ do
        for $i = k + 1$ to j do
            $v^i := v^i \oplus (a_k^i \odot v^k)$;
    for $i = 1$ to $j - 1$ do
        $a_j^i := (a_i^i)^* \odot v^i$;
    $a_j^j := v^j$;
    for $k = 1$ to $j - 1$ do
        for $i = j + 1$ to n do
            $a_j^i := a_j^i \oplus (a_k^i \odot v^k)$;
    $d = (v^j)^*$;
    for $i = j + 1$ to n do
        $a_j^i := a_j^i \odot d$; }

This algorithm requires $(2n^3 - 3n^2 + n)/6$ operations $\oplus$, $(2n^3 + 3n^2 - 5n)/6$ operations $\odot$, and $n(n+1)/2$ operations $*$. After its completion, the matrices L, D, and M are contained, respectively, in the lower triangle, on the diagonal, and in the upper triangle of the matrix C. In the case when A is symmetric about the principal diagonal and the semiring over which the matrix is defined is commutative, the algorithm can be modified in such a way that the number of operations is reduced approximately by a factor of two.

Other examples can be found in [10, 11, 22–25, 35, 36, 66, 67].

Remark 5. Note that to compute the matrices A^* and $A^* B$ it is convenient to solve the Bellman equation (2).

Some other interesting and important problems of linear algebra over semirings are examined, e.g., in [8, 9, 12, 18, 20, 22–25, 27, 57–60, 74–77].

6 The Idempotent Correspondence Principle

There is a nontrivial analogy between mathematics of semirings and quantum mechanics. For example, the field of real numbers can be treated as a "quantum object" with respect to idempotent semirings. So idempotent semirings can be treated as "classical" or "semi-classical" objects with respect to the field of real numbers.

Let $\mathbf{R}$ be the field of real numbers and $\mathbf{R}_+$ the subset of all non-negative numbers. Consider the following change of variables:

$$u \mapsto w = h \ln u,$$

where $u \in \mathbf{R}_+ \setminus \{0\}$, $h > 0$; thus $u = e^{w/h}$, $w \in \mathbf{R}$. Denote by $\mathbf{0}$ the additional element $-\infty$ and by S the extended real line $\mathbf{R} \cup \{\mathbf{0}\}$. The above change of variables has a natural extension D_h to the whole S by $D_h(0) = \mathbf{0}$; also, we denote $D_h(1) = 0 = \mathbf{1}$.

Denote by S_h the set S equipped with the two operations $\oplus_h$ (generalized addition) and $\odot_h$ (generalized multiplication) such that D_h is a homomorphism of $\{\mathbf{R}_+, +, \cdot\}$ to $\{S, \oplus_h, \odot_h\}$. This means that $D_h(u_1 + u_2) = D_h(u_1) \oplus_h D_h(u_2)$ and $D_h(u_1 \cdot u_2) = D_h(u_1) \odot_h D_h(u_2)$, i.e., $w_1 \odot_h w_2 = w_1 + w_2$ and $w_1 \oplus_h w_2 = h \ln(e^{w_1/h} + e^{w_2/h})$. It is easy to prove that $w_1 \oplus_h w_2 \to \max\{w_1, w_2\}$ as $h \to 0$.

Denote by $\mathbf{R}_{\max}$ the set $S = \mathbf{R} \cup \{\mathbf{0}\}$ equipped with operations $\oplus = \max$ and $\odot = +$, where $\mathbf{0} = -\infty$, $\mathbf{1} = 0$ as above. Algebraic structures in $\mathbf{R}_+$ and S_h are isomorphic; therefore $\mathbf{R}_{\max}$ is a result of a deformation of the structure in $\mathbf{R}_+$.

We stress the obvious analogy with the quantization procedure, where h is the analog of the Planck constant. In these terms, $\mathbf{R}_+$ (or $\mathbf{R}$) plays the part of a "quantum object" while $\mathbf{R}_{\max}$ acts as a "classical" or "semi-classical" object that arises as the result of a *dequantization* of this quantum object.

Likewise, denote by $\mathbf{R}_{\min}$ the set $\mathbf{R} \cup \{\mathbf{0}\}$ equipped with operations $\oplus = \min$ and $\odot = +$, where $\mathbf{0} = +\infty$ and $\mathbf{1} = 0$. Clearly, the corresponding dequantization procedure is generated by the change of variables $u \mapsto w = -h \ln u$.

Consider also the set $\mathbf{R} \cup \{\mathbf{0}, \mathbf{1}\}$, where $\mathbf{0} = -\infty$, $\mathbf{1} = +\infty$, together with the operations $\oplus = \max$ and $\odot = \min$. Obviously, it can be obtained as a result of a "second dequantization" of $\mathbf{R}$ or $\mathbf{R}_+$.

There is a natural transition from the field of real numbers or complex numbers to the semiring $\mathbf{R}_{\max}$ (or $\mathbf{R}_{\min}$). This is a composition of the mapping $x \mapsto |x|$ and the deformation described above.

In general an *idempotent dequantization* is a transition from a basic field to an idempotent semiring in mathematical concepts, constructions and results, see [38–40] for details.

For example, the basic object of the traditional calculus is a *function* defined on some set X and taking its values in the field $\mathbf{R}$ (or $\mathbf{C}$); its idempotent analog is a map $X \to S$, where X is some set and $S = \mathbf{R}_{\min}, \mathbf{R}_{\max}$, or another idempotent semiring. Let us show that redefinition of basic constructions of traditional calculus in terms

Universal Algorithms, Mathematics of Semirings and Parallel Computations

of idempotent mathematics can yield interesting and nontrivial results (see, e.g., [32, 37–43, 45, 48–53, 71, 72] for details, additional examples and generalizations).

Example 6 (Integration and measures). To define an idempotent analog of the Riemann integral, consider a Riemann sum for a function $\varphi(x)$, $x \in X = [a, b]$, and substitute semiring operations $\oplus$ and $\odot$ for operations $+$ and $\cdot$ (usual addition and multiplication) in its expression (for the sake of being definite, consider the semiring $\mathbf{R}_{\max}$):

$$\sum_{i=1}^{N} \varphi(x_i) \cdot \Delta_i \quad \mapsto \quad \bigoplus_{i=1}^{N} \varphi(x_i) \odot \Delta_i = \max_{i=1,\ldots,N} (\varphi(x_i) + \Delta_i),$$

where $a = x_0 < x_1 < \cdots < x_N = b$, $\Delta_i = x_i - x_{i-1}$, $i = 1, \ldots, N$. As $\max_i \Delta_i \to 0$, the integral sum tends to

$$\int_X^{\oplus} \varphi(x)\, dx = \sup_{x \in X} \varphi(x)$$

for any function $\varphi \colon X \to \mathbf{R}_{\max}$ that is bounded. In general, for any set X the set function

$$m_\varphi(B) = \sup_{x \in B} \varphi(x), \quad B \subset X,$$

is called an $\mathbf{R}_{\max}$-*measure* on X; since $m_\varphi(\bigcup_\alpha B_\alpha) = \sup_\alpha m_\varphi(B_\alpha)$, this measure is completely additive. An idempotent integral with respect to this measure is defined as

$$\int_X^{\oplus} \psi(x)\, dm_\varphi = \int_X^{\oplus} \psi(x) \odot \varphi(x)\, dx = \sup_{x \in X} (\psi(x) + \varphi(x)).$$

Using the standard partial order it is possible to generalize these definitions for the case of arbitrary idempotent semirings.

Example 7 (Fourier–Legendre transform). Consider the topological group $G = \mathbf{R}^n$. The usual Fourier–Laplace transform is defined as

$$\varphi(x) \mapsto \widetilde{\varphi}(\xi) = \int_G e^{i\xi \cdot x} \varphi(x)\, dx,$$

where $\exp(i\xi \cdot x)$ is a *character* of the group G, i.e., a solution of the following functional equation:

$$f(x + y) = f(x)f(y).$$

The idempotent analog of this equation is

$$f(x + y) = f(x) \odot f(y) = f(x) + f(y).$$

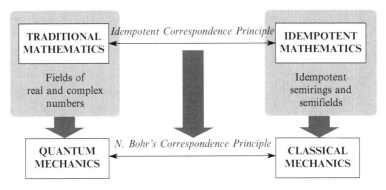

Fig. 1 Relations between idempotent and traditional mathematics

Hence natural "idempotent characters" of the group G are linear functions of the form $x \mapsto \xi \cdot x = \xi_1 x_1 + \cdots + \xi_n x_n$. Thus the Fourier–Laplace transform turns into

$$\varphi(x) \mapsto \widetilde{\varphi}(\xi) = \int_G^{\oplus} \xi \cdot x \odot \varphi(x)\,dx = \sup_{x \in G}(\xi \cdot x + \varphi(x)).$$

This is the well-known Legendre (or Fenchel) transform. Examples related to an important version of matrix algebra are discussed in section 4 above.

These examples suggest the following formulation of the idempotent correspondence principle [39, 40]:

> There exists a heuristic correspondence between interesting, useful, and important constructions and results over the field of real (or complex) numbers and similar constructions and results over idempotent semirings in the spirit of N. Bohr's correspondence principle in quantum mechanics.

So idempotent mathematics can be treated as a "classical shadow (or counterpart)" of the traditional Mathematics over fields.

A systematic application of this correspondence principle leads to a variety of theoretical and applied results, see, e.g., [37–43,45,46,55,71,72]. Relations between idempotent and traditional mathematics are presented in Fig 1. Relations to quantum physics are discussed in detail, e.g., in [38].

7 The Superposition Principle and Parallel Computing

In quantum mechanics the superposition principle means that the Schrödinger equation (which is basic for the theory) is linear. Similarly in idempotent mathematics the idempotent superposition principle means that some important and basic problems and equations (e.g., optimization problems, the Bellman equation and its versions and generalizations, the Hamilton–Jacobi equation) nonlinear in the usual sense can

Universal Algorithms, Mathematics of Semirings and Parallel Computations 81

be treated as linear over appropriate idempotent semirings (in a general form this superposition principle was formulated by Maslov), see [48–53].

Linearity of the Hamilton–Jacobi equation over $\mathbf{R}_{\min}$ (and $\mathbf{R}_{\max}$) can be deduced from the usual linearity (over $\mathbf{C}$) of the corresponding Schrödinger equation by means of the dequantization procedure described above (in Sect. 4). In this case the parameter h of this dequantization coincides with $i\hbar$, where $\hbar$ is the Planck constant; so in this case $\hbar$ must take imaginary values (because $h > 0$; see [38] for details). Of course, this is closely related to variational principles of mechanics.

The situation is similar for the differential Bellman equation (see, e.g., [32]) and the discrete version of the Bellman equations, see Sect. 4 above.

It is well known that linear problems and equations are especially convenient for parallelization, see, e.g., [73]. Standard methods (including the so-called block methods) constructed in the framework of the traditional mathematics can be extended to universal algorithms over semirings (the correspondence principle!). For example, formula (1) discussed in Sect. 3 leads to a simple block method for parallelization of the closure operations. Other standard methods of linear algebra [73] can be used in a similar way.

8 The Correspondence Principle for Computations

Of course, the idempotent correspondence principle is valid for algorithms as well as for their software and hardware implementations [39, 40, 42, 43]. Thus:

If we have an important and interesting numerical algorithm, then there is a good chance that its semiring analogs are important and interesting as well.

In particular, according to the superposition principle, analogs of linear algebra algorithms are especially important. Note that numerical algorithms for standard infinite-dimensional linear problems over idempotent semirings (i.e., for problems related to idempotent integration, integral operators and transformations, the Hamilton–Jacobi and generalized Bellman equations) deal with the corresponding finite-dimensional (or finite) "linear approximations." Nonlinear algorithms often can be approximated by linear ones. Thus the idempotent linear algebra is a basis for the idempotent numerical analysis.

Moreover, it is well-known that linear algebra algorithms easily lend themselves to parallel computation; their idempotent analogs admit parallelization as well. Thus we obtain a systematic way of applying parallel computing to optimization problems.

Basic algorithms of linear algebra (such as inner product of two vectors, matrix addition and multiplication, etc.) often do not depend on concrete semirings, as well as on the nature of domains containing the elements of vectors and matrices. Algorithms to construct the closure $A^* = \mathbf{1} \oplus A \oplus A^2 \oplus \cdots \oplus A^n \oplus \cdots = \bigoplus_{n=1}^{\infty} A^n$ of an idempotent matrix A can be derived from standard methods for calculating $(\mathbf{1} - A)^{-1}$. For the Gauss–Jordan elimination method (via LU-decomposition) this trick was used in [66], and the corresponding algorithm is universal and can be applied both to the Bellman equation and to computing the inverse of a real (or complex) matrix $(\mathbf{1} - A)$. Computation of A^{-1} can be derived from this universal algorithm with some obvious cosmetic transformations.

Thus it seems reasonable to develop universal algorithms that can deal equally well with initial data of different domains sharing the same basic structure [39, 40, 43].

9 The Correspondence Principle for Hardware Design

A systematic application of the correspondence principle to computer calculations leads to a unifying approach to software and hardware design.

The most important and standard numerical algorithms have many hardware realizations in the form of technical devices or special processors. *These devices often can be used as prototypes for new hardware units generated by substitution of the usual arithmetic operations for its semiring analogs and by addition tools for performing neutral elements* $\mathbf{0}$ *and* $\mathbf{1}$ (the latter usually is not difficult). Of course, the case of numerical semirings consisting of real numbers (maybe except neutral elements) and semirings of numerical intervals is the most simple and natural [38–43, 45, 46, 69]. Note that for semifields (including $\mathbf{R}_{\max}$ and $\mathbf{R}_{\min}$) the operation of division is also defined.

Good and efficient technical ideas and decisions can be transposed from prototypes into new hardware units. Thus the correspondence principle generated a regular heuristic method for hardware design. Note that to get a patent it is necessary to present the so-called "invention formula," that is to indicate a prototype for the suggested device and the difference between these devices.

Consider (as a typical example) the most popular and important algorithm of computing the scalar product of two vectors:

$$(x, y) = x_1 y_1 + x_2 y_2 + \cdots + x_n y_n. \tag{12}$$

The universal version of (12) for any semiring A is obvious:

$$(x, y) = (x_1 \odot y_1) \oplus (x_2 \odot y_2) \oplus \cdots \oplus (x_n \odot y_n). \tag{13}$$

In the case $A = \mathbf{R}_{\max}$ this formula turns into the following one:

$$(x, y) = \max\{x_1 + y_1, x_2 + y_2, \cdots, x_n + y_n\}. \tag{14}$$

This calculation is standard for many optimization algorithms, so it is useful to construct a hardware unit for computing (14). There are many different devices (and patents) for computing (12) and every such device can be used as a prototype to construct a new device for computing (14) and even (13). Many processors for matrix multiplication and for other algorithms of linear algebra are based on computing scalar products and on the corresponding "elementary" devices respectively, etc.

There are some methods to make these new devices more universal than their prototypes. There is a modest collection of possible operations for standard numerical semirings: max, min, and the usual arithmetic operations. So, it is easy to construct programmable hardware processors with variable basic operations. Using modern technologies it is possible to construct cheap special-purpose multi-processor chips implementing examined algorithms. The so-called systolic processors are especially convenient for this purpose. A systolic array is a "homogeneous" computing medium consisting of elementary processors, where the general scheme and processor connections are simple and regular. Every elementary processor pumps data in and out performing elementary operations in a such way that the corresponding data flow is kept up in the computing medium; there is an analogy with the blood circulation and this is a reason for the term "systolic," see e.g., [39, 40, 43, 52, 65–67].

Concrete systolic processors for the general algebraic path problem are presented in [65–67]. In particular, there is a systolic array of $n(n + 1)$ elementary processors which performs computations of the Gauss–Jordan elimination algorithm and can solve the algebraic path problem within $5n - 2$ time steps. Of course, hardware implementations for important and popular basic algorithms increase the speed of data processing.

The so-called GPGPU (General-Purpose computing on Graphics Processing Units) technique is another important field for applications of the correspondence principle. The thing is that graphic processing units (hidden in modern laptop and desktop computers) are potentially powerful processors for solving numerical problems. The recent tremendous progress in graphical processing hardware and software leads to new "open" programmable parallel computational devices (special processors), see, e.g., [78–80]. These devices are going to be standard for coming PC (personal computers) generations. Initially used for graphical processing only (at that time they were called GPUs), today they are used for various fields, including audio and video processing, computer simulation, and encryption. But this list can be considerably enlarged following the correspondence principle: the basic operations would be used as parameters. Using the technique described in this paper (see also our references), standard linear algebra algorithms can be used for solving different problems in different areas. In fact, the hardware supports all operations needed for the most important idempotent semirings: plus, times, min, max. The most popular linear algebra packages [ATLAS (Automatically Tuned Linear Algebra Software), LAPACK, PLASMA (Parallel Linear Algebra for Scalable Multicore Architectures)] can already use GPGPU, see [81–83]. We propose to make these tools more powerful by using parameterized algorithms.

Linear algebra over the most important numerical semirings generates solutions for many concrete problems in different areas, see, e.g., Sects. 4 and 5 above and references indicated in these sections.

Note that to be consistent with operations we have to redefine zero (0) and unit (1) elements (see above); comparison operations must be also redefined as it is described in Sect. 4.1. "Basic definitions." Once the operations are redefined, then the most of basic linear algebra algorithms, including back and forward substitution, Gauss elimination method, Jordan elimination method and others could be rewritten for new domains and data structures. Combined with the power of the new parallel hardware this approach could change PC from entertainment devices to power full instruments.

10   The Correspondence Principle for Software Design

Software implementations for universal semiring algorithms are not as efficient as hardware ones (with respect to the computation speed) but they are much more flexible. Program modules can deal with abstract (and variable) operations and data types. Concrete values for these operations and data types can be defined by the corresponding input data. In this case concrete operations and data types are generated by means of additional program modules. For programs written in this manner it is convenient to use special techniques of the so-called object oriented (and functional) design, see, e.g., [47, 62, 70]. Fortunately, powerful tools supporting the object-oriented software design have recently appeared including compilers for real and convenient programming languages (e.g., C^{++} and Java) and modern computer algebra systems.

Recently, this type of programming technique has been dubbed generic programming (see, e.g., [6, 62]). To help automate the generic programming, the so-called Standard Template Library (STL) was developed in the framework of C^{++} [62, 70]. However, high-level tools, such as STL, possess both obvious advantages and some disadvantages and must be used with caution.

It seems that it is natural to obtain an implementation of the correspondence principle approach to scientific calculations in the form of a powerful software system based on a collection of universal algorithms. This approach ensures a working time reduction for programmers and users because of the software unification. The arbitrary necessary accuracy and safety of numeric calculations can be ensured as well.

The system has to contain several levels (including programmer and user levels) and many modules.

Roughly speaking, it must be divided into three parts. The first part contains modules that implement domain modules (finite representations of basic mathematical objects). The second part implements universal (invariant) calculation methods. The third part contains modules implementing model dependent algorithms. These modules may be used in user programs written in C^{++}, Java, Maple, Mathlab etc.

The system has to contain the following modules:

- Domain modules:
  - Infinite precision integers;
  - Rational numbers;
  - Finite precision rational numbers (see [44]);
  - Finite precision complex rational numbers;
  - Fixed- and floating-slash rational numbers;
  - Complex rational numbers;
  - Arbitrary precision floating-point real numbers;
  - Arbitrary precision complex numbers;
  - p-adic numbers;
  - Interval numbers;
  - Ring of polynomials over different rings;
  - Idempotent semirings;
  - Interval idempotent semirings;
  - And others.

- Algorithms:
  - Linear algebra;
  - Numerical integration;
  - Roots of polynomials;
  - Spline interpolations and approximations;
  - Rational and polynomial interpolations and approximations;
  - Special functions calculation;
  - Differential equations;
  - Optimization and optimal control;
  - Idempotent functional analysis;
  - And others.

This software system may be especially useful for designers of algorithms, software engineers, students and mathematicians.

Note that there are some software systems oriented to calculations with idempotent semirings like $\mathbf{R}_{max}$; see, e.g., [64]. However these systems do not support universal algorithms.

11 Interval Analysis in Idempotent Mathematics

Traditional interval analysis is a nontrivial and popular mathematical area, see, e.g., [3, 20, 34, 54, 56, 59]. An "idempotent" version of interval analysis (and moreover interval analysis over positive semirings) appeared in [45, 46, 69]. Later appeared rather many publications on the subject, see, e.g., [12, 20, 27, 57, 58, 77]. Interval analysis over the positive semiring $\mathbf{R}_+$ was discussed in [8].

Let a set S be partially ordered by a relation $\preccurlyeq$. A *closed interval* in S is a subset of the form $\mathbf{x} = [\underline{\mathbf{x}}, \overline{\mathbf{x}}] = \{ x \in S \mid \underline{\mathbf{x}} \preccurlyeq x \preccurlyeq \overline{\mathbf{x}} \}$, where the elements $\underline{\mathbf{x}} \preccurlyeq \overline{\mathbf{x}}$ are called *lower* and *upper bounds* of the interval $\mathbf{x}$. The order $\preccurlyeq$ induces a partial ordering on the set of all closed intervals in S: $\mathbf{x} \preccurlyeq \mathbf{y}$ iff $\underline{\mathbf{x}} \preccurlyeq \underline{\mathbf{y}}$ and $\overline{\mathbf{x}} \preccurlyeq \overline{\mathbf{y}}$.

A *weak interval extension* $I(S)$ of a positive semiring S is the set of all closed intervals in S endowed with operations $\oplus$ and $\odot$ defined as $\mathbf{x} \oplus \mathbf{y} = [\underline{\mathbf{x}} \oplus \underline{\mathbf{y}}, \overline{\mathbf{x}} \oplus \overline{\mathbf{y}}]$, $\mathbf{x} \odot \mathbf{y} = [\underline{\mathbf{x}} \odot \underline{\mathbf{y}}, \overline{\mathbf{x}} \odot \overline{\mathbf{y}}]$ and a partial order induced by the order in S. The closure operation in $I(S)$ is defined by $\mathbf{x}^* = [\underline{\mathbf{x}}^*, \overline{\mathbf{x}}^*]$. There are some other interval extensions (including the so-called strong interval extension [46]) but the weak extension is more convenient.

The extension $I(S)$ is positive; $I(S)$ is idempotent if S is an idempotent semiring. A universal algorithm over S can be applied to $I(S)$ and we shall get an interval version of the initial algorithm. Usually both the versions have the same complexity. For the discrete stationary Bellman equation and the corresponding optimization problems on graphs interval analysis was examined in [45, 46] in details. Other problems of idempotent linear algebra were examined in [12, 27, 57, 58, 77].

Idempotent mathematics appears to be remarkably simpler than its traditional analog. For example, in traditional interval arithmetic, multiplication of intervals is not distributive with respect to addition of intervals, whereas in idempotent interval arithmetic this distributivity is preserved. Moreover, in traditional interval analysis the set of all square interval matrices of a given order does not form even a semigroup with respect to matrix multiplication: this operation is not associative since distributivity is lost in the traditional interval arithmetic. On the contrary, in the idempotent (and positive) case associativity is preserved. Finally, in traditional interval analysis some problems of linear algebra, such as solution of a linear system of interval equations, can be very difficult (generally speaking, they are NP-hard, see [14, 20, 33, 34] and references therein). It was noticed in [45, 46] that in the idempotent case solving an interval linear system requires a polynomial number of operations (similarly to the usual Gauss elimination algorithm). Two properties that make the idempotent interval arithmetic so simple are monotonicity of arithmetic operations and positivity of all elements of an idempotent semiring.

Usually interval estimates in idempotent mathematics are exact. In the traditional theory such estimates tend to be overly pessimistic.

Acknowledgements This work is partially supported by the RFBR grant 08-01-00601. The authors are grateful to A. G. Kushner for his kind help.

References

1. Aho, A.V., Ullman, J.D.: The Theory of Parsing, Translation and Compiling, Vol. 2: Compiling. Prentice-Hall, NJ (1973)
2. Aho, A.V., Hopcroft, J.E., Ullman, J.D.: The Design and Analysis of Computer Algorithms. Addison-Wesley, MA (1976)

Universal Algorithms, Mathematics of Semirings and Parallel Computations   87

3. Alefeld, G., Herzberger, J.: Introduction to Interval Computations. Academic Press, New York (1983)
4. Avdoshin, S.M., Belov, V.V., Maslov, V.P., Chebotarev, A.M.: Design of computational media: mathematical aspects. In: Maslov, V.P., Volosov, K.A. (eds.) Mathematical Aspects of Computer Engineering. MIR Publishers, Moscow (1988) 9–145
5. Baccelli, F., Cohen, G., Olsder, G.J., Quadrat, J.P.: Synchronization and Linearity: An Algebra for Discrete Event Systems. Wiley, New York (1992)
6. Backhouse, R., Jansson, P., Jeuring, J., Meertens, L.: Generic Programming – An Introduction. Lect. Notes Comput. Sci., 1608 (1999) 28–115
7. Backhouse, R.C., Carré, B.A.: Regular algebra applied to path-finding problems. J. Inst. Math. Appl **15** (1975) 161–186
8. Barth, W., Nuding, E., Optimale Lösung von Intervallgleichungsystemen. Computing **12** (1974) 117–125
9. Butkovič, P., Zimmermann, K.: A strongly polynomial algorithm for solving two-sided linear systems in max-algebra. Discrete Appl. Math. **154** (2006) 437–446
10. Carré, B.A.: An algebra for network routing problems. J. Inst. Appl. **7** (1971) 273–294
11. Carré, B.A.: Graphs and Networks. The Clarendon Press, Oxford (1979)
12. Cechlárová, K., Cuninghame-Green, R.A.: Interval systems of max-separable linear equations. Linear Algebra and its Applications **340** (2002) 215–224
13. Cohen, G., Gaubert, S., Quadrat, J.P.: Max-plus algebra and system theory: where we are and where to go now. Annu. Rev. Contr. **23** (1999) 207–219
14. Coxson, G.E.: Computing exact bounds on elements of an inverse interval matrix is NP-hard. Reliable Comput. **5** (1999) 137–142
15. Cuninghame-Green, R.A.: Minimax algebra. Lect. Notes in Economics and Mathematical Systems, 166. Springer, Berlin (1979)
16. Cuninghame-Green, R.A.: Minimax algebra and applications. Adv. Imag. Electron Phys. **90** (1995) 1–121
17. Cuningham-Green, R.A.: Minimax algebra and its applications. Fuzzy Set. Syst. **41** (1991) 251–267
18. Cuningham-Green, R.A., Butkovic, P.: The equation $a \otimes x = b \otimes y$ over (max, +). Theor. Comput. Sci. **293** (2003) 3–12
19. Del Moral, P.: A survey of Maslov optimization theory: optimality versus randomness. In: Kolokoltsov, V.N., Maslov, V.P. (eds.) Idempotent Analysis and Applications, pp. 243–302. Kluwer, Dordrecht (1997) (Appendix)
20. Fiedler, M., Nedoma, J., Ramik, J., Rohn, J., Zimmermann, K.: Linear Optimization Problems with Inexact Data. Springer, New York (2006)
21. Glazek, K.: A Guide to the Literature on Semirings and their Applications in Mathematics and Information Sciences: With Complete Bibliography. Kluwer, Dordrecht (2002)
22. Golan, J.S.: Semirings and their Applications. Kluwer, Dordrecht (1999)
23. Gondran, M.: Path algebra and algorithms. In: Roy, B. (ed.). Combinatorial Programming: Methods and Applications, NATO Adv. Study Inst. Ser., Ser. C. **19** (1975) 137–148
24. Gondran, M., Minoux, M.: Graphes et algorithmes. Editions Eyrolles, Paris (1979, 1988)
25. Gondran, M., Minoux, M.: Graphes, dioïdes et semi-anneaux. Editions TEC&DOC, Paris (2001)
26. Gunawardena, J. (ed.): Idempotency. Publ. of the Newton Institute 11, Cambridge University Press, Cambridge (1998)
27. Hardouin, L., Cottenceau, B., Lhommeau, M., Le Corronc, E.: Interval systems over idempotent semirings. Lin. Algebra Appl. **431** (2009) 855–862
28. Itenberg, I., Mikhalkin, G., Shustin, E.: Tropical Algebraic Geometry, Oberwolfach Seminars 35. Birkhäuser, Basel (2007)
29. Kleene, S.C.: Representation of events in nerve sets and finite automata. In: McCarthy, J., Shannon, C. (eds.). Automata Studies, 3–40. Princeton University Press, Princeton (1956)
30. Klement, E.P., Pap, E. (eds.), Mathematics of Fuzzy Systems. 25th Linz Seminar on Fuzzy Set Theory, Linz, Austria, Feb 3–7, 2004, Abstracts. J. Kepler University, Linz (2004)

31. Kolokoltsov, V.N.: Idempotency structures in optimization. J. Math. Sci. **104**(1) (2001) 847–880
32. Kolokoltsov, V., Maslov, V.: Idempotent analysis and applications. Kluwer, Dordrecht (1997)
33. Kreinovich, V., Lakeyev, A.V., Noskov, S.I.: Optimal solution of interval linear systems is intractable (*NP*-hard). Interval Computations **1** (1993) 6–14
34. Kreinovich, V., Lakeyev, A.V., Rohn, J., Kahl, P.: Computational Complexity and Feasibility of Data Processing and Interval Computations. Kluwer, Dordrecht (1998)
35. Kung, H.T.: Two-level pipelined systolic arrays for matrix multiplication, polynomial evaluation and discrete Fourier transformation. In: Demongeof, J. et al. (eds.). Dynamical Systems and Cellular Automata, 321–330. Academic Press, New York (1985)
36. Lehmann, D.J.: Algebraic structures for transitive closure. Theor. Comput. Sci. **4** (1977) 59–76
37. Litvinov, G.L.: Dequantization of mathematics, idempotent semirings and fuzzy sets. In: Klement, E.P., Pap, E. (eds.), Mathematics of Fuzzy Systems, 113–117. J. Kepler University, Linz (2004)
38. Litvinov, G.L.: The Maslov dequantization, idempotent and tropical mathematics: a brief introduction. J. Math. Sci. **140**(3) (2007) 426–444; arXiv:math.GM/0507014
39. Litvinov, G.L., Maslov, V.P.: Correspondence principle for idempotent calculus and some computer applications. (IHES/M/95/33), Institut des Hautes Etudes Scientifiques, Bures-sur-Yvette (1995); arXiv:math.GM/0101021
40. Litvinov, G.L., Maslov, V.P.: The correspondence principle for idempotent calculus and some computer applications. In: Gunawardena, J. (ed.). Idempotency. Publ. of the Newton Institute 11, 20–443. Cambridge University Press, Cambridge (1998)
41. Litvinov, G.L. Maslov, V.P. (eds.): Idempotent mathematics and mathematical physics. Contemporary Mathematics **377**, AMS, Providence, RI (2005)
42. Litvinov, G.L., Maslova, E.V.: Universal numerical algorithms and their software implementation. Programming and Computer Software **26**(5) (2000) 275–280; arXiv:math.SC/0102114
43. Litvinov, G.L., Maslov, V.P., Rodionov, A.Ya.: A unifying approach to software and hardware design for scientific calculations and idempotent mathematics. International Sophus Lie Centre, Moscow (2000); arXiv:math.SC/0101069
44. Litvinov, G.L., Rodionov, A.Ya., Tchourkin, A.V.: Approximate rational arithmetics and arbitrary precision computations for universal algorithms. Int. J. Pure Appl. Math. **45**(2) (2008) 193–204; arXiv:math.NA/0101152
45. Litvinov, G.L., Sobolevskiĭ, A.N.: Exact interval solutions of the discrete Bellman equation and polynomial complexity of problems in interval idempotent linear algebra. Dokl. Math. **62**(2) (2000) 199–201; arXiv:math.LA/0101041
46. Litvinov, G.L., Sobolevskiĭ, A.N.: Idempotent interval analysis and optimization problems. Reliable Comput. **7**(5) (2001) 353–377; arXiv:math.SC/0101080
47. Lorenz, M.: Object Oriented Software Development: A Practical Guide. Prentice Hall, NJ (1993)
48. Maslov, V.P.: New superposition principle for optimization problems. In: Seminaire sur les Equations aux Dérivées Partielles 1985/86. Centre Math. De l'Ecole Polytechnique, Palaiseau (1986) exposé 24
49. Maslov, V.P.: A new approach to generalized solutions of nonlinear systems. Soviet Math. Dokl. **42**(1) (1987) 29–33
50. Maslov, V.P.: On a new superposition principle for optimization problems. Uspekhi Mat. Nauk [Russian Math. Surveys] **42**(3) (1987) 39–48
51. Maslov, V.P.: Méthodes opératorielles. MIR, Moscow (1987)
52. Maslov, V.P. et al.: Mathematics of semirings and its applications. Technical report, Institute for New Technologies, Moscow (1991) (in Russian)
53. Maslov, V.P., Volosov, K.A. (eds.). Mathematical Aspects of Computer Engineering. MIR, Moscow (1988)
54. Matijasevich, Yu.V.: A posteriori version of interval analysis. In: Arató, M., Kátai, I., Varga, L. (eds.). Topics in the Theoretical Basis and Applications of Computer Sciences, 339–349. Proc. Fourth Hung. Computer Sci. Conf., Budapest, Acad. Kiado (1986)

55. Mikhalkin, G.: Tropical geometry and its applications. Proceedings of the ICM, Madrid, Spain **2** (2006) 827–852; arXiv:math.AG/0601041v2
56. Moore, R.E.: Methods and Applications of Interval Analysis. SIAM, Philadelphia (1979)
57. Myskova, H.: Interval systems of max-separable linear equations. Lin. Algebra Appl. **403** (2005) 263–272
58. Myskova, H.: Control solvability of interval systems of max-separable linear equations. Lin. Algebra Appl. **416** (2006) 215–223
59. Neumayer, A.: Interval Methods for Systems of Equations. Cambridge University Press, Cambridge (1990)
60. Pandit, S.N.N.: A new matrix calculus, SIAM J. Appl. Math. **9** (1961) 632–639
61. Pin, J.E.: Tropical semirings. In: Gunawardena, J. (ed.). Idempotency. Publ. of the Newton Institute 11, 50–60. Cambridge University Press, Cambridge (1998)
62. Pohl, I.: Object-Oriented Programming Using C^{++} (2nd edn). Addison-Wesley, MA (1997)
63. Quadrat, J.P.: Théorèmes asymptotiques en programmation dynamique. Comptes Rendus Acad. Sci. Paris **311** (1990) 745–748
64. Quadrat, J.P., Max-Plus working group: Maxplus Algebra Software. http://scilab.org/contrib; http://maxplus.org (2007)
65. Robert, Y., Tristram, D.: An orthogonal systolic array for the algebraic path problem. Computing **39** (1987) 187–199
66. Rote, G.: A systolic array algorithm for the algebraic path problem (shortest paths; matrix inversion). Computing **34** (1985) 191–219
67. Sedukhin, S.G.: Design and analysis of systolic algorithms for the algebraic path problem. Comput. Artif. Intell. **11**(3) (1992) 269–292
68. Simon, I.: Recognizable sets with multiplicities in the tropical semiring. Lecture Notes in Computer Science, 324 (1988) 107–120
69. Sobolevskiĭ, A.N.: nterval arithmetic and linear algebra over idempotent semirings. Doklady Akademii Nauk **369** (1999) 747–749 (in Russian)
70. Stepanov, A., Lee, M.: The standard template library. Hewlett-Packard, Palo Alto (1994)
71. Viro, O.: Dequantization of real algebraic geometry on a logarithmic paper. In: 3rd European Congress of Mathematics, Barcelona (2000); arXiv:math/0005163
72. Viro, O.: From the sixteenth Hilbert problem to tropical geometry. Japan. J. Math. **3** (2008) 1–30
73. Voevodin, V.V.: Mathematical foundation of parallel computings, 343. World Scientific, Singapore (1992)
74. Vorobjev, N.N.: The extremal matrix algebra. Soviet Math. Dokl. **4** (1963) 1220–1223
75. Vorobjev, N.N.: Extremal algebra of positive matrices. Elektronische Informationsverarbeitung und Kybernetik **3** (1967) 39–57 (in Russian)
76. Vorobjev, N.N.: Extremal algebra of nonnegative matrices. Elektronische Informationsverarbeitung und Kybernetik **6** (1970) 302–312 (in Russian).
77. Zimmermann, K.: Interval linear systems and optimization problems over max-algebras. In: Fiedler, M., Nedoma, J., Ramik, J., Rohn, J., Zimmermann, K. (eds.). Linear Optimization Problems with Inexact Data, 165–193. Springer, New York (2006)
78. Proceedings of the 23rd IEEE International Parallel and Distributed Processing Symposium (IPDPS'09), Rome, Italy May 23–May 29, IEEE Computer Society Press, 2009; ISBN: 978-1-4244-3751-1
79. Blithe, D. Rise of the graphics processor. Proc. of the IEEE **96**(5) (2008) 761–778
80. Owens, J.D. et al.: GPU computing. Proc. of the IEEE **96**(5) (2008) 879–899
81. ATLAS: http://math-atlas.sourceforge.net/
82. LAPACK: http://www.netlib.org/lapack/
83. PLASMA: http://icl.cs.utk.edu/plasma/

Scaling Invariant Interpolation for Singularly Perturbed Vector Fields (SPVF)

Viatcheslav Bykov, Vladimir Gol'dshtein, and Ulrich Maas

Abstract The problem of modelling, numerical simulations and interpretation of the simulations results of complex systems arising in reacting flows requires more and more sophisticated methods of qualitative system analysis. Recently, the concept of invariant, slow/fast, attractive manifolds has proven to be an efficient tool for such an analysis. In particular, it allows us to study main properties of detailed models describing the reacting flow by considering appropriate low dimensional manifolds, which appear naturally in the system state/composition space as a manifestation of a restricted number of real degrees of freedom exhibited by the system.

In order to answer the question of what are the minimal number of the real degrees of freedom (real system dimension) and to approximate low dimensional manifolds (i.e., reduced system's phase spaces) the concept of Singularly Perturbed Vector Fields (SPVF) has been suggested lately [1]. In the current work a scales invariant version of the SPVF will be presented and discussed.

1 Introduction

During the last 20 years, the rapid progress in computational ability of workstations, supercomputers and numerical integration software packages has considerably increased the role of accurate model formulation in the analysis of the dynamics of reacting flows. This is because simple models can not satisfy the needs of industry and science anymore. The amount of experimental data grows rapidly as well. Therefore, current models have to be adjusted to constantly growing data provided by experiment studies. As a result more and more detailed and sophisticated

---

V. Bykov (✉) and U. Maas
Karlsruher Institut fuer Technologie, Institut fuer Technische Thermodynamik,
Engelbert-Arndold-Strasse 4, Geb. 10.91, 76131 Karlsruhe, Germany
e-mail: bykov@itt.uni-karlsruhe.de, ulrich.maas@kit.edu

V. Gol'dshtein
Department of Mathematics, Ben Gurion University of the Negev, Beer-Sheva 84105, Israel
e-mail: vladimir@bgu.ac.il

A.N. Gorban and D. Roose (eds.), *Coping with Complexity: Model Reduction and Data Analysis*, Lecture Notes in Computational Science and Engineering 75, DOI 10.1007/978-3-642-14941-2_5, © Springer-Verlag Berlin Heidelberg 2011

descriptions of the system of reacting flows are in use, which leads to an increase of the system dimension and its complexity. This trend is crucial, particularly in modelling of reacting flows of combustion, biochemical processes and atmospheric chemistry.

At present, there are a lot of software packages for integration of system of governing kinetic equations (CHEMKIN, COSILAB, KINTECH, HOMREA etc., see e.g., [2–4]) as well as for development of detailed kinetic mechanisms, (see additionally [5–7] for more information). Nowadays, the development stage has become almost automatic. Therefore typical kinetic mechanisms provided by such packages of model development comprises of hundreds of chemically reacting species [8] and thousands of elementary reactions. Besides the problem of treatment of a high dimensional system, the detailed description of chemical reactions introduces a range of characteristic time scales that cover about ten to twelve orders of magnitude (from 1 to 10^{-12} s). The drastic disparity in time-scales results in a so-called stiffness of system of governing equations of a reacting flow and also causes scaling problems in space. These scaling problems together with the high dimensionality and the non-linearity of the chemical source term creates serious difficulties for numerical solution procedures and make the practical problems beyond the solving capacity of even today's workstations and super-computers.

There is strong need for algorithms that can handle automatically the enormous dimension, stiffness and non-linearity of reacting flow systems. It is of great interest to develop a tool that allows us to find a balance between complexity and required accuracy of the resulting model. There are two standard ways to deal with these problems. The physical model can be simplified first, meaning that problems of academic interest e.g., small length scales, simple geometries, symmetric flows etc., might be considered at this point. Alternatively, methods of model reduction can be developed and used to describe the chemical kinetics by means of reduced systems, while still retaining the essential dynamics of the detailed system without significant damage to the overall qualitative and quantitative properties of the reduced model by comparison with the original one [9, 10]. In this respect the current work is aimed to introduce a hierarchical system analysis and to present the recent progress in model reduction by "manifolds" methods. In particular, this work aims to single out the difference between the regular and singular perturbation analysis and the role of both approaches in the qualitative system analysis.

1.1 Multi-Scales Hierarchy and "Manifold" Methods

Practical reacting flow systems involve sub-processes with essentially different time scales defined as the reciprocal values of certain eigenvalues (see Fig. 1, where typical scales are sketched) this will be discussed in detail (see Sect. 2.3 and 3.2). As it was pointed out above, drastically different time scales cause the stiffness of the system of governing equations on one hand, but on the other hand they lead to the fact that there are some fast modes or processes which are quickly relaxed, some

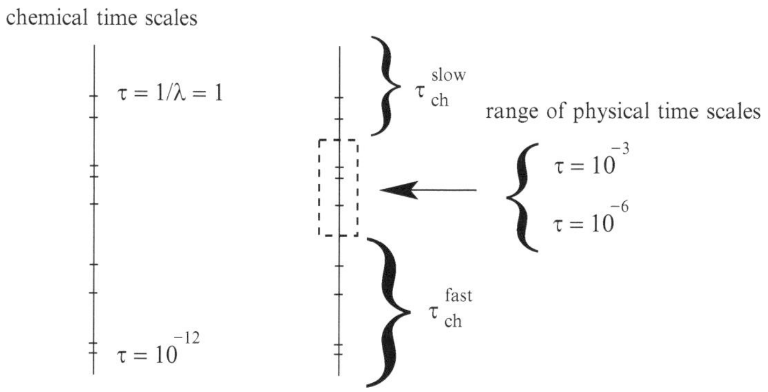

Scaling Invariant SPVF

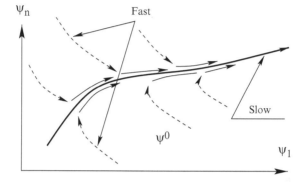

Fig. 1 Sketch of chemical and physical time scales

Fig. 2 State or phase space geometry shows that particular system solution trajectory can be decomposed into the fast motion represented by *dashed lines* and the slow motion (small *solid lines* with *arrows*) along the slow invariant attractor

stay stationary or quasi-stationary and finally only the slow processes govern the overall system dynamics. In this case, the long term system evolution is represented primarily by the dynamics of the slow processes along something that can be described or understood as a stable invariant geometrical attractor in the detailed state space (Fig. 2 illustrates typical phase portrait of such systems). This approach assumes the existence and uniqueness of the slow invariant manifold in the defined domain in the state space [11].

The main idea of any reduction methodology, thus, is in considering thermo-chemical states of the system to be confined only to this slow manifolds and the whole state vector can be redefined by using the manifolds. It is important to repeat that although the existence of multiple scales complicates the numerical treatment of the reacting flow system enormously it also allows to construct low-dimensional approximations of the detailed model describing the so-called short and longterm dynamics as well as the rate limiting processes accurately. Accordingly, finding of these manifolds along with a subset or combination of system parameters that consistently represents the full detailed system on this manifold in terms of a sufficiently small number of variables/parameters used to parametrize the manifold is the main purpose of any rigorous method devised to reduce the model. Furthermore, byproducts of such reduction methodology, i.e., reducing the dimension by removing the fast modes, is the decrease of the stiffness of the system of governing differential equations. This makes the existence of a multi-scales structure an important feature of chemical kinetic mechanisms and of reacting flow systems in general.

The main problem of approaches dealing with multi-scales system hierarchy lies within its implicit nature, i.e., for complex realistic models this structure is typically hidden in original mathematical model. Sometimes, it is possible to represent the multi-scales structure explicitly (even analytically) for comparatively simple models. One such model is discussed in detail in Sect. 2.2.1. In general, however, the transition from the original system with the hidden multi-scale structure to its explicit representation is not that trivial. As it will be shown below, in order to obtain an explicit decomposed form, deep understanding of physics, chemistry and properties of the mathematical model of the phenomenon under consideration is required. Moreover, this decomposition will always be confined to a particular asymptotic limit. Therefore, substantial amount of time, human resources and experience is needed to find a suitable system representation in a strictly decomposed form, where the system hierarchy is defined explicitly.

1.2 General Problems of Model Reduction

An automatic procedure for model reduction represents an alternative way to proceed with applications. Many methods performing automatic model reduction have been devised recently aiming to reduce stiffness, dimension, CPU time and memory storage for numerical treatments of different kinds of reacting flows (see e.g., [5, 7, 10, 12] for an overview of methods). However, in spite of many advantages these currently available automatic reduction methods there are principal and fundamental drawbacks of most approaches. The basic difficulty of model reduction is the lack of rigorous and united framework of model reduction by manifold's method that handles the most important theoretical problems with

1. Existence of fast/slow manifolds
2. Optimal dimension of fast/slow manifolds
3. Representation in the decomposed form
4. Stability of slow manifolds
5. Invariance of fast/slow manifolds.

Most methods solve only a part of arising problems, for instance, the Method of Invariant Manifolds/Grids (MIM/G) by Alexander Gorban and Ilya Karlin [13–15] is fully based on the invariance property while the decomposition (e.g., multiple time scales) is handled implicitly by implementing Newton iterations and the appropriate projection technique aiming to approximate the decomposition and the slow system manifold (the invariant manifold of slow motions) simultaneously. In particular, it means that the method has no problems with 5. Invariance (of the slow manifold) and 3. Decomposition, but problems with 1. Existence as well as with 2. Dimension (an optimal reduced model dimension) and 4. Stability of the slow manifold restricts the range of application of the MIM/G. Accordingly, in this case the system dynamics does not have a transient fast/slow behaviour one can safely use the designed reduced model on the basis of the MIM/G manifold [12,14]. However, once the system leaves the slow invariant manifold the reduced model based on the MIM/G would not be accurate anymore.

According to the method, one fixes first the dimension, defines initial guess for the manifold and then applies the so-called film equation to find the manifold (see e.g., [14] for more information). Therefore, the fact that a manifold of a fixed dimension with certain important properties e.g., attractive, stable and invariant exist everywhere in the whole domain of interest in the system state space as well as in the accessed region in the system parameter space is fundamental assumption of the method. Hence, in application of the MIM method this fact has to be thoroughly investigated before implementing the reduced model in real numerical simulation.

Another MIM – the Method of Integral Manifolds [16] suffers of a fundamental problem with the definition of the decomposition. It allows to study previously identified hierarchical structure of the system in detail, but it assumes a certain form of the system, namely, the system is considered in fully decomposed form as a Singularly Perturbed System (SPS) [17], i.e., the most complicated problems (2–3) are skipped. Obviously, this framework is extremely useful only when the major part of the system hierarchy analysis is already performed, and the important problem with identifying the dimension is already solved. It is clear that problems of the industrial level having high dimensionality and complexity cannot be decomposed that easily.

At present there is no method that allows to treat the problems above in a general and complete form. In the current work, however, we will further discuss an algorithm which overcomes most drawbacks of existing reduction methods concerning the problems above. In particular, it allows an automatic analysis of stability, dimension and existence of the slow manifolds, structure of fast motions/manifolds whenever specified conditions regarding the considered system are valid, i.e., the system defines the linearly decomposed Singularly Perturbed Vector Field (SPVF) [1, 18]. The word linear in this context signifies the existence of the linear transformation of the original coordinates leading to the representation of the system as the Singularly Perturbed System. It means that the fast manifolds of the original system are supposed to be linear.

This work aims to further develop and discuss a general mathematical background for model reduction, to fill a gap between model analysis and applications. In our view, only after the basics of model reduction have been established, understood and well studied, i.e., the language of model reduction is formulated; a straightforward and most efficient way of model reduction and its implementation can be found.

1.3 Mathematical Model of Reacting Flows

The strongly coupled description and interaction of reaction and transport processes [19] is modeled by the system of partial differential equations in most general form as the closed set of the standard conservation equations:

$$\frac{\partial \rho}{\partial t} + \text{div}\,(\rho\,\text{v}) = 0$$

$$\frac{\partial \rho_i}{\partial t} + \text{div}\,(\rho_i\,\text{v}) + \text{div}\,(\text{j}_i) = \rho\,M_i\,F_i$$

$$\frac{\partial\,(\rho\,\text{v})}{\partial t} + \text{div}\,(\rho\,\text{v} \otimes \text{v}) + \text{div}\,\left(\bar{\bar{\text{p}}}\right) = 0 \qquad (1)$$

$$\frac{\partial\,(\rho\,\text{u})}{\partial t} + \text{div}\,\left(\rho\,\text{u}\,\text{v} + \text{j}_q\right) + \bar{\bar{\text{p}}} : \text{grad}\,(\text{v}) = 0$$

$$p = (\rho,\,\text{u},\,\rho_1,\dots,\rho_{n_s})$$

where t denotes the time, v represents the velocity field, ρ the density, ρ_i the density of specie i, j_i the diffusion flux density of specie i, M_i the molar mass of specie of specie i, F_i the molar scale rate of formation of specie i, $\bar{\bar{\text{p}}}$ the pressure tensor, u the specific inner energy, j_q the conductive heat flux density, n_s number of different chemical species. The $(n = n_s + 2)$-dimensional vector $\psi = \left(h, p, \frac{w_1}{M_1}, \dots, \frac{w_{n_s}}{M_{n_s}}\right)^T$ can be used to shorten notations, where $w_1, \dots, w_{n_s}$ are the n_s species mass fractions $w_i = \frac{\rho_i}{\rho}$. The typical system of PDEs equation (1) for modelling the scalar field reacting flows in general symbolic vector notation can be written as

$$\frac{\partial \psi}{\partial t} = F\,(\psi) - \text{v} \cdot \text{grad}\,(\psi) - \frac{1}{\rho}\,\text{div}\,(D \cdot \text{grad}\,(\psi)) \equiv \Phi\,(\psi), \qquad (2)$$

here h denotes the enthalpy, p the pressure. D is the $(n$ by $n)$-dimensional matrix of the transport coefficients and $F(\psi)$ is the n-dimensional vector of thermo-chemical source terms [8]. Both are very complicated and non-linear functions of the thermo-kinetic state vector ψ (see e.g., [20–22]).

Because most problems with scaling and high dimension originate from the very stiff reaction source term (just compare typical time scales' range shown in Fig. 1) the analysis of the source term is a very important step of model reduction. Therefore, in the present work the main attention is paid on the following system of ODEs

$$\frac{d\psi}{dt} = F\,(\psi)\,. \qquad (3)$$

Questions of coupling with transport processes – with the other two terms in (2) are also under intensive investigations (see e.g., [23–26]), however, they are not discussed in this work.

At present there are several methods for an analysis of the system's (3) hierarchy. These methods have been improved constantly producing variations of methods designed for special tasks (see e.g., [5, 7, 10, 12]).

This study, both in the analysis and in the implementation of the proposed methods are based on one of them, i.e., on Intrinsic Low-Dimensional Manifolds method (ILDM), which is a kind of reference approach for this study (see e.g., [27] for definitions and detail). It can be considered as a main motivation of the authors' work in the field. It has a rigorous mathematical background established in a number of works [28–30], where it has been shown that the (ILDM) provides a good approximation of the invariant manifold of slow motions (up to the second order [29]). Furthermore, it combines both the system hierarchy analysis and a complete method of automatic reduced model generation, which provides an efficient subsequent implementation scheme into CFD codes [31, 32]. The reason for this choice is simple, at present, the ILDM method is an efficient and robust method of model reduction, moreover, it is realized in the FORTRAN codes HOMREA and INSFLA [33, 34].

2 Mathematics of Model Reduction

The basic concepts and problems of model reduction are discussed in this section. Because the use of multi-scale nature of different subsystems is almost universally used for modelling in physics and engineering, it is natural to start with introducing an original mathematical model for multi-scale phenomena, which is the standard Singularly Perturbed System (SPS) [16, 17]. The concept of the SPS is modified to include regular perturbation part that is relevant to model reduction. According to this concept, three sub-systems describe the so-called conserved, slow and fast motions of the system solution trajectory, moreover, discrepancies in time scales are controlled by defined small system parameters.

2.1 *SPS as a Mathematical Model of Multi-Scales Behavior*

A simple mathematical model for the decomposition of motions is the Singularly/Regularly Perturbed System (SPS).

$$
\begin{cases}
\frac{dU}{dt} = \delta F_c\,(U, V, W) \\
\frac{dV}{dt} = F_s\,(U, V, W) \\
\varepsilon \frac{dW}{dt} = F_f\,(U, V, W)
\end{cases}
\tag{4}
$$

Typically, the conserved part (the first subsystem with δ in (4) representing the first linear integrals of the system) is removed on the preliminary stage of the system analysis. However, for the method aiming to treat the original system in the most general form the conserved part, which is extended here to include a quasi-conserved part describing very slow processes, should be added to the SPS.

The system (4) describes the evolution of the so-called conserved or quasi-conserved variables

$$U = (U_1, \ldots, U_{m_c}),$$

slow variables

$$V = (V_1, \ldots, V_{m_s}),$$

which are assumed to change slowly compared to fast ones

$$W = (W_1, \ldots, W_{m_f}), \quad m_c + m_s + m_f = n.$$

Due to the presence of the system small parameters $0 < \varepsilon, \delta \ll 1$, and $F_c(U, V, W) \sim O(1)$, $F_s(U, V, W) \sim O(1)$ and $F_f(U, V, W) \sim O(1)$ the system (4) decomposes into three subsystems. A short outline of the main notations, definitions and concepts will be provided now. First of all, let us note that in the present form the main question of model reduction shrinks to the definition of the relation between system (3) and system (4). It is assumed in this work the transformation is linear, e.g., there is the constant block matrices

$$Z = \begin{pmatrix} Z_c & Z_s & Z_f \end{pmatrix} : \begin{pmatrix} Z_c & Z_s & Z_f \end{pmatrix} \cdot \begin{pmatrix} \tilde{Z}_c \\ \tilde{Z}_s \\ \tilde{Z}_f \end{pmatrix} = \begin{pmatrix} I_{m_c \times m_c} & 0 & 0 \\ 0 & I_{m_s \times m_s} & 0 \\ 0 & 0 & I_{m_f \times m_f} \end{pmatrix} \tag{5}$$

with $I_{m_i \times m_i}$ corresponding identity matrices, such that

$$\begin{pmatrix} U \\ V \\ W \end{pmatrix} = \begin{pmatrix} \tilde{Z}_c \\ \tilde{Z}_s \\ \tilde{Z}_f \end{pmatrix} \cdot \psi. \tag{6}$$

This transformation equation defines corresponding relations between F and F_c, F_s, F_f by

$$\begin{pmatrix} F_c \\ F_s \\ F_f \end{pmatrix} = \begin{pmatrix} \tilde{Z}_c \\ \tilde{Z}_s \\ \tilde{Z}_f \end{pmatrix} \cdot F. \tag{7}$$

Therefore, in the case of special representation of the system for a considered phenomenon is known and asymptotic limit specified precisely, one can apply the singular perturbation theory in order to obtain peculiarities of the reduced model.

According to the standard SPS theory, once the transformation equation (6) is found, one has the fast motions/manifolds estimated to the leading order of magnitude of the system small parameter ε describing the difference in time scales by

$$M_f^0 = \{(U, V, W) : U = U_0, V = V_0\}, \tag{8}$$

More detail on the theoretical background can be found in the literature (see e.g., [16, 17, 35]). It is obvious that the fast manifolds depend on the initial system state (U_0, V_0, W_0) because the slow as well as "conserved" variables during the fast transient period are estimated to be fixed or "frozen" (Fig. 3 shows typical phase space structure of the SPS).

The slow system manifold is then approximated to the leading order, when $\varepsilon \to 0$, $\delta \to 0$ by

$$M_s^0 = \{(U, V, W) : F_f(U_0, V, W) = 0\}, \qquad (9)$$

Geometrically, it means that in the ε vicinity of the slow manifold M_s^0 right hand sides (RHS) of (4) become comparable, i.e., the variables change with the same "speed." Note, however, that the higher order approximations of the invariant manifold of slow motions follow directly from the application of the SPS theory in a straightforward way [16]. For instance, the equation of the first order approximation in the implicit form is given by differentiation of the slow manifold equation (9) along the system (4) vector field $\Phi(U, V, W) = (F_c(U, V, W), F_s(U, V, W), F_f(U, V, W))^T$:

$$M_s^1 = \{(U, V, W) : D_{\{U,V,W\}}(F_f(U, V, W)) \Phi(U, V, W) = 0\}, \qquad (10)$$

where $D_{\{U,V,W\}}(F)$ denotes partial derivatives of the F by (U, V, W).

The stability analysis of the slow manifold is related to the following eigenvalue problem for points (U^*, V^*, W^*) on the slow manifold (9)

$$Re\left(\lambda\left(D_W F_f\left(U^*, V^*, W^*\right)\right)\right)|_{(U^*,V^*,W^*)\in M_s} < 0, \qquad (11)$$

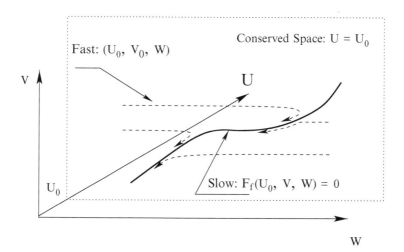

Fig. 3 Standard SPS phase space geometry

where $D_W F_f (U^*, V^*, W^*)$ is the matrix $m_f \times m_f$ of partial derivatives of the fast subsystem with respect to the fast variable and $\lambda (*)$ are eigenvalues of this matrix. It means, in particular, that knowledge about global fast manifolds is very important in the stability analysis of a reduced model described by a slow manifold.

Now, depending on the investigated dynamical regime one can use manifolds defined by (8) or (9) as a manifold equation and proceed with the reduction procedure.

In the present modified form the SPS theory can systematically be used as a natural mathematical construction for the system's hierarchy analysis and its subsequent reduction. Furthermore, the problem of model reduction with our interpretation reduces to the problem of identification of this special representation. Clearly, by assuming only a particular asymptotic limit and performing special non-dimensionalization, it might be possible to obtain the representation (4) of the system analytically. But, in the general case of the system (3), for more complex and practical systems (with large dimension, non-linearity, stiffness etc.) it is either very time/human/computational resources consuming or even prohibitive. Hence, if the original variables are not appropriate for this special form and the system is presented in the general form of (3), one has to deal with the following questions:

1. Is there a multi-scales hierarchy in the system (3)?
2. If yes, how to find it and define a minimal possible dimension of the reduced model (slow or fast manifold)?
3. How to transform the system to the explicit form of the SPS (3)?
4. How to define conserved, slow and fast manifolds as directions of fast relaxation processes?
5. How to study the properties of manifolds, i.e., existence, slowness/fastness, stability, attractiveness, domain of attraction etc.?
6. How to improve the slow stable manifold to an invariant manifold?

In our view any successful model reduction concept should be able to answer these questions, it should provide not only an approximation of the manifold of slow motions, but information of how the slow manifold is embedded into the state space is extremely important as well. Note that for the system in the form of the SPS (3) i.e., where first three questions are answered there are no principle difficulties with the other questions, therefore, one only needed to apply the singular perturbation theory in the manner outlined above.

2.2 Transformation to the SPS: Choice of New Coordinates

2.2.1 Semenov's Thermal Explosion

A very transparent and nevertheless meaningful example in combustion theory, which shows the importance of the features and concepts introduced above, is Semenov's classical thermal explosion model. It can be used as an illustration of

Scaling Invariant SPVF

the problem of transformation to the standard SPS. The non-dimensional form of Semenov's model reads

$$\gamma \frac{d\theta}{d\tau} = \eta \exp\left(\frac{\theta}{1 + \beta\theta}\right) - \alpha\theta$$
$$\frac{d\eta}{d\tau} = -\eta \exp\left(\frac{\theta}{1 + \beta\theta}\right) \tag{12}$$

where θ represents a non-dimensional quantity for the temperature, η is normalized by initial value of concentration of limiting chemical species. This system describes the homogeneous chemical reactor (given by Arrhenius type chemical kinetics – exponential term of highly exothermic chemical reaction) with heat losses are proportional to the difference of the reactor and ambient temperature (linear term).

This model has been very well studied in a number of works. However, in order to single out the meaning of the SPVF and indicate the relevance of standard techniques to suggested line of thinking it might be useful to repeat main stages of non-dimensionalization procedure implemented to obtain (12). First the chemical reaction is considered as a first order chemical reaction with an Arrhenius dependence of the rate coefficients

$$\frac{dC}{dt} = -C A \exp\left(-\frac{E}{RT}\right). \tag{13}$$

Typical characteristic time for isothermal chemical reaction (with constant temperature equals to the initial ambient temperature $T = T_0$) is defined as reciprocal of the constant $A \exp\left(-\frac{E}{RT_0}\right)$ of the linear equation for the species concentration C

$$t_R = A^{-1} \exp\left(\frac{E}{RT_0}\right), \tag{14}$$

here A is pre-exponential factor, E – activation energy, R – universal gas constant. Next, the energy equation is used in the form of reaction and heat loss balance

$$C_p V \frac{dT}{dt} = -Q \frac{dC}{dt} - \sigma S (T - T_0), \tag{15}$$

while characteristic time of the heat loss is defined as

$$t_H = \frac{C_p V}{\sigma S}, \tag{16}$$

where C_p is the heat capacity at constant pressure, V – volume, Q – heat of reaction, S – heat loss area, σ – heat loss coefficient. The desired form of the mathematical model (12) is obtained by performing in this way defined suitable non-dimensionalization following Frank–Kamenetskii approach of using the smallest characteristic time $\tau = t/t_R$ and non-dimensional system parameters

$$\eta = C/C_0, \ \theta = \frac{E}{RT_0} \frac{(T - T_0)}{T_0}$$
$$\alpha = \gamma \, t_H/t_R, \ \beta = \frac{RT_0}{E}, \ \gamma = \beta \frac{T_0 \, C_p}{Q \, C_0}. \tag{17}$$

The physical meaning of the parameters in (17) is clear because the model is simple while clever non-dimensionalization has been implemented. High exothermicity and activation energy of the chemical reaction typical for combustion models yield $\beta \ll 1, \gamma \ll 1, t_R \ll t_H$, i.e., explosive character of the system (12) is justified. In this way the system has become suitable for asymptotic analysis according to the standard SPS theory and different regimes have been thoroughly outlined [16, 36].

This brief description clearly shows the importance of physical experience in developing an accurate mathematical model appropriate for further specific asymptotic analysis. The knowledge about the system hierarchy has been already used on the modelling stage that allowed us to find the most suitable mathematical form of the model. It is clear, however, that for complex systems (with many species, reactions that have to be carefully studied by delineating particular characteristic times for the whole range of physical parameters) the procedure shown above cannot be implemented, thus, alternative way should be found to proceed "automatically" with such hierarchy analysis.

2.3 ILDM Method: Slow Manifold

One of the first method that has been devised to treat the model reduction in systematic way was the method of Intrinsic Low-Dimensional Manifolds (ILDM) [32]. The method not only assumes the existences of multiple scales, but it allows to define explicitly the slow manifold. Moreover, it suggests tabulation strategy and concept of generalized coordinates that render it as a very robust and attractive approach for implementing in the CFD codes.

Although the mathematical model of the ILDM method is described in details in [31, 32], a short repetition to outline the meaning of the ILDM method is needed here for completeness of the exposition.

In order to treat different time scales, the ILDM method proposes to analyze the (n by n)-dimensional Jacobi matrix F_ψ of the source term

$$\left(F_\psi \right)_{ij} = \frac{\partial F_i}{\partial \psi_j}.$$

It identifies locally (depending on the state ψ) fast/slow chemical processes by the Schur decomposition with subsequent solution of the Sylvester equation to obtain invariant subspaces of the Jacobian matrix as follows

$$F_\psi(\psi) = \left(Z_s(\psi)\; Z_f(\psi)\right) \cdot \begin{pmatrix} N_s(\psi) & 0 \\ 0 & N_f(\psi) \end{pmatrix} \cdot \begin{pmatrix} \tilde{Z}_s(\psi) \\ \tilde{Z}_f(\psi) \end{pmatrix}. \tag{18}$$

The matrices Z, $\tilde{Z}$ span up the right and the left invariant subspaces correspondingly,

$$\tilde{Z} = Z^{-1} = \left(Z_s\; Z_f\right)^{-1} = \begin{pmatrix} \tilde{Z}_s \\ \tilde{Z}_f \end{pmatrix}, \tag{19}$$

here Z_s is the (n by m_s)-dimensional matrix belonging to the invariant subspace of the m_s eigenvalues having the smallest real parts (defined by N_s), while Z_f is the (n by m_f)-dimensional matrix related to the m_f eigenvalues (defined by N_f) having the largest real parts respectively, where m_s and m_f denotes the number of eigenvalues according to slow and fast processes i.e.,

$$i = 1,\ldots, m_s \quad k = m_s + 1,\ldots, m_s + m_f \quad m_s + m_f = n \quad a, b > 0$$

$$\left| Re\left(\lambda_i\left(D_\psi F\right)\right) \right| < a \ll b < \left| Re\left(\lambda_k\left(D_\psi F\right)\right) \right|, \quad Re\left(\lambda_k\left(D_\psi F\right)\right) < 0 \tag{20}$$

$$0 < \epsilon = a/b \ll 1$$

Note that within the ILDM method the conserved subspace (which is linear, because of elements and mass conservation relations define first linear integrals of the kinetic mechanism models) is handled within the slow subsystem. The general assumption is that the fast processes have already relaxed define a m_s-dimensional manifold in the state space. This manifold is composed of points where the reaction rates in direction of the m_f fastest processes vanish, therefore, they can be expressed by

$$M_s^{ILDM} = \{\tilde{Z}_f(\psi)\, F(\psi) = 0\}, \tag{21}$$

where $\tilde{Z}_f$ is the (m_f by n)-dimensional matrix of left Schur vectors corresponding to the fast relaxing processes. One can see easily the connection between (21) and (9).

If conditions (20) are valid, then one has the local transformation to the standard SPS form which is defined similar to (6) by invariant subspaces of $D_\psi F$ with $\varepsilon = \frac{a}{b} \ll 1$

$$\begin{cases} U = \tilde{Z}_s(\psi_0)\, \psi \\ V = \tilde{Z}_f(\psi_0)\, \psi \end{cases}, \tag{22}$$

unfortunately, the transformation is valid only in the vicinity of the slow manifold and cannot be used to find out how the slow manifold is embedded into the state space, even the stability is assumed only locally due to (20). In particular, it means that the locally varying characteristic time scales, although they allow to find the slow manifold up to the second order of the accuracy, they are not suitable for identification of the global transformation of the system (3) to the form (4) and, therefore, dimensions, stability and existence problems cannot be treated by local type approaches (which are based on the Jacobian matrix analysis).

Thus, a global tool for system analysis that answers the main questions above and provides with additional information should be constructed. For this reason the framework of Singularly Perturbed Vector Fields (SPVF) has been developed and will be discussed below.

3 Coordinate Free Singular Perturbations: SPVF

The asymptotic analysis of Semenov's model and its complications [16] lead us to the following conclusions:

- In order to define characteristic time scales and to justify the SPS analysis a kind of linearization is used and needed.
- Original coordinates are not suitable for the asymptotic analysis – it is a quite rare situation that the original coordinates are appropriate.
- A coordinate free concept has to be developed.
- Scaling invariant linear interpolation/approximation problem over the interval/ domain is the first step to proceed with the transformation to the standard SPS.
- Improvements of the transformation of the model to the SPS [36] hints on a possible solution in the general case.

As the result of different attempts to generalize the non-dimensionalization procedure, the concept of Singularly Perturbed Vector Fields has been emerged [37]. Roughly speaking both concepts of the SPS and SPVF are equivalent but locally only. The latter can be understood as an analogue of the decomposed form (4), but valid only locally if an appropriate coordinate system is specified, while keeping same dimension structure all over the domain (see Fig. 4).

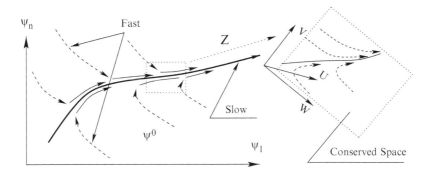

Fig. 4 Standard SPS vs. SPVF phase space geometry

In order to account for changed structure of fast manifolds (fast dynamics of relaxation) and perform the global analysis the so-called fiber bundles are used, which is a natural substitute to a non-linear coordinate system [1]. Now, everything depends on whether the fiber structure can be defined in advance or not. Figure 4 illustrates the relation between the standard SPS and SPVF cases, once the transformation is found the rest is almost trivial, the standard SPS theory can be exclusively applied (see Sect. 2.1).

It has been shown, in a number of works [1, 18, 30], how the structure of fiber bundles of fast motions can be used to define local coordinates explicitly decomposing the system's motions. In many cases, however, a simple structure, e.g., rotation of the original coordinate system or linear change of coordinates, might be sufficient for the explicit decomposition of motions (see Fig. 4). This figure and the last remark present the essence of the suggested and discussed in the current work approach.

The standard existing strategies can be revised within the SPVF method as follows. First, one tries to use certain scaling or normalization procedures in the original coordinate system such that it transforms to the standard SPS explicit form written in the original coordinates. The well known Quasi-Steady States Approach (QSSA) is a typical representative of the initial stage of system hierarchy analysis [35, 38, 39]. Of course an assumption that the original variables already appropriate for the decomposition (4) (for description of the system in the SPS form) is a very strong one. It limits considerably both an application range and the achieved reduced dimension of such methods.

Next, local analysis can be performed (ILDM, CSP [40,41], sensitivity study etc.) which is used to approximate efficiently the slow manifold. It is suggested, however, that the fast relaxation motions do not follow the hyper-planes of the constant original coordinates, but they are not identified and assumed to be "unimportant" for the reduced model formulation. It means that the slow manifold is found, while the way it is embedded into the state space through the fast relaxation is still unknown. Obviously, the information on a slow manifold only cannot be efficiently used for construction of model reduction technique, especially for modelling of the transient system behavior and for reproducing of the accurate dynamics on the slow manifold.

The final step would be a complete description of the fast manifolds structure and all advantages of the representation (4) might be employed at this stage. However, in general, it is a very complicated task. At present, an algorithm to treat the case of linear fast manifolds has been suggested. Although the very strong assumption about linearity of the coordinate transformation has been used it is definitely less restrictive compare to conventional methods such as for instance QSSA approach since linear combinations of original coordinates are allowed as well.

3.1 Linearly Decomposed Vector Fields: GQL

Since linearly decomposed vector field is assumed, it is natural to use a kind of linearization (linear interpolation) of the original vector field F defined by the RHS of (3). This linear approximation will provide with the transformation and with the fast manifolds structure. In order to define the linear approximation of the vector function the so-called Global Quasi-Linearization (GQL) procedure has been suggested in [18]. The main steps of the original algorithm are outlined shortly as following.

- Suppose $F(\psi)$ defines a SPVF that depends on a "hidden" small parameter ε.
- Choose n linearly independent points $\psi_1, \ldots, \psi_n$ to form a reference set $\Psi = \{\psi_1, \ldots, \psi_n\}$
- Calculate images of the states from the reference set by the vector field $\{F(\psi_1), \ldots, F(\psi_n)\}$ and use them as rows for matrices in order to define a linearization matrix T as follows

$$T := [F(\psi_1), \ldots, F(\psi_n)] [\psi_1, \ldots, \psi_n]^{-1}. \tag{23}$$

This matrix has a simple geometrical interpretation. It is the matrix of linear mapping that transforms points $\psi_1, \ldots, \psi_n$ into $F(\psi_1), \ldots, F(\psi_n)$. In this way the problem of transformation is reduced to the problem of the reference set definition.

- Calculate invariant eigenspaces of T and eigenvalues. It will answer main questions of model reduction and describe the linear hierarchy of the system.

It is obvious that the crucial point of the algorithm is the second step, it must be adapted for every particular model. Practical recommendations for the choice are the following. Points $\psi_1, \ldots, \psi_n$ in the state space should not be "close" one to another, because it can lead to a degeneration of the matrix $[\psi_1, \ldots, \psi_n]^{-1}$ as well as the vector field $F(\psi)$ should have essentially "different" behavior for different points $\psi_1, \ldots, \psi_n$.

It is clear that the problem with the choice of the reference sequence of the GQL approximation T should be solved by introducing a scaling invariant procedure of the choice [18, 36]. The property to be scaling invariant is required because the whole algorithm should not be dependent on additional knowledge of the processes involved and must be applicable automatically to wide range of complex systems describing reacting flows.

3.2 Scaling Invariant Form

Scaling invariance is a very important feature of the reduction method, it is a basic property to be self-consistent. Below we present the scaling invariant sorting procedure to identify the reference. In order to make the method independent of additional dimension analysis some geometrical invariants of the original vector field define the GQL matrix T and, consequently, the chosen reference set.

First of all a stochastic set of uniformly distributed points in the domain of interest are located. Then, by taking subsets of this sequence, a set of GQLs T_k, $k = 1, \ldots, N$ is calculated.

In order to pick up the best one from the set of GQLs, the following invariants of the vector field based on an integral of the characteristic polynomial coefficients of the differential $D_\psi (F(\psi))$ of the vector field $F(\psi)$ are studied and compared to those of T_k.

$$I_j (F; \Omega) = \left| \int_\Omega a_j (\psi; F) \, d\psi \right|, \ j = 1, \ldots, n, \tag{24}$$

where $a_i (\psi; F): f_{D_\psi F(\psi)}(t) = t^n + a_1 (\psi; F) t^{n-1} + \cdots + a_n (\psi; F)$.

The same invariants for the vector field produced by the GQL procedure are the following:

$$I_j (T; \Omega) = |a_j (T)| \, |\Omega|, \ j = 1, \ldots, n, \tag{25}$$

where $a_i (T): f_T(t) = t^n + a_1 (T) t^{n-1} + \cdots + a_n (T)$.

There is a clear interpretation of the integral of the last invariant giving the volume of the image of the domain of interest Ω: $|Im (F(\Omega))| \equiv I_n (F; \Omega)$.

Other integrals do not have such clear geometrical interpretations, but certainly have connection to the geometrical structure of the vector field. Hence, the final choice of the GQL is performed on a basis of the following functional

$$\Pi (T; \Omega) = \sum_j \frac{|I_j (T; \Omega) - I_j (F; \Omega)|}{|I_j (T; \Omega)| + |I_j (F; \Omega)|}. \tag{26}$$

The chosen GQL minimizes the functional above over the reference sequence

$$T^*: \ \min_T (\Pi (T; \Omega)) = \Pi (T^*; \Omega). \tag{27}$$

The rest is standard, the spectrum of T^* answers the question about existing hierarchy, a gap condition can be used for definition of the decomposition dimension and of the small system parameter, similar to (18)

$$T^* = \left(Z_c \ Z_s \ Z_f \right) \cdot \begin{pmatrix} N_c & 0 & 0 \\ 0 & N_s & 0 \\ 0 & 0 & N_f \end{pmatrix} \cdot \begin{pmatrix} \tilde{Z}_c \\ \tilde{Z}_s \\ \tilde{Z}_f \end{pmatrix}, \tag{28}$$

where dimensions and small parameters ϵ and δ of the system representation (4) are defined simultaneously on the basis of the gap condition of the GQL matrix T similar to (20),

$$\delta = \frac{\max |\lambda_i (N_c)|}{\min |\lambda_i (N_s)|}, \ \epsilon = \frac{\max |\lambda_i (N_s)|}{\min |\lambda_i (N_f)|} \tag{29}$$

while invariant subspaces of groups of eigenvalues define the quasi-conserved, slow and fast invariant subspaces together with projections, for instance

$$Pr_c = Z_c \, \tilde{Z}_c,$$
$$Pr_s = Z_s \, \tilde{Z}_s,$$

projection onto the conserved and slow subspace respectively. The first projector can be used to split off conserved subspace. The most important fast invariant subspace of the linear interpolation equation (27) provides with

$$Pr_f = Z_f \, \tilde{Z}_f$$

and defines the equation of the slow manifold which in the original coordinates reads

$$M_s^0 (\psi_0) = \{ \psi : \tilde{Z}_f \, F(\psi) = 0, \, \tilde{Z}_c \psi = \tilde{Z}_c \, \psi_0 \}, \tag{30}$$

moreover, it can be used to study the stability of the slow manifold equation (30)

$$Re \left(\lambda \left(\tilde{Z}_f \, D_\psi F \left(\psi^* \right) Z_f \right) \right) |_{\psi^* \in M_s^0} < 0, \tag{31}$$

In a view of defined projectors the system (3) is decomposed explicitly, i.e., can be represented in form (4) and the SPS theory can be applied for analysis and reduction. Finally, the fast manifold or the fast relaxation dynamics of the system can be estimated due to the slow invariant subspace

$$M_f^0 (\psi_0) = \{ \psi : \tilde{Z}_s \, \psi = \tilde{Z}_s \, \psi_0, \, \tilde{Z}_c \psi = \tilde{Z}_c \, \psi_0 \}. \tag{32}$$

3.3 Discussion

The most important message of the current work concerns the need to find or construct the functional that can be efficiently used to compare the global properties of non-linear functions. The functional allows to chose the best linear approximation of the non-linear function with respect to the system hierarchy. By studying the properties of the linear approximation the questions of model reduction are answered in a complete form.

In particular, it allows simultaneously:

- To check whether the system defines a linearly decomposed SPVF or not (29)
- To find appropriate dimensions of different scale's ranges (28) and (29)
- To transform the given system to the standard SPS form as defined by (5), (6) and (28), which in turn permits
- Using the SPS theory for detailed study of the properties of the manifolds needed for model reduction purposes (Sect. 2.1)
- Improving reduced models and projecting the detailed system on the slow manifold consistently with asymptotically decomposed representation of the system.

The version above represents scaling invariant criteria for the choice of the optimal GQL together with the reference set. The GQL algorithm has been tested on the very important problem of auto-ignition of combustion systems [42]. In this case the method is adopted for the purpose of definition of the best performing reduced model with respect to the ignition delay time estimation.

Although considerable progress in the understanding of main problems of model reduction, there are still many difficult technical aspects in implementation of the GQL method which, however, can be treated without principal difficulties.

4 Conclusions

In the present work, the need for an automatic methods of hierarchy system analysis and reduction of detailed kinetics mechanisms of reacting flows was discussed. The problem of model reduction has been treated in a quite general formulation. Important problems that have to be solved by any successful reduction methods were discussed and outlined as well. A further developments and a discussion of the coordinate free singular perturbations approach was the main part of the paper. Accordingly, manifolds based approaches for model reduction were discussed. In the case of linearly decomposed vector fields the method of hierarchy analysis and model reduction that follows the main ideas of the standard ILDM technique has been presented. Finally, the suggested scaling invariant modification overcomes most of the problems of implementation with respect to the system hierarchy investigation.

Acknowledgements This research was supported by the Deutsche Forschungsgemeinschaft (DFG). Bykov thanks the Centre for Advanced Studies in Mathematics at the Ben-Gurion University of the Negev (BGU) for financial support of his stay at the BGU during spring 2009.

References

1. Bykov, V., Goldfarb, I., Gol'dshtein, V.: Singularly perturbed vector fields. J. Phys. Conf. Ser. **55** (2006) 28–44
2. Available at www.ca.sandia.gov/chemkin/
3. Green, W.H., Barton, P.I., Bhattacharjee, B., Matheu, D.M., Schwer, D.A., Song, J., Sumathi, R., Carstensen, H.-H., Dean, A.M., Grenda, J.M.: Computer construction of detailed chemical kinetic models for gas-phase reactors. Ind. Eng. Chem. Res. **40(23)** (2001) 5362–5370
4. Miyoshi, KUCRS software library, version May 2005 beta, available from the author. See the web: www.frad.t.u-tokyo.ac.jp/~miyoshi/KUCRS/
5. Griffiths, J.F.: Reduced kinetic models and their applications to practical combustion systems. Prog. Energ. Combust. Sci. **21** (1995) 25–107
6. Chevalier, C., Pitz, W.J., Warnatz, J., Westbrook, C.K.: Hydrocarbon ignition: automatic generation of reaction mechanisms and applications to modelling of engine knock. Proc. Comb. Inst. **24** (1992) 93–101

7. Tomlin, A.S., Turanyi, T., Pilling, M.J.: Mathematical tools for the construction, investigation and reduction of combustion mechanisms. In: Comprehensive Chemical Kinetics **35**: Low-temperature Combustion and Autoignition, M.J. Pilling (eds.). Elsevier, Amsterdam (1997)
8. Warnatz, J., Maas, U., Dibble, R.W.: Combustion, 4th edn. Springer, Berlin (2004)
9. Peters, N., Rogg, B.: Reduced kinetics mechanisms for application in combustion systems. Springer, Berlin (1993)
10. Okino M.S., Mavrovouniotis, M.L.: Simplification of mathematical models of chemical reaction systems. Chem. Rev. **98(2)** (1998) 391–408
11. Strygin, B.B., Sobolev, V.A.: Decomposition of Motions by the Integral Manifolds Method. Moscow, Nauka (1988) (in Russian)
12. Gorban, A.N., Karlin, I.V., Zinovyev, A.Yu.: Constructive methods of invariant manifolds for kinetic problems. Phys. Rep. **396(4–6)** (2004) 197–403
13. Gorban, A.N., Karlin, I.V.: Method of invariant manifold for chemical kinetics. Chem. Eng. Sci. **58** (2003) 4751–4768
14. Gorban, A.N., Karlin, I.V., Zinovyev, A.Yu.: Invariant grids for reaction kinetics. Physica A **333** (2004) 106–154
15. Gorban, A.N., Karlin, I.V.: Constructive methods of invariant manifolds for physical and chemical kinetics. Lecture Notes in Physics 660, p. 498. Springer, Berlin (2005)
16. Gol'dshtein, V., Sobolev, V.: Qualitative analysis of singularly perturbed systems of chemical kinetics. In: Singularity theory and some problems of functional analysis. American Mathematical Society, Translations, S.G. Gindikin (ed.) **153(2)** (1992) 73–92
17. Fenichel, N.: Geometric singular perturbation theory for ordinary differential equations. J. Differ. Equat. **31** (1979) 53–98
18. Bykov, V., Gol'dshtein, V., Maas, U.: Simple global reduction technique based on decomposition approach. Combust. Theor. Model. **12(2)** (2008) 389–405
19. Maas, U., Bykov, V., Rybakov, A., Stauch, R.: Hierarchical modelling of combustion processes. Proc. Teraflop WS (2009) 111–128
20. Hirschfelder, J., Curtiss, C.: Molecular theory of gases and liquids. Wiley, New York (1964)
21. Bird, R., Stewart, W., Lightfoot, E.: Transport phenomena. Wiley Interscience, New York (1960)
22. Ern, A., Giovangigli, V.: Multicomponent transport algorithms. Lecture Notes in Physics. Springer, Berlin (1994)
23. Bykov, V., Maas, U.: The extension of the ILDM concept to reaction-diffusion Manifolds. Proc. Comb. Inst. **31** (2007) 465–472
24. Bykov, V., Maas, U.: Problem adapted reduced models based on reaction-diffusion nanifolds (REDIMs). Proc. Comb. Inst. **32(1)** (2009) 561–568
25. Singer, M.A., Pope S.B., Najm, H.N.: Operator-splitting with ISAT to model reacting flow with detailed chemistry. Combust. Theor. Model. **10(2)** (2006) 199–217
26. Ren, Z., Pope, S.B., Vladimirsky A., Guckenheimer, J.M.: Application of the ICE-PIC method for the dimension reduction of chemical kinetics coupled with transport. Proc. Comb. Inst. **31** (2007) 473–481
27. Maas, U.: Mathematische Modellierung instationaerer Verbrennungsprozesse unter Verwenung detaillieter Reaktionsmechanismen. PhD thesis, Naturwissenschaftlich-Mathematische Gesamtfakultaet. Ruprecht-Karls-Universitaet, Heidelberg (1988)
28. Rhodes, C., Morari, M., Wiggins, S.: Identification of low order manifolds: validating the algorithm of Maas and Pope. Chaos **9** (1999) 108–123
29. Kaper, H.G., Kaper, T.J.: Asymptotic analysis of two reduction methods for systems of chemical reactions. Argonne National Lab, preprint ANL/MCS-P912-1001 (2001)
30. Bykov, V., Goldfarb, I., Gol'dshtein, V., Maas, U.: On a modified version of ILDM approach: asymptotical analysis based on integral manifolds method. IMA J. Appl. Math. **71(3)** (2006) 359–382
31. Maas U., Pope, S.B.: Simplifying chemical kinetics: intrinsic low-dimensional manifolds in composition space. Combust. Flame **88** (1992) 239–264
32. Maas, U.: Efficient calculation of intrinsic low-dimensional manifolds for the simplification of chemical kinetics. Comput. Visual. Sci. **1** (1998) 69–82

33. Maas U., Warnatz, J.: Ignition processes in hydrogen-oxygen mixtures. Combust. Flame **74** (1988) 53–69
34. Maas, U.: Coupling of chemical reaction with flow and molecular transport. Appl. Math. **3** (1995) 249–266
35. Bowen, J.R., Acrivos, A., Oppenheim, A.K.: Singular perturbation refinement to quasi-steady state approximation in chemical kinetics. Chem. Eng. Sci. **18** (1963) 177–188
36. Bykov, V., Gol'dshtein, V.: On a decomposition of motions and model reduction. J. Phy. Conf. Ser. **138** (2008) 012003
37. Bykov, V., Goldfarb, I., Gol'dshtein, V.: Novel numerical decomposition approaches for multiscale combustion and kinetic models. J. Phys. Conf. Ser. **22** (2005) 1–29
38. Bodenstein, M., Lind, S.C.: Geschwindigkeit der Bildung des Bromwasserstoffs aus seinen Elementen. Z. Phys. Chem. **27** (1906) 168–175
39. Williams, F.A.: Combustion theory, the fundamental theory of chemically reacting systems, 2nd edn. Benjamin/Cummings, California (1985)
40. Lam, S.H., Goussis, D.M.: The GSP method for simplifying kinetics. Int. J. Chem. Kinet. **26** (1994) 461–486
41. Lam, S.H.: Reduced chemistry-diffusion coupling. Combust. Sci. Tech. **179** (2007) 767–786
42. Bykov, V., Maas, U.: Investigation of the hierarchical structure of kinetic models in ignition problems. Z. Phys. Chem. **223(4–5)** (2009) 461–479

Think Globally, Move Locally: Coarse Graining of Effective Free Energy Surfaces

Payel Das, Thomas A. Frewen, Ioannis G. Kevrekidis, and Cecilia Clementi

Abstract We present a multi-scale simulation methodology, based on data-mining tools for the extraction of low-dimensional *reduction coordinates*, to explore dynamically a protein model on its underlying effective folding free energy landscape. In practice, the averaged coarse-grained description of the local protein dynamics is extracted in terms of a few reduction coordinates from multiple, relatively short molecular dynamics trajectories. By exploiting the information collected from the fast relaxation dynamics of the system, the reduction coordinates are extrapolated "backward-in-time" to map globally the underlying low-dimensional free energy landscape. We demonstrate that the proposed method correctly identifies the transition state region on the reconstructed two-dimensional free energy surface of a model protein folding transition.

1 Introduction

Many complex physical systems exhibit conformational transitions between free energy minima separated by high barriers. Examples encompass nucleation processes, nanocluster assembly, macromolecular dynamics, and protein folding [32]. Transitions across high free energy barriers typically occur on a time scale several orders of magnitude slower than the time scale associated with the fast motions (e.g., atomic vibrations) of the system. This wide separation of time scales severely

---

P. Das and C. Clementi
Department of Chemistry, Rice University, Houston, TX 77005, USA
e-mail: daspa@us.ibm.com, cecilia@rice.edu

I.G. Kevrekidis (✉)
Department of Chemical Engineering, Princeton University, Princeton, NJ 08544, USA
e-mail: yannis@princeton.edu
and
PACM and Mathematics, Princeton University, Princeton, NJ 08544, USA

T.A. Frewen
Department of Chemical Engineering, Princeton University, Princeton, NJ 08544, USA
e-mail: tfrewen@gmail.com

A.N. Gorban and D. Roose (eds.), *Coping with Complexity: Model Reduction and Data Analysis*, Lecture Notes in Computational Science and Engineering 75, DOI 10.1007/978-3-642-14941-2_6, © Springer-Verlag Berlin Heidelberg 2011

restricts the use of all-atom Molecular Dynamics (MD) simulation to study such processes. One promising way to overcome this limitation focuses the simulation on the study of a few "collective" degrees of freedom that are important for the long time scale dynamics of the system. Toward this goal, several approaches [2, 6, 9, 23–25, 34] have been proposed, which aim to explore the effective free energy surface associated with a complex dynamical process in terms of a few collective coordinates. These techniques are usually based on the following approximation: a set of collective degrees of freedom (slow variables, effective coordinates, or "reduction coordinates") exists that exhibits significant variation during the considered transition, while fluctuations along the remaining degrees of freedom can be treated as thermal noise. Under this assumption, the long time scale dynamics of the system can be described by an effective coarse-grained Fokker–Planck equation (FPE) representing the time evolution of the probability distribution function of the slow variables. However, an effective FPE may not be available in closed form for the complex process under investigation. The recently proposed Coarse Molecular Dynamics (CMD) approach [1, 17, 20, 21, 30] bypasses the explicit formulation of such an equation, and only needs to assume its existence to dynamically explore the behavior of the collective coordinates. In this approach, the coarse-grained information about the slow dynamics of the system is extracted on-the-fly (in terms of the reduction coordinates) from multiple short microscopic simulations, allowing for estimation of the (local) coefficients of an effective FPE for different values of the reduction coordinates.

The only a priori information required to perform this task is the definition of an appropriate set of reduction coordinates, effectively capturing the slow time scale evolution of the system dynamics. However, for complex processes such a definition can be very challenging. Nonlinear dimensionality reduction approaches (e.g., [6, 27]) have been proposed to automatically extract a few optimal reduction coordinates from the molecular configurations generated during MD (or other atomistic, such as Monte Carlo (MC)) simulations. Such approaches provide a discrete mapping of the microscopic configurations of the system to the coarse space spanned by the reduction coordinates, but can not produce – at least in their current implementation – the explicit functional form of this mapping, hindering their direct use in active simulation/exploration tasks. As an alternative, a set of empirically defined collective variables – for which there is an easy, explicit mapping to the system Cartesian coordinates – can be used in the CMD formulation; the results can then be validated a posteriori by means of dimensionality reduction analysis (see Sect. 4). As a first step in this direction, we present here a multi-scale simulation procedure within the framework of CMD to efficiently explore/characterize the effective free energy surface associated with a protein folding transition. The averaged coarse description of the protein folding free energy (in terms of a few reduction coordinates) is extracted locally from the fast relaxation dynamics of the system, by means of multiple short ($\sim$ps), suitably initialized, conventional MD trajectories. The average relaxation dynamics yields information on the local time evolution of the reduction coordinates (in a sense to be clarified below) that is then extrapolated "backward-in-time" to climb up free energy barriers. This strategy enables

Coarse Graining of Effective Free Energy Surfaces

the system to rapidly escape from free energy minima and reach the transition state region, efficiently mapping the overall low-dimensional folding free energy landscape. Comparisons with the results obtained from traditional, long-time, MD simulation shows that this coarse reverse integration approach correctly identifies the folding transition state region, with the advantage of a significant reduction in the simulation time.

2 Coarse Reverse Integration

The coarse reverse integration procedure was previously introduced in the context of "simple" reactions that can be effectively described by a single reaction coordinate [1, 17]. Here it is extended to study the behavior of a larger and more complex system, in which more than one reaction coordinate may be required to successfully describe the process. For this purpose, in our two-coarse-dimension case, the system is explored through the evolution of a one-parameter family of points defining a closed curve in the coarse state space [12]. For a deterministic gradient problem, the reverse evolution dynamics occur along the (negative) gradient of the deterministic potential. For a stochastic problem with simple state-independent noise (when the diffusion matrix in the effective FPE has identical, constant diagonal entries) this gradient is proportional to the (negative of the) drift coefficients; for such a "scalar" diffusion case, the deterministic evolution using these drift coefficients corresponds to an "upward" evolution in the effective potential, and can be useful in efficiently exploring the effective free energy surface. Even for more complex stochastic problems, with state-dependent noise, if an effective potential exists (under appropriate "potential conditions" [29]), a reverse coarse integration using the drift equations may be useful in exploring the effective potential surface. It is important to note, however, that in such a case evaluating the changes in the effective potential associated with a step in our drift-based dynamics requires estimates of *both* the local drift *and* the local diffusion coefficients. In the remainder of the paper "coarse integration" is always performed for the deterministic dynamics based on the local FPE drift coefficients.

This coarse reverse integration can be summarized in terms of the following four steps (illustrated in Fig. 1, and described in detail below):

1. Prescribe the coarse variables X, Y providing a good low-dimensional description of the system; then construct (a discretization of) the initial ring $\Phi(\alpha, t = 0)$ as in Fig. 1, enclosing a free energy minimum in this space;
2. *Lift* the system "on-the-fly" from every discretization point on the current ring to produce an ensemble of high-dimensional configurations consistent with the current coarse ring;
3. *Run* multiple, short MD simulations to locally characterize the fast relaxation dynamics of the system at these coarse ring discretization points;
4. Obtain the local effective FPE drift (the forward time derivatives of our coarse evolution) from the MD runs, corresponding to the vector d in Fig. 1; decompose it into tangential and normal components (with respect to the curve $\Phi(\alpha, t)$); use

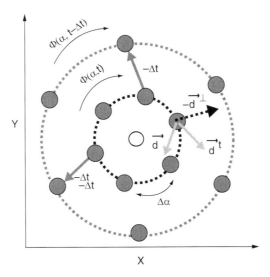

Fig. 1 Schematic of the coarse reverse evolution of a ring in the two-dimensional space of the reduction coordinates X and Y. The ring $\Phi(\alpha, t)$ encircles a free energy minimum. The position of the minimum is marked by an *open circle*. At any time, t, during the reverse integration the ring is parametrized by its normalized arc-length α. The local drift of the effective FPE associated with each point $(X(t), Y(t))$ on the curve $\Phi(\alpha, t)$ (*filled circles*) at a given time t, is related to the local effective free energy gradient, and is obtained by averaging over many short MD simulations initialized at $(X(t), Y(t))$. This local drift vector $\boldsymbol{d}$ is decomposed into components tangential, $\boldsymbol{d}^t$, and normal, $\boldsymbol{d}^\perp$, to the ring. An outer integration step (here a simple forward Euler step) is then used to extrapolate ("project") the position of each point $(X(t), Y(t))$ in the direction $-\boldsymbol{d}^\perp$ (shown as a *dashed* vector in the figure). The backward extrapolation yields the curve $\Phi(\alpha, t - \Delta t)$, corresponding to the ring at time $t - \Delta t$ in the deterministic dynamics described by the drift equations

the normal component to evolve ("project," here using a simple explicit Euler step) the current ring "backward-in-time."

The basic initialization and the remaining three steps are performed in an iterative manner starting from the neighborhood of different minima on the coarse free energy surface; the procedure will stagnate in the neighborhood of saddle points. For problems with "scalar" state-independent noise, the "dynamic" saddles of the drift equations correspond to the effective potential saddles; in this case our dynamic evolution will idle at the top of the free energy barrier (i.e., in the transition state region). The "backward-in-time" step (4) is simply related to a step "up" in effective free energy (see below).

For more complex problems with state dependent effective diffusion coefficients the local values of these coefficients need to be taken into account in order to locate effective potential barriers; this information can be used to evolve an initial ring in terms of equal "upward" effective potential steps. Yet "navigation" using only the drift dynamics may still be useful in efficient exploration of the effective free energy surface.

3 Exploration of the Two-Dimensional Folding Free Energy Surface for a SH3 Model Protein

We test the performance of the proposed method by applying it to explore the two-dimensional folding free energy surface of a simplified model of the src-SH3 domain, a 57 residue protein. The native coordinates for src-SH3 were taken from residues 84–140 of the structure from the Protein Data Bank (1FMK.pdb) [33].

Previous experimental and simulation studies have revealed a two-state folding mechanism for this protein in which a single free energy barrier separates the two minima, i.e., the folded and the unfolded states [1, 3, 5, 13]. The protein is modeled using an off-lattice C_α representation associated with a realistic, although simplified, minimally frustrated Hamiltonian, as described in [5]. The use of a simplified protein model effectively reduces the computational time required to sample the conformational space by several orders of magnitude compared to all-atom protein models, while retaining the relevant physical features of the folding landscape. Therefore, the complete folding free energy landscape obtained from the analysis of long MD trajectories of this simplified protein model can be used to validate the results of the coarse reverse integration.

3.1 Choice of Reduction Coordinates

The *initialization* of the system for the first integration step can be accomplished by identifying putative free energy minima in terms of a few a priori defined reduction coordinates. An initial guess on the position of such minima can be obtained by using a number of different strategies; details on the minima identification are discussed in the next subsection for the specific case of the protein folding process of the SH3 model presented here. Each free energy minimum is represented by a closed space curve, or "ring" on the coarse space parametrized by the chosen reduction coordinates, say X and Y (see Fig. 1). At any (pseudo-)time t during the reverse integration a discrete representation of the ring is provided by a finite number of points $(X(t), Y(t))$, or "nodes." The curve representing the ring is parametrized by its normalized arc-length α and is therefore represented as $\boldsymbol{\Phi}(\alpha, t)$. Appropriate extensions of such a parametrized curve (e.g., an initial ellipsoid in three dimensions) enables the generalization of the reverse integration approach for d-dimensional free energy surfaces, where $d > 2$; more sophisticated algorithms do exist for the exploration of higher-dimensional manifolds in the computation of invariant manifolds in the context of dynamical systems [15, 16].

Non-linear dimensionality reduction of a large set of folding simulation data obtained using the SH3 protein model reveals that its folding landscape is intrinsically low-dimensional [6]. As such, this system represents an ideal case to test the validity of the proposed method. Previous results have shown that this folding process can be effectively characterized by two empirically defined structure-based reduction coordinates [5]: the number of native contacts formed, Q, and the number of non-native contacts formed, A.[1] In the following we use Q and A to map the underlying effective folding free energy surface of SH3. The validity of these coordinates is verified a posteriori by means of nonlinear dimensionality reduction, as discussed in Sect. 4.

The folding free energy surface as a function of Q and A is shown in Fig. 2, as obtained from a standard statistical mechanical analysis [11] of equilibrium conformational sampling generated using a $1.2\,\mu s$ long MD simulation at the folding temperature T_f (for this model, with the choice of the parameters as in [5], $k_B T_f \simeq \langle \epsilon \rangle$, where $\langle \epsilon \rangle \simeq 1\,\text{Kcal mol}^{-1}$ is the average energy per native contact).

The entire span of the Q axis is considered for the exploration of the folding free energy surface using our coarse reverse integration. On the other hand, a cutoff for the highest possible value of A is set such that $A_{max} = 40$, if $0 \le Q \le 40$, otherwise, $A_{max} = 50 - 0.25 \times Q$ (as shown in Fig. 2). Setting such a cutoff value for A allows us to restrict the reverse integration of rings within the region of (Q, A) space that is relevant for folding. An a priori criterion for the selection of this region can be obtained by using polymer physics arguments: by comparing results obtained in the thermodynamic limit with simulation data on finite systems, it was shown in [3] that most relevant configurations visited during folding are confined in the region $Q + A \le Q_{folded}$. With this choice the rings are not allowed to visit unphysical protein conformations with a very high number of simultaneously formed native and non-native contacts. Clearly, such an assumption needs to be validated a posteriori. In the following sections we show that the main folding reaction occurs in a region with smaller values of A, confirming that this choice of A_{max} does not affect the results.

3.2 Definition of the Initial Rings: A Hybrid MC-MD Scheme

The location of the minima on the two-dimensional folding free energy surface (as a function of Q and A) can be obtained from equilibrium MD simulations. As mentioned in the discussion above, the slow *drift* dynamics in the reduction coordinates space are simply related to the effective potential gradient when the diffusion coefficients can be approximated as "scalar" [12]. Preliminary computations on the system considered here show that the off-diagonal elements of the diffusion matrix

---

[1] A residue pair (i, j) is considered to form a *native contact* if in the native structure there is at least a heavy atom from residue i closer than a distance cutoff (chosen to be $4.5\,\text{Å}$) from a heavy atom of residue j. By using this definition, there are 124 native contacts for the src-SH3 native structure. Among all the possible pairs of residues not involved in native contacts, we define a given pair to be a *non-native contact* if – in addition to it not belonging to the list of native contacts – the two constituting residues are not observed to be in contact (that is, closer than a distance cutoff) with a probability larger than 0.1 during a large sample of simulated dynamics trajectories at the folding temperature (see [3] for details).

Coarse Graining of Effective Free Energy Surfaces

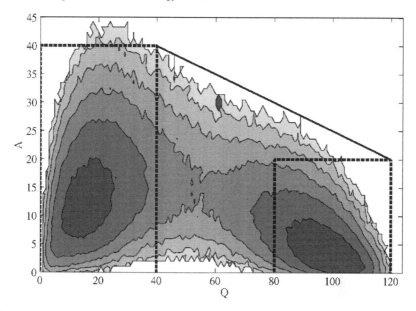

Fig. 2 Free energy surface of coarse-grained SH3 folding as a function of the empirical coordinates Q and A. The configuration sampling used to estimate the folding free energy was obtained from a long MD simulation run performed at $T = T_f$. The higher free energy regions are shown in lighter shades of *gray*. Each contour represents $1RT_f$ of free energy difference. The folded and unfolded minima are located in the regions ($Q \sim 85 - 110$, $A \sim 0 - 10$) and ($Q \sim 8 - 24$, $A \sim 6 - 20$), respectively. All the points inside the regions defined by the *black dashed lines* are used to identify the initial rings around the free energy minima using the hybrid MC-MD scheme described in the text. The value for A_{max} bounding the region explored is shown, as a function of Q, as a solid *black line*

are at least an order of magnitude smaller than the diagonal ones (data not shown). In the following we assume that the scalar approximation for the diffusivity approximately holds for the SH3 protein model with the selected reduction coordinates. In particular, we exploit the fact that the potential minima are (for "scalar" diffusion systems) attractors of the drift dynamics parametrized by Q and A; even in the non-scalar diffusion case, one can locate these minima as roots of simultaneous equations in the coarse variables; see [30] for such a computation.

Two broad regions on the (Q, A) surface are initially selected as the possible locations of the minima: ($0 \leq Q \leq 40$, $0 \leq A \leq 40$) for the unfolded state, and ($80 \leq Q \leq 120$, $0 \leq A \leq 20$) for the folded state (see Fig. 2). These choices are based on the definition of Q and A: The folded state ensemble is more likely to have a higher number of native contacts formed in contrast to the unfolded state ensemble. The variation along the A axis is chosen to be larger for the unfolded state, as configurations corresponding to the unfolded minimum can have a higher number of non-native contacts formed compared to the folded configurations. We *initialize* the system at equidistributed locations in the above-defined regions. In principle, an infinite number of high-dimensional states can be projected to the same point in the low dimensional space (X, Y); here $(X, Y) = (Q, A)$. In practice, we are interested in obtaining a good sampling of the physically reasonable ones among all of them. For every coarse initial condition of the system, about 250 high-dimensional configurations, or "replicas," are generated by means of a Simulated Annealing Monte Carlo (SA-MC) scheme, where a bias potential E_{bias} is added to the potential energy of the system, in a harmonic form:

$$\frac{E_{bias}(\Gamma)}{RT_f} = K((X_\Gamma - X_t)^2 + (Y_\Gamma - Y_t)^2). \tag{1}$$

In (1), K is a stiff spring constant, and (X_Γ, Y_Γ) represent the coarse description of the system associated with any molecular configuration Γ. The use of such a biasing potential restricts the configurational sampling close to the target values of reduction coordinates (X_t, Y_t).

For the specific case of the protein folding reaction described here, detailed configurations consistent with any fixed $(X(t), Y(t))$ point are created starting from the native state configuration of the protein as follows: A randomly selected short segment $(i - j)$, $4 \leq |i - j| \leq 6$, of the protein chain is rotated by a random angle $\theta \in (0, \pi)$. The new configuration is accepted with a probability P according to the Metropolis criterion, $P = \min\left(1, e^{-\beta_1(E_{new} - E_{old})}\right)$, where $E_{old/new} = E(\Gamma_{old/new}) + E_{bias}(\Gamma_{old/new})$, and $E(\Gamma)$ is the potential energy associated with the molecular configuration Γ. The SA-MC simulation is initialized with an annealing temperature $\beta_1 \ll \beta_f = (RT_f)^{-1}$, where T_f is the folding temperature of the protein (that we assume is known a priori). The temperature of the system is then gradually reduced until $\beta_1 \simeq \beta_f$.

At the end of each SA-MC simulation, the distribution of the potential energy (without the bias E_{bias}) is constructed using all sampled conformations. Multiple high-dimensional copies, or "replicas," consistent with the instantaneous coarse description of the protein system, i.e., $(X(t), Y(t))$, are selected with energy corresponding to the peak of the distribution. Considering only configurations around the most probable energy allows us to filter out rare and/or unphysical states of the systems.

Multiple ($\sim$10) independent, short ($\sim$ps) MD simulations are then initiated from each of the selected individual replica, as described in more detail below. Figure 3 shows that the majority of these trajectories collapse within 1.5 ps simulation time inside well defined regions on the (Q, A) space (highlighted in red). These regions reproduce reasonably well the two minima on the folding free energy surface obtained using equilibrium MD simulation. The initial rings of our procedure are defined through the density contour of the most densely populated regions (black solid rings in Fig. 3). Each ring is described by means of a set of connected nodes (black dots in Fig. 3).

3.3 Climbing Over a Free Energy "Mountain"

The initialization described above is used to locate an initial ring of nodes in terms of the reduction coordinates Q and A, at the time $t = 0$. At any time, t, during

Coarse Graining of Effective Free Energy Surfaces

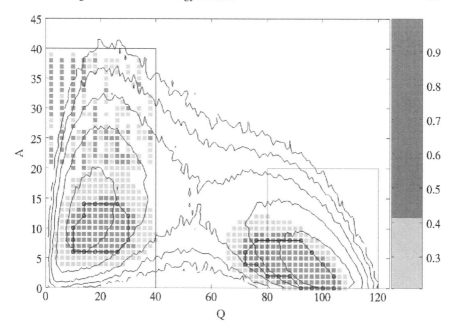

Fig. 3 End point distribution on the (Q, A) plane estimated from short (1.5 ps) bursts of simulations. The short simulations are initialized at equidistributed points on the regions shown in Fig. 2. The local density of configurations on the coarse space is shaded according to the shademap. The *black* contours represent the folding free energy surface obtained from the analysis of long MD simulation data (as in Fig. 2). The initial rings selected are shown by *black dots* and *lines* (see text)

the reverse integration procedure the ring is parametrized by its normalized arclength α; each node in the ring at time t is identified as $(Q(\alpha, t), A(\alpha, t))$. Starting from the initial ring of nodes $(Q(\alpha, t=0), A(\alpha, t=0))$, the system is *lifted* on-the-fly from each particular node (that is, for each value of the variable α along the ring) by creating approximately 50 replicas consistent with each of the following five initial coarse conditions, i.e., $(Q(\alpha, t), A(\alpha, t))$, $(Q(\alpha, t)+1, A(\alpha, t)+1)$, $(Q(\alpha, t)+1, A(\alpha, t)-1)$, $(Q(\alpha, t)-1, A(\alpha, t)+1)$, and $(Q(\alpha, t)-1, A(\alpha, t)-1)$. The replicas are created by the SA-MC procedure described in the previous section. Clearly, one needs to accommodate the integer nature of the coarse variables in the continuum coarse integration/ring description framework used here. This is accomplished through interpolation from the stencil of adjacent integer lattice points above.

From each replica, several ($\sim$100), very short, independent MD simulations at folding temperature T_f are *run* with initial velocities randomly assigned from a Maxwell–Boltzmann distribution. The timespan δt of these MD simulations is given by the typical relaxation time scale of the system, τ_{fast} (for the system considered here, $\tau_{fast} \leq 0.05$ ps). Coarse evolution time derivatives at time t may be estimated by averaging over all the short MD trajectories, over all the replicas initialized at the point $X(t) = (Q(t); A(t))$:

$$\dot{X}(t) = \begin{pmatrix} v_Q(t) \\ v_A(t) \end{pmatrix} \approx \begin{pmatrix} (\overline{Q}(t + \delta t; Q(t)) - Q(t))/\delta t \\ (\overline{A}(t + \delta t; A(t)) - A(t))/\delta t \end{pmatrix} \tag{2}$$

where $(\overline{Q}(t + \delta t; Q(t)), \overline{A}(t + \delta t; A(t)))$ is the average end-point of the MD trajectories starting from $(Q(t), A(t))$. We reiterate that this deterministic dynamic evolution using equations of motion with the local drift vector $d(t) = (v_Q(t), v_A(t))$ as right-hand-sides is physically meaningful for deterministic gradient problems, yet it can still be used to fruitfully explore effective free energy surfaces for stochastic problems. The relationship between the estimated "time" derivatives appearing in the above equation and the reverse ("upward") steps in an effective potential is determined by the form of the effective diffusion tensor for a given problem [12, 29]. If the diffusion tensor entries are diagonal and identical (referred to briefly as "scalar") our deterministic evolution is meaningful in terms of the effective potential – with a stepsize scaling depending on the diffusion constant D. When the diffusion coefficients are also estimated, differences in effective potential (free energy) between a reference state at the node $(Q(\alpha, t), A(\alpha, t))$ and the state $(Q(\alpha, t + \delta t), A(\alpha, t + \delta t)$ can be directly computed from the following line integral [29]:

$$\beta \Delta E^{\text{eff}} = -D^{-1} \left(\int_{Q(\alpha,t)}^{Q(\alpha,t+\delta t)} v_Q(Q', A(\alpha, t)) dQ' \right.$$

$$\left. + \int_{A(\alpha,t)}^{A(\alpha,t+\delta t)} v_A(Q(\alpha, t + \delta t), A') dA' \right). \tag{3}$$

Here, we do not require the stepsize scaling to coarsely identify the transition state region: just following the drifts "in reverse" suffices. The above analysis is based on the assumption of the existence of an effective potential – one can numerically test whether the relevant *potential conditions* for the system hold to confirm this (see [29] where the case of a more general diffusion matrix $D(X)$ with all entries possibly state-dependent is also discussed). In order to estimate the forward time derivatives of our drift-based dynamics for a particular node, a Gaussian smoothing is performed over the time derivatives estimated from the entire ensemble of replica MD runs initialized on the five-point (Q, A) stencil representing that particular node. The rings are then projected "backward-in-time" with a time step Δt. Given the coordinates $(X(t), Y(t))$ (that is, $(Q(t), A(t))$ for the specific protein system considered here) of each node on a ring at time t, and the corresponding drift vector $d(t)$, the components normal to the ring $d^{\perp}$ can be computed as:

$$d^{\perp} = d - (\hat{t} \cdot d)\hat{t}, \tag{4}$$

where $\hat{t}$ is the tangent vector to the curve $\Phi(\alpha, t)$. Following the discussion above, if the diffusion coefficients are "scalar," evolution of the ring backward to time

Coarse Graining of Effective Free Energy Surfaces 123

$t-\Delta t$ using explicit Euler integration normal to the ring [7, 8, 14, 18] at each node $(X(t), Y(t))$, that is:

$$X(t - \Delta t) = X(t) - \Delta t d_X^\perp \tag{5}$$

$$Y(t - \Delta t) = Y(t) - \Delta t d_Y^\perp \tag{6}$$

corresponds to an "upward" step in effective potential. The final step in the computational procedure involves a reparametrization of the ring to obtain node equidistribution at $\Phi(\alpha, t-\Delta t)$; it is possible to combine the evolution and redistribution steps through a Lagrange multiplier formulation [7], which, however, we do not implement here, avoiding the estimation of further partial derivatives from simulation data.

3.4 Reconstructed Folding Free Energy Landscape of SH3 Protein

Figure 4 shows the "backward-in-time" coarse evolution of the rings, starting from different minima, for the first few consecutive reverse integration steps with a time step $\Delta t = 0.2\,\mathrm{ps}$.[2] The final location of the nodes in (Q, A) space are obtained by averaging the results obtained from 10 independent runs. Figure 4 shows that the system starts climbing uphill from both minima within a single coarse reverse integration step. As evident from Fig. 4, the rings quickly become elongated (i.e., they "move" slower in the direction of the barrier). Within four to five reverse integration steps the rings starting from different minima converge[3] at the top of the free energy barrier (white circled region in Fig. 4, corresponding to $Q \sim (54 - 55)$ and $A \sim (10 - 22)$). The location of the "top" of the folding free energy barrier on the (Q, A) surface, as approximated by the coarse reverse integration, is in good agreement with the transition state location obtained from the analysis of long, equilibrium MD simulations.

4 Test of the Validity of the Reduction Coordinates Q and A

From a mathematical point of view, the problem of rigorous definition of collective coordinates in macromolecular systems becomes the problem of identifying a low dimensional surface (or, more generally, a *manifold*) that usefully describes

---

[2] The results presented in this study are robust against the choice of the coarse projective step Δt, if selected from a range of values from 0.1 to 0.25 ps. Within this interval, we choose to use a sufficiently large projective step size, that allows us to map the full landscape with reasonably accuracy in a relatively small number of reverse integration steps.

[3] We consider that "opposite" rings meet if the distance $dist = \sqrt{((Q_i - Q_j)^2 + (A_i - A_j)^2)} < 1$, between two nodes (Q_i, A_i) and (Q_j, A_j) from opposite rings.

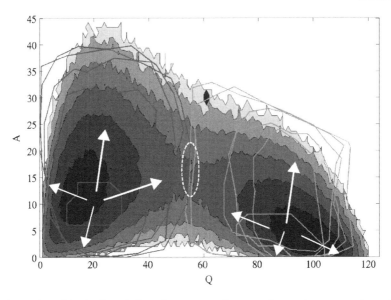

Fig. 4 "Backward-in-time" evolution of our coarse rings superimposed on grayscale contours of the folding free energy surface in terms of Q and A, as obtained from the analysis of long MD simulation data. *Closed grey curves* represent the propagation of nodes starting from rings enclosing the unfolded and the folded state, evolving in the direction indicated by the *arrows*. Five reverse coarse projective integration steps of our local-drift-based dynamics are performed. The region where rings representing different minima eventually "meet" (*dashed white circled* region) matches well with the transition state region identified on the equilibrium folding free energy surface

thesystem motion, and is non-trivially embedded into the much higher dimensional space of the degrees of freedom of the detailed simulation code. Although several non-linear dimensionality reduction techniques have been proposed, the development of new such methods is still an active area of research. Nonlinear dimensionality reduction techniques have been recently applied to infer appropriate coarse variables to describe complex macromolecular motions over long time scales. These techniques can be used a posteriori to test the validity of *empirical* coarse coordinate-based CMD computations. The discrete (point-based) nature of such descriptions necessitates the use of "out-of-sample" extension algorithms before they can be integrated in our CMD formulation [10, 22].

We briefly describe here two nonlinear dimensionality reduction techniques independently used by our groups, and show how they can be used to assess the suitability of our empirical coordinates Q and A.

4.1 The Scalable Isomap (ScIMap) Approach

We have recently adapted a *nonlinear* dimensionality reduction method originally developed to address image-recognition problems to map the "embedding" of a protein folding landscape [6, 28]. In particular, our approach is based upon the Isomap algorithm [31]. Let us assume that a reliable sampling is available for the configurations populated by the system of interest on a given time-scale. We assume that a low-dimensional embedded manifold exists that encompasses all the points of the sampling. We do not have an explicit mathematical formula for this manifold; rather, we have a discrete set of points sampled from it. The goal is to find a minimal set of variables that can parametrize the same configurational space.

Differential geometry tells us that the essential information on a manifold metric is contained in the definition of the *geodesic distance*, that is, the shortest path between any pair of points, when the path is confined to the manifold. By exploiting this fact, if we find a way to measure geodesics on the manifold, we can obtain a low dimensional embedding that preserves as best as possible geodesic distances between all pairs of data points in the sample under consideration [31]. The idea at the base of the Isomap algorithm is to estimate the geodesic distance between any pair of points on the manifold. To this effect, proximity relations are defined for each high dimensional configuration, that is represented as a vertex in a graph. Each vertex is connected to the k closest configurations according to a distance measure (typically, the least Root Mean Square Deviation, lRMSD, that is the root mean square deviation computed on pairs of optimally aligned configurations [19]), that are referred to as the k nearest neighbors. The emerging graph captures the connectivity of the data set. Each edge of the graph is associated with the lRMSD distance between the configurations that it connects. The distance between any pair of configurations is estimated as the length of the shortest path between them in the graph, where the path length is obtained by adding up the lRMSD distances associated with the edges of the path. The reduction coordinates are then computed as a function of the distance matrix whose entries represent shortest-path distances between a significant portion of the configurations in the data set.

A detailed description of the ScIMap implementation, the approximations involved, and the results obtained on the application to protein folding processes can be found in [6, 28].

4.2 The Diffusion Map (DMap) Approach

As in ScIMap, in the diffusion map approach [4, 26, 27] each high-dimensional configuration in the available sample is described as a vertex in a graph. The main difference from ScIMap is that in the diffusion map approach the connection "strength" between all pairs of vertices is given by a (Gaussian) kernel matrix with entries of the form $K(\mathbf{x}, \mathbf{y}) = \exp\left(-\|\mathbf{x} - \mathbf{y}\|^2 / \sigma^2\right)$ where $\mathbf{x}$ and $\mathbf{y}$ are high-dimensional vectors containing the coordinates of atoms in a configuration, $\|\cdot\|$ is the lRMSD distance measure (as defined above), and σ is a parameter that embodies the extent of the local neighborhood around each high-dimensional point. The parameter σ may be interpreted as a length scale for which the Euclidean distance is a reliable measure of similarity between $\mathbf{x}$ and $\mathbf{y}$.

We define a diagonal $N \times N$ normalization matrix $\mathbf{D}$ with entries given by $D_{ii} = \sum_{k=1}^{N} K_{ik}$ and construct a Markov matrix M (describing transition probabilities between configurations) using $\mathbf{M} = \mathbf{D}^{-1}\mathbf{K}$. The leading eigenvectors ψ_j obtained by solution of the eigenvalue problem

$$\mathbf{M}\psi_j = \lambda_j \psi_j, \qquad \text{for} \quad j = 1, 2, \ldots, N \qquad (7)$$

especially in the presence of a gap in the eigenvalues λ_j, provide "automated" candidate reduction coordinates for the data [27].

4.3 Analysis of the Results

Coarse coordinates can be associated to a set of high-dimensional protein configurations both by means of a SciMap and of a diffusion map analysis. The results obtained by SciMap for the folding reaction of the SH3 model described in the paper have been described in a previous publication [6]. The results of the corresponding diffusion-map based analysis of the same data are qualitatively very similar. Figure 5 plots the components of the sampled MD simulation data in the second and third eigenvectors (ψ_2 and ψ_3) of the Markov matrix $\mathbf{M}$ defined above (i.e., the two leading diffusion map coordinates); the first eigenvector ψ_1 with eigenvalue 1 is trivial and does not provide useful information. On average, points at the maximum (minimum) values of coordinate ψ_2 in Fig. 5 correspond to configurations in the folded (unfolded) well. The ordering of points in the coordinate ψ_2 appears similar to their ordering in coordinate Q (from left to right in Fig. 5 the shade changes gradually from white to black). A similar result holds when the first embedded dimension x_1 emerging from SciMap is used as coarse coordinate (see Fig. 4 in [6]). In particular, Fig. 6 indicates that both the diffusion map coordinate ψ_2 and the SciMap coordinate x_1 reflect the large variation of Q around the transition region, while the range of values of both ψ_2 and x_1 is small in the minimum free energy wells.

The eigenspectrum of the diffusion map shows a large gap between the first significant eigenvalue (corresponding to ψ_2) and the following ones. A similar result emerges from the analysis of the residual variance $\sigma_r(d)$ of SciMap as a function of the number of dimensions d considered in approximating the data.[4] As discussed in [6], the embedded landscape associated with the folding simulations of the coarse-grained model of SH3 used in the paper has extremely low residual variance, even when only a single dimension is considered, suggesting that the folding landscape of SH3 can be essentially described by one reaction coordinate, in agreement with results from previous work [1, 5]. The main coarse coordinate emerging

---

[4] The residual variance $\sigma_r(d)$ provides an estimate of the distortion introduced when the first d dimensions emerging from the SciMap analysis are used as coarse coordinates to describe the data (see [6, 31] for details).

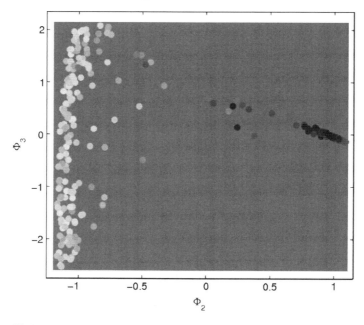

Fig. 5 Diffusion map representation of data sampled from long time MD simulation; x-axis (y-axis) indicates the component of each data point (a configuration) in the first (second) significant eigenvector Φ_2 (Φ_3). Points are shaded from *white* to *black* according to the corresponding value of Q, from 0 to 1

from dimensionality reduction (either ψ_2, or x_1) parametrizes the process occurring over the slowest time scale – the transition between the unfolded and folded well. The one-to-one mapping between ψ_2 or x_1 and the empirical coordinate Q indicates that Q also suitably parametrizes this transition. Figure 5 also suggests that, while the next coarse diffusion map coordinate, ψ_3, varies very little in the neighborhood of the folded state, it exhibits significant variation in the neighborhood of the unfolded state. This is true also for the second ScIMap coordinate, x_2 (see Fig. 4 in [6]), and it suggests that ψ_3 or x_2 may be a good coarse variable to parametrize the (faster) dynamics within the unfolded well. Figure 7 plots the values of A versus the corresponding values of ψ_3 and x_2 on the sampled simulation data. Two comparisons are made, considering only the data corresponding to the folded state (left panel), as well as only those corresponding to the unfolded state (right panel): the variation of x_2 correlates with A *within the unfolded well*, while ψ_3 appears to (much less strongly) correlate with A *within the folded well*. In light of these a posteriori observations, one might argue that the empirical coarse variables (Q,A) can be considered a good choice to parametrize the local relaxation dynamics: Q signals the folding/unfolding global transition, while A appears to relate to dynamics inside individual wells.

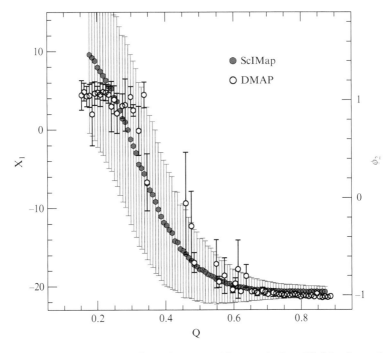

Fig. 6 The diffusion map top significant eigenvector (Φ_2, *open dots*), and SciMap first embedding dimension (x_1, *solid dots*) are plotted against the empirical reduction coordinate Q, defined as the fraction of native contacts formed. The main coordinate extracted with both dimensionality reduction methods captures the variability of Q around the transition region. The error bars illustrate the spread around the average of Φ_2 and x_1 corresponding to different values of Q

5 Conclusions

A multi-scale hybrid simulation procedure that couples the low- and high-dimensional descriptions of a complex system has been presented and illustrated. This approach can provide an efficient tool to locate regions of critical importance on the underlying effective free energy surface of a macromolecular dynamic process characterized by a distinct time scale separation. The only a priori knowledge required about the system is the definition of a few reasonable reduction coordinates. The great advantage of this method is that the MD simulation does not need to directly access the long time-scale motions of the system. Instead, multiple short conventional MD trajectories are used to explore the longer time system evolution in terms of the coarse variables, which makes this approach eminently suitable for parallelization.[5] The results presented in this study suggest that this procedure may

---

[5] The coarse reverse integration approach required about 5 CPU hours (Intel Xeon 2.2 GHz) to fully explore the folding free energy surface, while the traditional long MD simulation used to significantly sample the protein conformational space requires > 100 CPU hours. We expect the

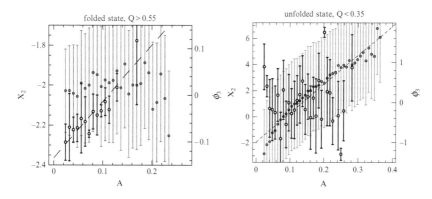

Fig. 7 The diffusion map second significant eigenvector (Φ_3, *open dots*), and SciMap second embedding dimension (x_2, *solid dots*) are plotted against the empirical reduction coordinate A, defined as the fraction of non-native contacts formed, for data corresponding to the folded state (*left panel*), and data corresponding to the unfolded state (*right panel*)

provide an efficient, alternative approach to the study of complex dynamic processes occurring over long time scales, inaccessible to conventional molecular simulation techniques.

Acknowledgements The authors have planned and started their collaboration on this project during the program on "Bridging Time and Length Scales in Materials Science and Bio-Physics" that was held at the NSF-funded Institute for Pure and Applied Mathematics (IPAM) at UCLA in September-December 2005. This work has been supported in part by grants from NSF (C.C. Carccr CHE-0349303, CCF-0523908, and CNS-0454333), and the Robert A. Welch Foundation (C.C. Norman Hackermann Young Investigator award, and grant C-1570). The Rice Cray XD1 Cluster ADA used for the calculations is funded by NSF under grant CNS-0421109, and a partnership between Rice University, AMD and Cray. T.A.F and I.G.K are partially supported by NSF and DARPA.

References

1. Amat, M. A., Kevrekidis, I. G., Maroudas, D.: Coarse molecular dynamics applied to the study of structural transitions in condensed matter. Phys. Rev. B **74** (2006) 132201
2. Bolhuis, P. G., Chandler, D., Dellago, C, Geisseler, P. L.: Transition path sampling: Throwing ropes over rough mountain passes, in the dark. Ann. Rev. Phys. Chem. **53** (2002) 291–318
3. Clementi, C., Plotkin, S. S.: The effects of nonnative interactions on protein folding rates: Theory and simulation. Protein Sci. **13** (2004) 1750–1766
4. Coifman, R., Lafon, S., Lee, A., Maggioni, M., Nadler, B., Warner, F., Zucker, S.: Geometric diffusions as a tool for harmonic analysis and structure definition of data. part i: Diffusion maps. Proc. Natl. Acad. Sci. USA **102** (2005) 7426–7431

---

speed-up to be dramatically more significant for systems exhibiting slower coarse dynamics (i.e., a larger time scale separation) compared to the case considered here.

5. Das, P., Matysiak, S., Clementi, C.: Balancing energy and entropy: A minimalist model for the characterization of protein folding landscapes. Proc. Natl. Acad. Sci. USA **102** (2005) 10141–10146

6. Das, P., Moll, M., Stamati, H., Kavraki, L. E., Clementi, C.: Low-dimensional, free-energy landscapes of protein-folding reactions by nonlinear dimensionality reduction. Proc. Natl. Acad. Sci. USA **103** (2006) 9885–9890

7. E. W., Ren, W., Vanden-Eijnden, E.: String method for the study of rare events. Phys. Rev. B **66** (2002) 052301

8. E. W., Ren, W., Vanden-Eijnden, E.: Finite temperature string method for the study of rare events. J. Phys. Chem. B **109** (2005) 6688–6693

9. Elber, R., Meller, J., Olender, R.: Stochastic path approach to compute atomically detailed trajectories: Application to the folding of c peptide. J. Phys. B **103** (1999) 899–911

10. Erban, R., Frewen, T. A., Wang, X., Elston, T., Coifman, R., Nadler, B., Kevrekidis, I. G.: Variable-free exploration of stochastic models: a gene regulatory network example. J. Chem. Phys. **126** (2007) 155103

11. Ferrenberg, A. M., Swendsen, R.H.: Optimized monte carlo data analysis. Phys. Rev. Lett. **63** (1989) 1185–1198

12. Frewen, T. A., Kevrekidis, I. G., Hummer, G.: Exploration of effective potential landscapes using coarse reverse integration. J. Chem. Phys **131** (2009) 134104

13. Grantcharova, V. P., Baker, D.: Folding dynamics of the src sh3 domain. Biochemistry **36** (1997) 15685–15692

14. Guckenheimer, J., Worfolk, P.: Bifurcations and Periodic Orbits of Vector Fields. Kluwer, Dordrecht (1993)

15. Henderson, M. E.: Multiple parameter continuation: Computing implicitly defined k-manifolds. Int. J. Bifurcat. Chaos **12** (2002) 451–476

16. Henderson, M. E.: Computing invariant manifolds by integrating fat trajectories. SIAM J. Appl. Dyn. Syst. **4** (2005) 832–882

17. Hummer, G., Kevrekidis, I. G.: Coarse molecular dynamics of a peptide fragment: Free energy, kinetics, and long-time dynamics computations. J. Chem. Phys. **118** (2003) 10762–10773

18. Johnson, M. E., Jolly, M. S., Kevrekidis, I. G.: Two-dimensional invariant manifolds and global bifurcations: Some approximation and visualization studies. Num. Alg. **14** (1997) 125–140

19. Kabsch, W.: Discussion of solution for best rotation to relate 2 sets of vectors. Acta. Cryst. **A34** (1978) 827–828

20. Kevrekidis, I. G., Gear, C. W., Hyman, J. M., Kevrekidis, P. G., Runborg, O., Theodoropoulos, C.: Equation-Free Multiscale Computation: enabling microscopic simulators to perform system-level tasks. Comm. Math. Sci. **1** (2003) 715–762

21. Kopelevich, D. I., Panagiotopoulos, Z. A., Kevrekidis, I. G.: Coarse-grained kinetic computations for rare events: Application to micelle formation. J. Chem. Phys. **122** (2005) 044908

22. Laing, C. R., Frewen, T. A., Kevrekidis, I. G.: Coarse-grained dynamics of an activity bump in a neural field model. Nonlinearity **20** (2007) 2127–2146

23. Laio, A., Parrinello, M.: Escaping free-energy minima. Proc. Natl. Acad. Sci. USA **99** (2002) 12562–12566

24. Micheletti, C., Bussi, G., Laio, A.: Optimal langevin modeling of out-of-equilibrium molecular dynamics simulations. J. Chem. Phys. **129** (2008) 074105

25. Mossa, A., Clementi, C.: Supersymmetric langevin equation to explore free energy landscapes. Phys. Rev. E **75** (2007) 046707

26. Nadler, B., Lafon, S., Coifman, R. R., Kevrekidis, I. G.: Diffusion maps, spectral clustering and eigenfunctions of fokker-planck operators. Volume 18 of Advances in Neural Information Processing Systems, page 955 (2005). MIT, MA

27. Nadler, B., Lafon, S., Coifman, R. R., Kevrekidis, I. G.: Diffusion maps, spectral clustering and reaction coordinates of dynamical systems. Appl. Comput. Harmon. A. **21** (2006) 113–127

28. Plaku, E., Stamati, H., Clementi, C., Kavraki, L. E.: Fast and reliable analysis of molecular motion using proximity and dimensionality reduction. Proteins Struct. Funct. Bioinformatics **67** (2007) 897–907

29. Risken, H.: The Fokker-Planck Equation: Methods of Solutions and Applications. Springer Series in Synergetics, 2nd edition. Springer, Berlin (1996)
30. Sriraman, S., Kevrekidis, I. G., Hummer, G.: Coarse nonlinear dynamics and metastability of filling-emptying transitions: Water in carbon nanotubes. Phys. Rev. Lett. **95** (2005) 130603
31. Tenenbaum, J. B., de Silva, V., Langford, J.C.: A global geometric framework for nonlinear dimensionality reduction. Science **290** (2000) 2319–2323
32. Wales, D. J.: Energy Landscapes. Cambridge University Press, Cambridge (2003)
33. Xu, W. , Harrison, S. C., Eck, M. J.: Three-dimensional structure of the tyrosine kinase c-Src. Nature **385** (1997) 595–602
34. Voter, A. F.: Hyperdynamics: Accelerated molecular dynamics of infrequent events. Phys. Rev. Lett. **78** (1997) 3908–3911

Extracting Functional Dependence from Sparse Data Using Dimensionality Reduction: Application to Potential Energy Surface Construction

Sergei Manzhos, Koichi Yamashita, and Tucker Carrington

Abstract To construct a continuous D-dimensional function that represents sparse data, we use either a single function depending on $d < D$ coordinates or a sum of L such *component* functions. Neural networks (NN) are used as building blocks that approximate the component functions. For a given density of data, only component functions with $d < d_{max}$ can reliably be recovered (or only coupling up to a certain order can be recovered if original coordinates are used). This justifies using functions of lower dimensionality. We use functions of $d < D$ new, adapted coordinates. The coordinate transformation is done by a NN and is adjusted during the fit for the best approximation error. When $Ld < D$, dimensionality reduction is achieved. We discuss the usefulness of representations via low-dimensional functions for the reconstruction of potential energy surfaces of molecular and reactive systems from sparse ab initio data.

1 Introduction

In this chapter, we discuss a neural network based method for constructing multivariate functions from sparse data and how it incorporates dimensionality reduction, and present a quantum chemistry application – construction of a potential energy function. The task of constructing a smooth functional form in a high-dimensional space is ubiquitous in science [1], but the dimensionality problem is especially severe in quantum chemistry [2, 3]. An important goal of theoretical chemistry is to compute molecular or material properties and reaction outcomes. This almost always involves solving the Schrödinger equation. In principle, the dimensionality

---

S. Manzhos (✉) and K. Yamashita
Department of Chemical System Engineering, School of Engineering, University of Tokyo, 7-3-1, Hongo, Bunkyo-ku, Tokyo 113-8656, Japan
e-mail: sergei@tcl.t.u-tokyo.ac.jp, yamasita@chemsys.t.u-tokyo.ac.jp

T. Carrington
Department of Chemistry, Queens University, Kingston, ON, Canada K7L 3N6
e-mail: Tucker.Carrington@chem.queensu.ca

A.N. Gorban and D. Roose (eds.), *Coping with Complexity: Model Reduction and Data Analysis*, Lecture Notes in Computational Science and Engineering 75, DOI 10.1007/978-3-642-14941-2_7, © Springer-Verlag Berlin Heidelberg 2011

of the Schrödinger equation is $3(N_e + N_n)$, where N_e is the number of electrons, and N_n is the number of nuclei. Solution of the Schrödinger equation is simplified by [4] separating it into an electronic problem for frozen nuclei and a nuclear problem. From the electronic Schrödinger equation,

$$\hat{H}_{el}\psi_{el}(r) = E_{el}(R)\psi_{el}(r)$$
$$\hat{H}_{el} = -\frac{h^2}{8\pi^2}\frac{1}{m_e}\sum_k \nabla_k^2 + \frac{1}{4\pi\epsilon_0}\left(-\sum_{iI}\frac{Z_I e^2}{\Delta r_{iI}}\right.$$
$$\left. +\sum_{i<j}\frac{e^2}{\Delta r_{ij}} + \sum_{I<J}\frac{Z_I Z_J e^2}{\Delta R_{IJ}}\right). \tag{1}$$

Here R denotes all coordinates of all nuclei. One can compute the electronic energy E_{el} and the wavefunction $\psi_{el}(r)$ which depends on the positions of all electrons in the molecule or reactive system. For even a simple molecule like methane, CH_4, there are 10 electrons, and the electronic Schrödinger equation is a 30-dimensional problem. Often one wishes to solve the electronic Schrödinger equation for more complicated systems, e.g., molecular networks on surfaces or nanotubes [5]. The dimensionality of the electronic Schrödinger equation is a serious bottleneck for molecular and material design [2] but is not the subject of this chapter.

The dimensionality issue that we address in this chapter arises when, after solving the electronic problem (1), one (using the Born–Oppenheimer approximation [4]) wishes to calculate rate constants, spectra etc. by solving the nuclear Schrödinger equation,

$$\left(\hat{T}_v + V\right)\psi_v(R) = E_v\psi_v(R)$$
$$V(R) = E_{el}(R) \tag{2}$$

The potential term V – the potential energy surface (PES) – is the electronic energy. PES's govern reaction dynamics and are extremely important in chemistry [6–9]. The nuclear Schrödinger equation can be solved if it is possible to compute $V(R)$ at every point at which it is needed. One option is to solve the electronic Schrödinger equation every time a potential value is required to solve the nuclear Schrödinger equation. Although this is frequently done [10], it necessitates making severe approximations to solve the electronic problem; otherwise, the procedure is too costly. It would be much better to have a good means of constructing $V(R)$ from a relatively small number of solutions of the electronic Schrödinger equation. This is the goal of this chapter. Such a procedure would greatly simplify molecular modelling of processes like the dynamics of polyatomic molecules on surfaces, the modelling of which is important for catalysis [11–13]. For an isolated molecule, the dimensionality of a PES is $D = 3N_n - 6$, for a molecule on a surface it is $3N_n$. These D coordinates span the so-called configuration space. Even though it is too costly to accurately solve the electronic Schrödinger equation at all (millions or billions) R points needed to solve the nuclear problem, it is possible to sample V by solving the electronic problem (1) at $\sim 10^4$ points. The problem we address in this chapter is the determination of V from $\sim 10^4$ values. Once the potential has been obtained, one solves the nuclear problem. It is possible to deal with hundreds of degrees of freedom if the nuclear problem is solved classically [14]. If it is solved quantum mechanically, even $D \geq 6$ is difficult. For example, the computation of the vibrational spectrum of the four-atom molecule hydrogen peroxide (a six-dimensional configuration space) requires a quadrature grid with as many as 30 million points [2, 15]. Our purpose is to develop a method for constructing a smooth potential function from a very sparse (as low as, in a direct product sense, ~ 2–3 data per dimension) set of solutions of the electronic equation. Points computed from the electronic solution will be referred to as ab initio samples. This is essentially a fitting problem. It is difficult because we assume no knowledge of the function and because the dimensionality is large. It is therefore important to be able to reduce the dimensionality by introducing new coordinates. In what follows, we describe a method to deal with the dimensionality of the potential term V in (2). However, not every method to map the coordinate space onto a low-dimensional manifold is suitable. When the nuclear Schrödinger equation is solved, it is advantageous to have a potential that can be written as a sum of products [16–18]. If the new coordinates are linear combinations of the old coordinates, this is possible. It is in this context that we address the problem of the representation of the potential energy surface in this chapter. In Sect. 2, we describe the representation of a multivariate function with lower-dimensional functions and a corresponding algorithm with a dimensionality reduction capability. This reviews the material of [19–22]. In Sect. 3, we present applications of the method to fitting potential energy surfaces by using multiple low-dimensional terms as well as dimensionality reduction. Finally, in Sect. 4, we place the method in the larger context of existing D-reduction approaches and discuss the applicability of dimensionality reduction techniques to problems of theoretical chemistry.

2 Method

We construct a continuous function from sparse data by representing it with low-dimensional terms. Such representations have been used for a long time; in particular, in chemistry, the so-called Multimode expansion in the orders of coupling of the *original* coordinates is used [23–25]:

$$
f(x_1, x_2, \ldots, x_D) = f_0 + \sum_{n=1}^{D} f_i^{(1)}(x_i) + \sum_{i<j=1}^{C_D^2} f_{ij}^{(2)}(x_i, x_j) + \ldots
$$

$$
+ \sum_{i_1, i_2, \ldots, i_n}^{C_D^n} f_{i_1, i_2, \ldots, i_n}^{(n)}(x_{i_1}, x_{i_2}, \ldots, x_{i_n}) + \ldots + f_{12 \ldots D}^{(D)}(x_1, x_2, \ldots, x_D) \quad (3)
$$

For physical reasons, it often converges after 3rd or 4th order ($n = 3, 4$) [23–26]. The f functions are called component functions. We have shown that only coupling (component functions) up to a certain order can be recovered for a given sampling density [19], which favors using this sort of expansion. The Multimode component functions are defined as hyper-surfaces of increasing dimensionality passing through an expansion center y:

$$f_0 = f(y)$$
$$f_i(x_i) = f(x_i|y) - f_0$$
$$f_{ij}(x_i, x_j) = f(x_i, x_j|y) - f_i(x_i) - f_j(x_j) - f_0$$
$$\ldots \qquad (4)$$

where the notation $(x_i, x_j, \ldots | y)$ means that all components of x are set to those of the expansion center y except for $x_i, x_j, \ldots$, which are allowed to vary. This definition of the component functions has two important problems: each function must be fitted to its own data set, and the number of terms grows combinatorially with both the dimensionality of x and the order of expansion n. With this choice of the component functions, the expansion is also good, by construction, only close to the sub-dimensional surfaces and not in all space.

The Multimode expansion, however, is a particular case of the so-called high-dimensional model representation, HDMR [27, 28], which aims to minimize the error functional $\int_{K^D} \left[f(x) - f^{HDMR}(x) \right]^2 d\mu(x)$ with respect to the measure $d\mu$. It is obtained from the more general approach by replacing $d\mu$ with a product of δ-functions: $d\mu(x) = \prod_{i=1}^{D} \delta(x_i - y_i) dx_i$. With an HDMR, it is possible to use *one* set of data distributed according to a density w and minimize the error in *all* space. If one sets $d\mu(x) = w(x) \prod_{i=1}^{D} dx_i$ and imposes constraints on the component functions, it is possible to derive closed-form expressions for the component functions [29], but they are in terms of high-dimensional integrals which are very costly for $D > 6$.

In [19], we showed that it was possible to use a general weight function without doing costly high-dimensional integrals by using neural networks to represent the component functions. Despite the advantages of this approach, it is still limited by the combinatorial explosion of the number of component functions with dimension in the expansion (3).

One can prove [20] that there exists a linear coordinate transformation to new coordinates, with which it is possible to achieve arbitrary accuracy with an HDMR type expression of any order k:

$$\exists \hat{A}, b : y = \hat{A}x + b$$

$$f^{HDMR}(y) = \left[f_0 + \sum_i f_i \left(y_i^{(1)} \right) + \sum_{i_1, i_2} f_{i_1, i_2} \left(y_{i_1}^{(2)}, y_{i_2}^{(2)} \right) + \ldots \right]$$

$$\sum_{i_1,i_2,\ldots,i_k} f_{i_1,i_2,\ldots,i_k} \left(y_{i_1}^{(k)}, y_{i_2}^{(k)}, \ldots, y_{i_k}^{(k)} \right)$$

$$\forall \epsilon > 0, \left| f^{HDMR}(y) - f(x) \right| < \epsilon, \forall k \geq 1 \tag{5}$$

In (5), the lower-order terms (shown in brackets) are not mandatory – they can be lumped into the highest-order terms. The low-dimensional component functions $f_{i_1,i_2,\ldots,i_k}$ are easier to construct than the original D-dimensional function f: fitting methods which are not practical for the high-dimensional function f could still work for the component functions. In particular, the effective density of sampling is increased in the subspace of y-coordinates, making the construction of a reliable continuous function feasible.

For all calculations in this chapter, we lump the low-order terms and minimize the error between the resulting expansion and the function f:

$$f(y(x)) \approx \sum_{i}^{L} f_i^{NN} \left(y_{i_1}(x), y_{i_2}(x), \ldots, y_{i_{d<D}}(x) \right)$$

$$\min \int_{K^D} \left[f(x) - \sum_{i}^{L} f_i^{NN} \left(y_{i_1}(x), y_{i_2}(x), \ldots, y_{i_{d<D}}(x) \right) \right] w(y(x)) \, dx \tag{6}$$

We represent the component functions f_i^{NN} with neural networks [30], which results in a general method. Neural networks are used both to find the best coordinate transformation in terms of the fitting error and the number of terms, and to fit the terms in new coordinates. For more details, see [20–22].

2.1 Neural Networks

Neural networks (NN) are representations with univariate functions called neurons, whose arguments are linear combinations of coordinates x_i. A common single-hidden layer network [15, 31, 32] is shown schematically in Fig. 1 and is expressed as

$$f^{NN}(x) + d = \sum_{q=1}^{S} c_q \sigma \left(w_q x + d_q \right) \tag{7}$$

The parameters c_i, d_i, w_{ij}, b_i are adjusted to achieve a good fit. NN's are universal approximators [33, 34] and have a good dimensionality scaling. Any non-linear function can be a neuron [35], and the use of exponential neurons, for instance, results in a sum of products of one-dimensional terms: easy to integrate and differentiate [36] (if only one non-linear layer of neurons is used). This is important,

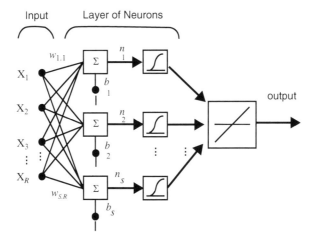

Fig. 1 The architecture of a typical single hidden layer neural network

as the cost of multidimensional quadratures is one of the biggest problems in the modelling of big molecular and reactive systems.

In high dimension and with sparse data, neural networks eventually do suffer from the same set of problems as any black-box method [37], but we use them as building blocks to construct low-dimensional component functions of (6), thereby making the most of their advantages and minimizing the disadvantages.

2.2 Fitting Algorithm

The algorithm based on (6) fits a sum of small, low-dimensional in y, neural networks to the data. Neural networks are sequentially fitted or trained to the difference of the data and the sum of all other neural networks [22]. For example, if two-dimensional terms are used,

$$f_i^{NN}\left(y_{i_1}(x^n), y_{i_2}(x^n)\right) = V(x^n) - \sum_{j \neq i} f_j^{NN}\left(y_{j_1}(x^n), y_{j_2}(x^n)\right) \quad (8)$$

where $V(x^n) = V_n$ are samples of the function. Moreover, it is possible to use only a subset of data for training a partial NN. The subsets are randomly redrawn from the full set during neural network training [21]. Because the non-linear part of the fit is done in low-dimensional y-coordinates, there is a cost advantage. In principle, this permits the fitting of arbitrarily large datasets of arbitrarily large dimensionality. A Matlab code for this algorithm is now available for general use through Computer Physics Communications' programs library [22].

3 Applications

In the following, we present two applications of the algorithm described in Sect. 2, one showing the advantage of the coordinate transformation in terms of the required number and dimensionality of the functions and another that explicitly uses dimensionality reduction.

3.1 Reduction of the Number and Dimensionality of Terms with the Help of Optimized Coordinates

In [21], we fitted a 12-dimensional potential function of vinyl bromide, Fig. 2, and tested the role of coordinates y in achieving an acceptable quality of interpolation despite sparse sampling.

A particular configuration of atoms can be specified by the bond lengths 1C4C, 1C3H, 1C2H, 4C5H, 4C6Br; the angles 3H1C4C, 5H4C1C, 2H1C4C, 6Br4C1C; and the dihedral angles between the planes 3H1C4C and 5H4C1C, the planes 2H1C4C and 5H4C1C, and the planes 3H1C4C and 6Br4C1C (Fig. 2). We sampled the energy of this molecule from a reference potential surface of [38]. Many thousands of points, some of which correspond to geometries at which the molecule begins to break into fragments, are included, i.e., we have a complicated energy landscape. The range of the data is about 70,000 cm^{-1}. The distribution of function values is as shown in Fig. 1 of [21]. We monitor the error on a separate, test, set of

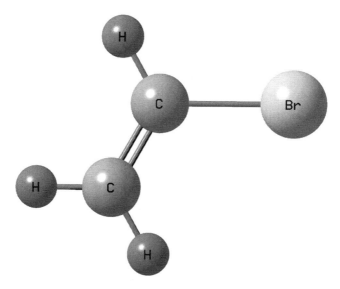

Fig. 2 The structure of vinyl bromide

Table 1 The rmse error (in cm^{-1}) at the test points when fitting the PES of vinyl bromide in 12 dimensions in the original bond coordinates and in NN-optimized coordinates

No. of 2D terms	Original coordinates		Optimized coordinates	
	20,000 data	50,000 data	20,000 data	50,000 data
6	n/a	n/a	1,723	1,626
12	n/a	n/a	434	371
66	2,137	1,670		

data of equal size. General methods like neural networks can always reduce the error at the fitted data by using more parameters. Unless the fitting function has special properties that ensure the quality of the fit (between the training points), the quality of an approximant that is not tested at points not used to produce the fit is uncertain. The desired error for this sort of system is about $300\,cm^{-1}$, which constitutes the so-called chemical accuracy. It is limited by the density of sampling [19].

If one fits the data with low-dimensional combinations of the original coordinates, the root mean square error (rmse) at the test points is large. Fitting 50,000 data points with 66 2nd order terms gives a rmse of $1,670\,cm^{-1}$ (see Table II of [21]). The most obvious way to improve the fit is to also use 3rd order terms, but the number of 3rd order terms (C_{12}^3) and the number of data points required to determine all the NN parameters make such a fit simply impractical. A much better fit can be determined by using new, adapted coordinates. With adapted coordinates, 20,000 points are enough to determine the dozen terms required to reduce the error to about $400\,cm^{-1}$ (Table 1). Moreover, with adapted coordinates, it is possible to use higher-order component functions, because their number remains small. This enables one to reduce the rmse further. The 3-, 4-, and 5-D terms remain well-determined with a data set of about 20,000 points (Fig. 3).

3.2 Dimensionality Reduction

If the product of the number of terms and their dimensionality is smaller than the dimensionality of the space ($Ld < D$), then we have achieved dimensionality reduction. The simplest case is that of a fit with one term ($L = 1$) depending on a small number of new coordinates obtained from the original coordinates by a linear transformation:

$$f\left(\boldsymbol{y}(\boldsymbol{x})\right) \approx f^{NN}\left(y_1(\boldsymbol{x}), y_2(\boldsymbol{x}), \ldots, y_{d<D}(\boldsymbol{x})\right)$$
$$\boldsymbol{y} = \hat{A}\boldsymbol{x} + \boldsymbol{b} \tag{9}$$

where the coordinates y are optimized by a NN to minimize the fitting error. We illustrate the dimensionality reduction capability by fitting the potential energy function of a molecule-surface reaction.

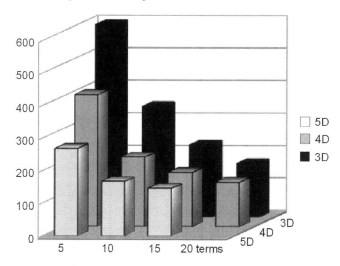

Fig. 3 The rmse (in cm^{-1}) at the test set of points when fitting the 12-dimensional PES of vinyl bromide with different numbers of terms of different dimensionality using 20,000 test and fit data

Molecule-surface, specifically, catalytic, reactions are key processes which permit technological advances in areas such as environmental protection and new energy sources – fuel cells, for example [39]. Hence the importance of theoretical modelling of reactions at surfaces. Availability of a potential energy function would greatly simplify such modelling. Unfortunately, such functions have so far been calculated only for reactions of diatomic molecules with surfaces [40–43], partly because of the difficulties described above of constructing a multidimensional function from sparse data.

The model system we consider is nitrous oxide on copper. N_2O is an ozone depletion agent and a greenhouse gas. It is a useful oxygen atom donor. N_2O is also a by-product of industrial chemistry and of the catalytic removal of NO_x [44]. We aim to build a PES describing the dissociative adsorption of N_2O on Cu(100). The energy landscape is sampled with electronic structure calculations using density functional theory (DFT) [45]. The Cu surface is modelled with a slab in a supercell containing 16 Cu atoms in four layers, as shown in Fig. 4, which is replicated ad infinum in x and y directions, the z axis being perpendicular to the surface. Cu atoms are kept frozen, and the slabs are separated by about 20 Å of vacuum. More details of the calculation scheme are given in [22].

The configuration space of the molecule on the surface can be characterized by at least nine coordinates, for example, the Cartesian coordinates of the N and O atoms, or the internal coordinates used in [22]. Those coordinates, however, may not be the best coordinates for representing the underlying physics. For example, if the PES is to describe dissociation, asymptotic conditions are important. Black-box fitting methods [46–50] do not incorporate asymptotic conditions; specifically, neural networks do not have good extrapolating properties. One could instead choose

Fig. 4 The slab model of the Cu(100) surface and the N_2O molecule. Cu atoms are shown in *grey*, N atoms in *black*, and the O atom in *light grey*

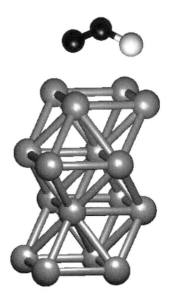

coordinates which do incorporate them. A potential function must also be invariant under permutations of identical atoms and behave properly when some coordinate become large. These conditions can be easily satisfied by using 15 inverse interparticle distances R_{ij}^{-1}: three interatomic distances of N_2O and twelve distances between the top four Cu atoms and the N and O atoms are used. The usefulness of R_{ij}^{-1} coordinates has been recognized in other fitting methods [43, 51, 52]. In particular, the method of Bowman and Braams [51, 52] uses symmetry-adapted polynomials of functions of interatomic distances. Increasing the number of coordinates from 9 to 15 has the important disadvantage that it causes a significant drop in the density of data and an increase in the cost of the non-linear fitting of NN parameters. The disadvantage of adding extra dimensions due to the use of interatomic distances becomes worse as the size of the system increases. We use D-reduction to build a function in these physically-motivated, but computationally inconvenient coordinates. The configuration space was sampled from a uniform distribution of N and O atomic positions within the vacuum region of the supercell. The only restriction was imposed on the length of the NN-bond: it was sampled from a distribution $p(r_{NN}) \propto 1 - \left(1 - e^{-\beta_0(r_{NN}-r_e)}\right)$, where $\beta_0 = 2.69 \text{Å}^{-1}$, $r_e = 1.098 \text{Å}$ [53]. This prevents sampling high-energy configurations corresponding to N–N bond breaking without introducing any bias into the dynamics of the studied reaction $N_2O^{gas} \rightarrow N_2^{gas} + O_{ads}$. We calculated about 6,000 ab initio energies in a 40,000 cm^{-1} range (higher energy points were discarded as they would be inaccessible during the reaction with energies of interest). 4,300 data were used to fit and 1,845 data to test. Again, it is only by computing the error for a set of test points that one can confirm the quality of a fit. The density of sampling by the fitted points is only 1.75 data per dimension. Before fitting, the data set was augmented by adding symmetry equivalent points obtained by permuting like atoms.

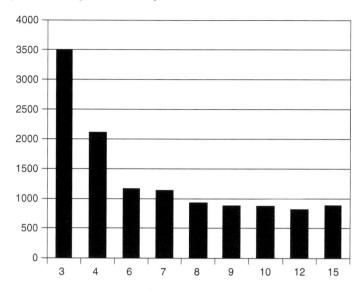

Fig. 5 The mean absolute error (in cm^{-1}) at the test points as a function of the dimensionality of the neural network when fitting the PES for N$_2$O/Cu(100) in 15D. The results are optimized with respect to the number of neurons

In Fig. 5, the mean absolute error at the test points is plotted vs. the dimensionality of the neural network in new, adapted coordinates. It quickly drops until $d = 6$, and goes below 1,000 cm^{-1} (which is a value we deem acceptable) at $d = 8$ with only marginal improvement thereafter. The intrinsic dimensionality of this data set is therefore about 8. This means that the information contained in the data can be represented on an eight-dimensional manifold with almost no loss of accuracy. A data set with a higher density of sampling or sampling different regions of the configuration space might have a different intrinsic dimensionality. We were able to use physically-motivated coordinates that expand the dimensionality from 9 to 15 while collapsing the dimensionality back to 8 for the purpose of the non-linear fit. This approach allowed us to build reliably a continuous function from extremely sparse data.

4 Discussion

Dimensionality reduction is an active and developed field of research. Many D-reduction techniques have been proposed, see Refs. [54, 55] for a review. Most known methods can be organized in the diagram of Fig. 6. The methods are separated into nonlinear and linear depending on the type of coordinate transformation.

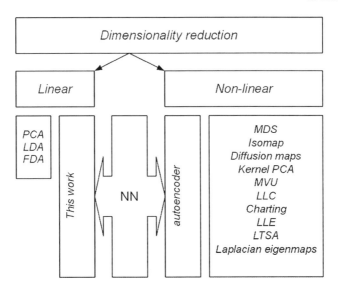

Fig. 6 Classification of dimensionality reduction techniques

The familiar principal component analysis (PCA) can be viewed as a simple linear D-reduction technique. Most methods involve non-linear manipulations of data which cannot even be strictly presented as a coordinate transformation in a closed form. However, if there is a mapping, it must in principle be learnable to arbitrary accuracy by a neural network. In that sense, the neural network autoencoder [56] can be considered the ultimate non-linear dimensionality reduction technique (Fig. 6).

Like the autoencoder, we also use a NN, but in our case, the NN finds the best linear combinations of variables as the function is being fit in the new variables (Fig. 7). A difference between this method and the PCA is that no constraints are imposed on the possible transformations, which increases the efficiency of the embedding into the low dimensional space.

If we use non-linear neurons in the coordinate transformation layer or many layers, we obtain the autoencoder. The D-reduction can then be more significant, but the cost explodes, and one also has to train a mirror NN to learn the inverse mapping [56]. We, on the contrary, reduced the cost of the fit with a simpler scheme.

There is recent interest in dimensionality reduction for the description of molecular systems. See, for example, a comparison of the performance of different dimensionality reduction techniques when using different coordinates to describe the structure of 1,2,4-trifluorocyclooctane [57]. From Fig. 2 of [57], it is clear that, indeed, the autoencoder performs as well as any other method, while PCA results are more coordinate-dependent. While the error of mapping with non-linear methods drops steeply towards the effective dimensionality as the dimensionality is increased, that of PCA decreases only slowly with dimension and is higher.

In [22], we performed a linear D-reduction for the same $N_2O/Cu(100)$ system considered here, but in internal coordinates (and with a different sampling scheme).

Extracting Functional Dependence from Sparse Data

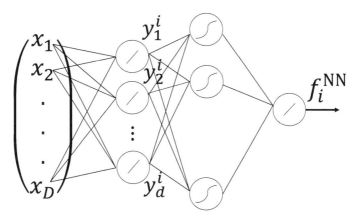

Fig. 7 Linear dimensionality reduction with a neural network

Even though tested on a different system than that of [57], our linear method shows a steep drop in error vs. dimension in both sets of coordinates, and to a similar intrinsic dimensionality, a behaviour more characteristic of non-linear mappings. The underperformance of PCA appears therefore to be due to a large extent to the restrictive nature of the coordinate transformation (the new coordinates are eigenvectors of the covariance matrix of the zero-mean data) and not only to the fact that it is linear. Relaxing restrictions on possible transformations leads to more effective D-reduction. This gives us hope that the linear method used here can perform well in many settings.

An even more important question than *how significant is the D-reduction?* is *is it useful?* The D-reduction enables one to obtain a lower-dimensional function that represents the potential, but is it expensive to use it to compute dynamical properties? If dynamical properties are computed by solving the nuclear Schrödinger equation quantum mechanically, then one represents wavefunctions as linear combinations of basis functions and computes integrals of products of basis functions and the potential. The basis functions are products of low-dimensional (often univariate) functions. In the univariate case, one must compute integrals of the form

$$\int \phi_{i,1}(x_1), \phi_{i,2}(x_2), \ldots, \phi_{i,D}(x_D) f(y) \phi_{i,1}^*(x_1), \phi_{i,2}^*(x_2), \ldots, \phi_{i,D}^*(x_D) dx \quad (10)$$

where $*$ denotes the complex conjugate [2]. The cost of these integrals is a major contribution to the overall cost of a quantum dynamics calculation. If a non-linear mapping $x \in K^D \to y \in k^{d<D}$ is used, the integrals are full-dimensional. With a linear mapping, on the contrary, it is possible, by choosing the right neurons, to factor the integrals. Component functions can be fitted with a neural network with exponential neurons [36] (alternatively, they can be refitted in the y coordinates using either an exponential neuron NN or other methods to give a sum-of-products form [58–61]), and then the potential assumes the form of a sum of products of

one-dimensional functions in new as well as in original coordinates. For example, a second-order term of (6) in new coordinates will be (cf. (7)):

$$f_i^{NN}\left(y_{i_1}, y_{i_2}\right) = \sum_{q=1}^{S} c_q e^{\left(w_q y_i + d_q\right)} = \sum_{q=1}^{S} \tilde{c}_q \prod_{p=1}^{D} e^{\tilde{w}_{qp} x_p} \tag{11}$$

and the integral of (10) will factor. This illustrates the importance of linear D-reduction for applications in physics and chemistry.

5 Conclusion

The number of nuclear coordinates causes a severe dimensionality problem in the modelling of chemical reactions. It makes both the representation of potentials and their use difficult. In this chapter, we discussed new ideas that facilitate both fitting potentials to sparse data and also their use (via factorizability of integrals). These new ideas are all based on the use of approximations using terms that depend on fewer coordinates. Of course, the simplest way to reduce the dimensionality of the problem is simply to fix some coordinates assumed to be less important, but that is often not adequate [62]. We propose a more accurate and less drastic approximation. Using dimension reduction to fit potentials should have important applications, because it enables one to obtain good PESs from a relatively small number of samples. PESs make efficient molecular modelling possible.

We have developed a neural network based method that represents multivariate functions with a small number of low-dimensional terms in new, adapted coordinates obtained by a linear transformation. It also allows for dimensionality reduction. Because the transformation is linear, the use of such functions is simplified, in particular, for the application of integro-differential operators. The dimensionality problem is particularly severe in the case of the solution of the Schrödinger equation.

Acknowledgements Sergei Manzhos is supported by a GCOE grant from Japan Society for the Promotion of Science. Tucker Carrington is grateful for the support of the Canadian Natural Science and Engineering Research Council.

References

1. Handbook of applied mathematics: selected results and methods. Pearson, C.E. (ed.), Van Nostrand Reinhold, NY (1983)
2. Encyclopedia of computational chemistry. Schleyer, P. von Ragué (ed.), Wiley, New York (1998)
3. Bowman, J.M., Carrington, Jr., T., and Meyer, H.-D.: Quantum approaches for computing vibrational spectra of polyatomic molecules. Mol. Phys. **106** (2008) 2145–2182

4. Levine, I.N.: Quantum chemistry, 6th ed., Prentice Hall, NJ (2009)
5. Setyowati, K., Piao, M.J., Chen, J., and Liu, H.: Carbon nanotube surface attenuated infrared absorption. Appl. Phys. Lett. **92** (2008) 043105
6. Schinke, R.: Photodissociation dynamics, Cambridge University Press, Cambridge (1993)
7. Schatz, G.C.: The analytical representation of electronic potential-energy surfaces. Rev. Mod. Phys. **61** (1989) 669–688
8. Truhlar, D.G., Steckler, R., and Gordon, M.S.: Potential energy surfaces for polyatomic reaction dynamics. Chem. Rev. **87** (1987) 217–236
9. Hirst, D.M.: Potential energy surfaces, Taylor and Francis, London (1985)
10. Hase, W.L., Song, K., and Gordon, M.S.: Direct dynamics simulations. Comput. Sci. Eng. **5** (2003) 36–44
11. Kroes, G.-J.: Frontiers in surface scattering simulations. Science **321** (2008) 794–797
12. Thomas, J.M.: Heterogeneous catalysis: Enigmas, illusions, challenges, realities, and emergent strategies of design. J. Chem. Phys. **128** (2008) 182502
13. Carter, E.: Challenges in modeling materials properties without experimental input. Science **321** (2008) 800–803
14. Pulay, P. and Paizs, B.: Newtonian molecular dynamics in general curvilinear internal coordinates. Chem. Phys. Lett. **353** (2002) 400–406
15. Manzhos, S., Wang, X., Dawes, R., and Carrington, Jr., T.: A Nested molecule-independent neural network approach for high-quality potential fits. J. Phys. Chem. A **110** (2006) 5295–5304
16. Beck, M.H., Jaeckle, A., Worth, G.A., and Meyer, H.-D.: The multiconfiguration time-dependent Hartree (MCTDH) method: a highly efficient algorithm for propagating wavepackets. Phys. Rep. **324** (2000) 1–105
17. Dawes, R. and Carrington, Jr., T.: Using simultaneous diagonalization and trace minimization to make an efficient and simple multidimensional basis for solving the vibrational Schrödinger equation. J. Chem. Phys. **124** (2006) 054102
18. Cooper, J. and Carrington, Jr., T.: Computing vibrational energy levels by using mappings to fully exploit the structure of a pruned product basis. J. Chem. Phys. **130** (2009) 214110
19. Manzhos, S. and Carrington, Jr., T.: A random-sampling high dimensional model representation neural network for building potential energy surfaces. J. Chem. Phys. **125** (2006) 084109
20. Manzhos, S. and Carrington, Jr., T.: Using redundant coordinates to represent potential energy surfaces with lower-dimensional functions. J. Chem. Phys. **127** (2007) 014103
21. Manzhos, S. and Carrington, Jr., T.: Using neural networks, optimized coordinates, and high-dimensional model representations to obtain a vinyl bromide potential surface. J. Chem. Phys. **129** (2008) 224104
22. Manzhos, S., Yamashita, K., and Carrington, Jr., T.: Fitting sparse multidimensional data with low-dimensional terms. Comput. Phys. Comm. **180** (2009) 2002–2012
23. Carter, S., Bowman, J.M., and Harding, L.B.: Ab initio calculations of force fields for H_2CN and C1HCN and vibrational energies of H_2CN. Sprectrochimica Acta A **53** (1997) 1179–1188
24. Carter, S., Culik, S.J., and Bowman, J.M.: Vibrational self-consistent field method for many-mode systems: a new approach and application to the vibrations of CO adsorbed on Cu(100). J. Chem. Phys. **107** (1997) 10458
25. Carter, S. and Handy, N.C.: On the representation of potential energy surfaces of polyatomic molecules in normal coordinates. Chem. Phys. Lett. **352** (2002) 1–7
26. Li, G., Rosenthal, C., and Rabitz, H.: High dimensional model representations. J. Phys. Chem. A **105** (2001) 7765–7777
27. Sobol, I.M.: Sensitivity analysis for non-linear mathematical models. Math. Model. Comput. Exp. **1** (1993) 407–414
28. Rabitz, H. and Alis, O.F.: General foundations of high-dimensional model representations. J. Math. Chem **25** (1999) 197–233
29. Wang, S.-W., Georgopoulos, P.G., Li, G., and Rabitz, H.: Random Sampling-High Dimensional Model Representation (RS-HDMR) with nonuniformly distributed variables: application to an integrated multimedia/multipathway exposure and dose model for trichloroethylene. J. Phys. Chem. A **107** (2003) 4707–4716

30. Hassoun, M.H.: Fundamentals of artificial neural networks, MIT, MA (1995)
31. Malshe, M., Pukrittayakamee, A., Raff, L.M., Hagan, M., Sukkapatnam, S., and Komanduri, R.: Accurate prediction of higher-level electronic structure energies for large databases using neural networks, Hartree-Fock energies, and small subsets of the database. J. Chem. Phys. **131** (2009) 124127
32. Sumpter, B.G., Getino, C., Noid, D.W.: Theory and applications of neural computing in chemical science. Annu. Rev. Phys. Chem. **45** (1994) 439–481
33. Hornik, K., Stinchcombe, M., and White, H.: Multilayer feedforward networks are universal approximators. Neural Network. **2** (1989) 359–366
34. Hornik, K.: Approximation capabilities of multilayer feedforward networks. Neural Networks **4** (1991) 251–257
35. Gorban, A.N.: Approximation of continuous functions of several variables by an arbitrary nonlinear continuous function of one variable, linear functions, and their superpositions. Appl. Math. Lett. **11** (1998) 45–49
36. Manzhos, S. and Carrington, Jr., T.: Using neural networks to represent potential surfaces as sums of products. J. Chem. Phys. **125** (2006) 194105
37. Donoho, D.L.: High-Dimensional Data Analysis: The curses and blessings of dimensionality, Aide-memoire of the invited lecture at the conference *Mathematical Challenges of the 21st Century*, AMS, CA (2000)
38. Malshe, M., Raff, L.M., Rockey, M.G., Hagan, M.T., Agrawal, P.A., and Komanduri, R.: Theoretical investigation of the dissociation dynamics of vibrationally excited vinyl bromide on an ab initio potential-energy surface obtained using modified novelty sampling and feedforward neural networks. II. Numerical application of the method. J. Chem. Phys. **127** (2007) 134105
39. Watanabe, T., Ehara, M., Kuramoto, K., and Nakatsuji, H.: Possible reaction pathway in methanol dehydrogenation on Pt and Ag surfaces/clusters starting from O-H scission: dipped adcluster model study. Surf. Sci. **603** (2009) 641–646
40. Kroes, G.-J., Pijper, E., and Salin, A.: Dissociative chemisorption of H_2 on the Cu(110) surface: a quantum and quasiclassical dynamical study. J. Chem. Phys **127** (2007) 164722
41. Diaz, C., Perrier, A., and Kroes, G.-J.: Associative desorption of N_2 from Ru(0 0 0 1): a computational study. Chem. Phys. Lett. **434** (2007) 231–236
42. Kroes, G.-J. and Meyer, H.-D.: Using n-mode potentials for reactive scattering: application to the 6D H_2 + Pt(1 1 1) problem. Chem. Phys. Lett. **440** (2007) 334–340
43. Crespos, C., Collins, M.A., Pijper, E., and Kroes, G.-J.: Multi-dimensional potential energy surface determination by modified Shepard interpolation for a molecule-surface reaction: H_2 + Pt(1 1 1). Chem. Phys. Lett. **376** (2003) 566–575
44. Kapteijn, F., Rodrigez-Mirasol, J., and Moulijn, J.A.: Heterogeneous catalytic decomposition of nitrous oxide. Appl. Catal. B **9** (1996) 25–64
45. Parr, R.G. and Weitao, Y.: Density-functional theory of atoms and molecules, Oxford University Press, Oxford (1994)
46. Lorentz, G.G., Chui, C.K., and Shumaker, L.L.: Approximation theory, vol. II, Academic Press, New York (1976)
47. Singh, S.P., Barry, J.H.W., and Watson, B.: Approximation theory and spline functions, Reidel, Dordrecht, (1984)
48. Maisuradze, G.G., Thompson, D.L., Wagner, A.F., and Minkoff, M.J.: Interpolating moving least-squares methods for fitting potential energy surfaces: detailed analysis of one-dimensional applications. J. Chem. Phys. **119** (2003) 10002–10014
49. Hollebeek, T., Ho, T.-S., and Rabitz, H.: Constructing multidimensional molecular potential energy surfaces from ab initio data. Annu. Rev. Phys. Chem. **50** (1999) 537–570
50. Szalay, V.: Iterative and direct methods employing distributed approximating functionals for the reconstruction of a potential energy surface from its sampled values. J. Chem. Phys. **111** (1999) 8804–8818
51. Sharma, A.R., Braams, B.J., Carter, S., Shepler, B.C., and Bowman, J.M.: Full-dimensional ab initio potential energy surface and vibrational configuration interaction calculations for vinyl. J. Chem. Phys. **130** (2009) 174301

Extracting Functional Dependence from Sparse Data 149

52. Czako, G., Shepler, B., Braams, B.J., and Bowman, J.M., Accurate ab initio potential energy surface, dynamics, and thermochemistry of the $F+CH_4 \rightarrow HF+CH_3$ reaction. J. Chem. Phys. **130** (2009) 084301

53. Lee, A.R., Kalotas, T.M., and Adams, N.A.: Modified Morse potential for diatomic molecules. J. Mol. Spectrosc. **191** (1998) 137–141

54. van der Maaten, L.J.P., Postma, E.O., and van den Herik, H.J.: Dimensionality reduction: A comparative review (Technical Report TiCC-TR 2009-005), Tilburg University, The Netherlands (2009)

55. Ravisekar, B.: A comparative analysis of dimensionality reduction techniques (research report), Georgia Institute of Technology, GA (2006)

56. Hinton, G.E. and Salakhutdinov, R. R.: Reducing the dimensionality of data with neural networks. Science **313** (2006) 504–507

57. Brown, W.M., Martin, S., Pollock, S.N., Coutsias, E.A., and Watson, J.-P.: Algorithmic dimensionality reduction for molecular structure analysis. J. Chem. Phys. **129** (2008) 064118

58. Sobol, I.M.: The use of Haar series in estimating the error in the computation of infinite-dimensional integrals. Soviet Math. Dokl. **8** (1967) 810–813

59. Sobol, I.M.: Functions of many variables with rapidly convergent Haar series. Soviet Math. Dokl. **1** (1960) 655–658

60. Jaeckle, A. and Meyer, H.-D.: Product representation of potential energy surfaces. J. Chem. Phys. **104** (1996) 7974–7984

61. Jaeckle, A. and Meyer, H.-D.: Product representation of potential energy surfaces. II. J. Chem. Phys. **109** (1998) 3772–3779

62. Nave, S. and Jackson, B.: Methane dissociation on Ni(111): the effects of lattice motion and relaxation on reactivity. J. Chem. Phys. **127** (2007) 224702

A Multilevel Algorithm to Compute Steady States of Lattice Boltzmann Models

Giovanni Samaey, Christophe Vandekerckhove, and Wim Vanroose

Abstract We present a multilevel algorithm to compute steady states of lattice Boltzmann models directly as fixed points of a time-stepper. At the fine scale, we use a Richardson iteration for the fixed point equation, which amounts to time-stepping towards equilibrium. This fine-scale iteration is accelerated by transferring the error to a coarse level. At this coarse level, one directly solves for the density (the zeroth moment of the lattice Boltzmann distributions), for which a coarse-level equation is known in some appropriate limit. The algorithm closely resembles the classical multigrid algorithm in spirit, structure and convergence behaviour. In this paper, we discuss the formulation of this algorithm. We give an intuitive explanation of its convergence behaviour and illustrate with numerical experiments.

1 Introduction

For a broad class of systems, in applications ranging from physics, fluid flow [9, 10] and biology [2] to traffic flow [4], a macroscopic (coarse) partial differential equation (PDE) that models density evolution in a space-time domain, is often insufficient to accurately describe physical interactions between individual particles (atoms, molecules, bacteria, vehicles). For instance, the dynamics of a system of colliding particles with interactions that depend sensitively on the relative particle velocities can, in general, not be modeled completely by a reaction-diffusion equation for the particle density [13]. In such cases, one needs to resort to a more microscopic (fine-scale) description, such as a kinetic equation that models

---

G. Samaey (✉) and C. Vandekerckhove
Scientific Computing, Department of Computer Science, K.U. Leuven, Celestijnenlaan 200A, 3001 Leuven, Belgium
e-mail: giovanni.samaey@cs.kuleuven.be, christophe.vandekerckhove@gmail.com

W. Vanroose
Department of Mathematics and Computer Science, Universiteit Antwerpen, Middelheimlaan 1, 2020 Antwerpen, Belgium
e-mail: wim.vanroose@ua.ac.be

A.N. Gorban and D. Roose (eds.), *Coping with Complexity: Model Reduction and Data Analysis*, Lecture Notes in Computational Science and Engineering 75, DOI 10.1007/978-3-642-14941-2_8, © Springer-Verlag Berlin Heidelberg 2011

151

the evolution of the velocity-position phase space distribution density of particles as a combination of advection according the current velocity and collisions that redistribute velocities [25].

A lattice Boltzmann model (LBM) is a simple and effective space-time-velocity discretization of a fine-scale kinetic equation [20]. Only a low number of velocities are considered, and the discrete velocities are related to the lattice spacing and time step in such a way that the corresponding distributions are simply shifted over an integer number of lattice sites over one time step. The advantage is that one can decompose evolution in two separated steps: first, one simply advects each distribution function according to its corresponding velocity; subsequently, one performs a collision step in which the velocities are redistributed. A more precise mathematical description of LBMs will be presented in Sect. 2.

In the appropriate diffusion limit [15, e.g.], the above-described fine-scale kinetic models collapse onto a limiting coarse partial differential equation. However, their associated computational cost is much higher, due to the increased dimension. For such cases, several approaches have been developed that exploit the link between the coarse and fine-scale levels of descriptions to significantly accelerate computations. Of particular importance in this work is the equation-free framework that was proposed by Kevrekidis et al. [12, 21]. The framework introduces *lifting* and *restriction* operators to map a coarse state to a fine-scale state and vice versa. These operators are then used to construct a *coarse time-stepper* for the unavailable coarse equation as a three step procedure: (1) lifting, i.e. the creation of appropriate initial conditions for the fine-scale model, conditioned upon the coarse state at time t^*; (2) simulation, using the fine-scale model, over the time interval $[t^*, t^* + \Delta t]$; and (3) restriction, i.e. the extraction of the coarse state at time $t^* + \Delta t$. The resulting coarse time-Δt map can then be used in conjunction with projective integration methods [6] to accelerate time integration, or with a matrix-free algorithm to directly compute coarse steady states [17–19]. We refer to [11] for a recent review and references to related methods.

In this paper, we borrow important concepts from the equation-free framework, and use them to construct a new algorithm that directly computes *fine-scale steady states* of an LBM as fixed points of the time-stepper. The proposed algorithm bears many similarities with standard multigrid [1, 22]. Multigrid is a sophisticated iterative method for the computation of a steady state of a PDE. It is most easily understood by realizing that the error in any initial guess consists of high and low wavenumber modes. Basic iterative methods (such as a Jacobi or Gauss–Seidel iteration, or Richardson's method, which corresponds to time-stepping) are very efficient in removing high-frequency error components; such methods are therefore called *smoothers*. In multigrid, one uses the fact that the smoother yields errors (and residuals) that can be accurately represented on a coarser mesh (since they are smooth). One then transfers the residuals to the coarser grid (*restriction*) and solves for the error. The resulting error is then transferred back to the finer grid (*prolongation*). This sequence is called *coarse grid correction*, and the combination of smoothing and coarse grid correction yields a very powerful algorithm. At the coarse grid, the same procedure can be applied. One then performs again a few

A Multilevel Algorithm to Compute States of Lattice Boltzmann Models 153

smoothing steps and a coarse grid correction on an even coarser grid, leading to a hierarchy of grids – hence the name *multigrid*.

The multilevel algorithm that we propose here follow the same strategy. At the fine scale, we consider the computation of a fixed point of a lattice Boltzmann time-stepper; the basic iteration (smoother) on this level is a Richardson iteration, which amounts to time-stepping towards equilibrium. Instead of moving to a coarser grid, as in multigrid, however, here we move to a *coarser level of description*, i.e. the level of the coarse PDE. Note that at this level, we can only use the PDE that is known in the diffusion limit, which yields only an approximation to the evolution of the density of the LBM. The transfer operators that connect the coarse and fine-scale levels of description are exactly the lifting and restriction operators that have been introduced in the equation-free framework; these then replace the prolongation and restriction operators that are used in multigrid.

The remainder of the paper is organized as follows. In Section 2 we define our model problems. Section 3 describes the behaviour of Richardson iteration for this problem. We then proceed to formulate our multilevel algorithm in Sect. 4. Section 5 contains an intuitive explanation of the convergence behaviour. We conclude in Sect. 6.

2 Model Problem

2.1 Lattice Boltzmann Models

Throughout this paper, we consider reaction-diffusion lattice Boltzmann models (LBMs) [3,20] in one space dimension as the illustrative example. We briefly review the principles of LBMs and describe the specific model problem of this paper.

We define a discrete number of particle distribution functions $f_i(x_j, t_n)$, with $1 \leq j \leq N$, $-q \leq i \leq q$ and $n \geq 0$, with grid spacing δx in space and δt in time, and choose the velocities v_i such that they correspond to a movement over an integer number of lattice points during one time-step,

$$
v_i = c_i \frac{\delta x}{\delta t}, \qquad c_i = -q, \ldots, -1, 0, 1, \ldots, q.
$$

Furthermore, we introduce equilibrium distributions $f_i^{eq}(x, t) = w_i^{eq} \rho(x, t)$, which correspond to local diffusive equilibrium. Here, the coefficients $w_i^{eq} = w_{-i}^{eq}$ depend on the number of speeds $2q + 1$.

Remark 1 (Notation). To avoid potential confusion, we emphasize that, in this text, we will use the symbol δt (and, for notational consistency, δx) to denote the parameters of the fine-scale (lattice Boltzmann) time-stepper, whereas the symbol Δt is used to indicate the size of a coarse time step.

With these definitions, we can write an evolution law for the distributions $f(x,t) = (f_i(x,t))_{i=-q}^{q}$ as

$$f_i(x+c_i\delta x, t+\delta t) = (1-\omega) f_i(x,t) - \omega f_i^{eq}(x,t) + \beta \sum_j V_{ij} f_j(x,t) + R_i(x,t),$$

(1)

which we denote in compact form as

$$\mathbf{f}^{n+1} = \phi(\mathbf{f}^n),$$

(2)

where $\mathbf{f} = (f_i(\mathbf{x}))_{i=-q}^{q}$, $\mathbf{x} = (x_j)_{j=1}^{N}$ and the superscript n denotes the time instance $t = n\delta t$. Equation (1) decomposes evolution into a collision phase (the right-hand side) and a streaming phase, during which the post-collision values are propagated to a neighboring site. The first two terms of the right-hand side represent a BGK relaxation to local diffusive equilibrium on a characteristic time scale $\tau = 1/\omega$. The third term is an external force, which is discretized as proposed by Luo [14], and the terms $R_i(x,t)$ model reactive collisions. Note that, for pure diffusion, a lattice Boltzmann time-stepper is stable for $\omega \in [0, 2]$.

We now define the first few (non-dimensional) moments of the distribution function as

$$\rho(x,t) = \sum_{i=-q}^{q} f_i(x,t),$$

$$\varphi(x,t) = \sum_{i=-q}^{q} c_i f_i(x,t),$$

(3)

$$\xi(x,t) = \frac{1}{2} \sum_{i=-q}^{q} c_i^2 f_i(x,t),$$

which represent density, momentum and energy, respectively. The lattice Boltzmann equation can be written equivalently as a set of coupled PDEs for the $2q + 1$ moments of the individual distribution functions. A Chapman–Enskog expansion shows that, when the density is sufficiently smooth, the lattice Boltzmann model (1) (or the set of coupled equations for the moments) can be approximated by a single reaction-advection-diffusion equation only depending on the density [5, 16],

$$\partial_t \rho(x,t) = \partial_x(D\partial_x\rho(x,t)) + \beta\partial_x\rho(x,t) + r(\rho(x,t)),$$

$$D = \frac{2-\omega}{2\omega} \frac{\delta x^2}{\delta t} \sum_i c_i^2 w_i^{eq},$$

(4)

where the transport coefficients D and β and the reaction rate $r(\rho)$ depend on the relaxation, external force and collision terms in the lattice Boltzmann model.

A Multilevel Algorithm to Compute States of Lattice Boltzmann Models   155

When the reaction terms R_i can be written as functions of the density, $R_i(x,t) = w_i^{eq}\delta t R(\rho(x,t))$, we obtain $r(\rho) = R(\rho)$. In general, however, if $R_i(x,t)$ depend on the individual distribution functions, it is cumbersome to eliminate the velocity dependence and derive an explicit formula for $r(\rho)$.

From the assumption that a coarse model for $\rho(x,t)$ exists, it follows that the higher order moments $\varphi(x,t)$ and $\xi(x,t)$ can be written as functionals of $\rho(x,t)$. (We call this *slaving*.) For a 3-speed lattice Boltzmann reaction-diffusion model, i.e. equation (1) with $q = 1$ and $\beta = 0$, these *slaving relations* can be written down analytically as an asymptotic expansion in $1/\omega$. Up to third order, we have [23],

$$\varphi(x,t) = -\frac{2\delta x}{3\,\omega}\partial_x\rho + \frac{\delta x\delta t}{3\omega^2}\frac{(4\omega-2)}{(\omega-2)}\partial_x r(\rho) + \frac{\delta x\delta t}{3\omega^2}\frac{(-2\omega^2+2\omega-2)}{(\omega-2)}\partial_{xt}^2\rho(x,t)$$

$$\xi(x,t) = \frac{1}{3}\rho(x,t) - \frac{\delta t}{6\omega}\left(r\left(\rho(x,t)\right) - \partial_t\rho(x,t)\right).$$

$$(5)$$

These expansions can alternatively be written down in terms of $\rho(x,t)$ and its spatial derivatives only, by making use of (4).

2.2   Fixed Point Formulation

We are interested in the fixed points of (2), i.e. solutions of the nonlinear equation

$$\mathbf{f}^* - \phi(\mathbf{f}^*) = 0. \tag{6}$$

This nonlinear problem can be solved using Newton's method. We will apply a multilevel algorithm on the linear systems arising in each Newton iteration, which can be written as

$$\left(I - J_\phi(\bar{\mathbf{f}})\right)\Delta\mathbf{f} = -r(\bar{\mathbf{f}}), \tag{7}$$

or, for short,

$$A_f\mathbf{x} = b. \tag{8}$$

Here, we use $\mathbf{x} = \Delta\mathbf{f}$ and $J_\phi(\bar{\mathbf{f}}) = (\partial\phi/\partial\mathbf{f})(\bar{\mathbf{f}})$ and $r(\bar{\mathbf{f}}) = \bar{\mathbf{f}} - \phi(\bar{\mathbf{f}})$ denote the Jacobian of ϕ and the residual of (6), respectively, evaluated at a current guess $\bar{\mathbf{f}}$. For notational convenience, we have eliminated the dependence on $\bar{\mathbf{f}}$ in A_f and b. The Jacobian J_ϕ is not available in closed form, since we can only evaluate the time-stepper, but Jacobian-vector products can be estimated as

$$J_\phi(\bar{\mathbf{f}})\mathbf{v} \approx \frac{\phi(\bar{\mathbf{f}}+\epsilon\mathbf{v}) - \phi(\bar{\mathbf{f}})}{\epsilon},$$

with ϵ appropriately small. This enables the use of methods that require only matrix-vector products.

3 Richardson Iteration

We first consider a basic iterative method, namely Richardson iteration,

$$\mathbf{x}^{(k+1)} = S(\mathbf{x}^{(k)}, A_f, b) = \mathbf{x}^{(k)} + (A_f \mathbf{x}^{(k)} - b), \tag{9}$$

which is selected over alternatives such as Jacobi or Gauss–Seidel, because it does not involve a matrix splitting of A_f, and hence requires only matrix-vector products. This method can be seen to correspond to time-stepping for the error equation of the linearized system. To see this, we elaborate

$$\begin{aligned}
\mathbf{x}^{(k+1)} &= \mathbf{x}^{(k)} + (A_f \mathbf{x}^{(k)} - b) \\
&= \mathbf{x}^{(k)} + \left((I - J_\phi)\mathbf{x}^{(k)} - b \right) \\
&= J_\phi \mathbf{x}^{(k)} - b.
\end{aligned}$$

The last equation shows that the Richardson iteration corresponds to time-stepping with the linearized time-stepper, modified by the righthand side of the linear system (the residual of the nonlinear equation). Hence, the spectral properties of the smoothing operator will be those of the linearized time-stepper.

Recall that we assume that an equation exists that closes at the level of the density $\rho(x,t)$ alone. As a consequence, this assumption implies that the higher order moments become functionals of $\rho(x,t)$ on time-scales which are fast compared to the overall system evolution (*slaving*). This behaviour is reflected in the spectral properties of the smoother: Richardson iterations quickly reduce errors in the higher order moments, whereas errors in modes that correspond to the density are not significantly affected by the smoother. This observation is illustrated by a numerical experiment.

We consider the pure diffusion 3-speed lattice Boltzmann problem, i.e. equation (1) with $q = 1$, $\beta = 0$ and $R_i \equiv 0$ using periodic boundary conditions. The domain is $[0, 1]$, and the model parameters are chosen to be $D = 1$, $\omega = 1.25$ and $\delta x = 1/128$. The time step then follows from (4). A fixed point of this time-stepper can be found as the solution of the linear system

$$A_f \mathbf{x} = 0. \tag{10}$$

We now look into the behaviour of Richardson iteration for the linear system (10).

Remark 2. Because of the periodic boundary conditions, the time-stepper has a trivial eigenvalue 1, and consequently, the fixed point problem is singular. This singularity can be removed by adding a phase condition and an artificial parameter, see e.g. [19]. In this experiment, we perform a spectral analysis of Richardson iteration; for this purpose, simply considering Richardson iteration on the singular system is sufficient.

Because of the homogeneous righthand side, Richardson iterations reduce to time-stepping with the original linear time-stepper. For the 3-speed model, there is a linear 1-to-1 relation between the distribution functions and the moments (3). It can easily be verified (for instance using Maple) that, when working in the equivalent moment description, the eigenvectors have the form $(\rho, \varphi, \xi) = (A \sin(k \cdot x), B \cos(k \cdot x), C \sin(k \cdot x))$, with $1 \leq k \leq N - 1$. For each frequency, there are three eigenmodes. However, the analytic determination of the constants, as well as the corresponding eigenvalues, does not yield a workable closed formula. Therefore, we compute the spectrum numerically. The results are depicted in Fig. 1 (top left). It is seen that the spectrum breaks down into two sets of eigenvalues: a set of $N - 1$ real and positive eigenvalues, and a set of $2N - 2$ complex eigenvalues with negative real part. Figure 1 (right) shows an eigenvector corresponding to a particular mode in each of these two sets, in the moment representation. One can visually verify that the eigenvectors corresponding to the rightmost set of real eigenvalues appear to satisfy the slaving relations (5); we call these the *slaved modes*. The slaving relations are not satisfied for the eigenvectors corresponding to the set of complex eigenvalues (non-slaved modes). Clearly, the slower a mode is damped, the better the slaving relations are satisfied.

We now look at the behaviour of the Richardson iteration for each of these eigenmodes individually. Starting from a normalized eigenmode that satisfies $\max_l V_l = 1$ (with $1 \leq l \leq 3N - 3$ denoting the components), we perform a single Richardson iteration and look at the amplification factor. Figure 1 (bottom left) shows the maximal value of the resulting vector as a function of the wave number. As could be expected from the spectrum of the time-stepper, we can draw two main conclusions:

- Regardless of the wave number, modes that corresponds to non-slaved states are damped quickly by the time-stepper; only modes that corresponds to slaved states can persist over longer time scales.
- For the slaved modes, we observe that low wavenumber modes are damped much more slowly than high wavenumber waves.

Based on these observations, we conclude that it would be beneficial to combine Richardson iterations with a method to directly reduce the errors in the density. To this end, we now turn to a multilevel algorithm.

4 Multilevel Algorithm

To solve the linear system (8) more efficiently, we make use of the link between the fine-scale model (1) and the approximate coarse description (4). We first describe lifting and restriction operators that map coarse to fine-scale states, and vice versa. We then proceed to formulate the complete algorithm. An intuitive explanation on it convergence properties is deferred to Sect. 5.

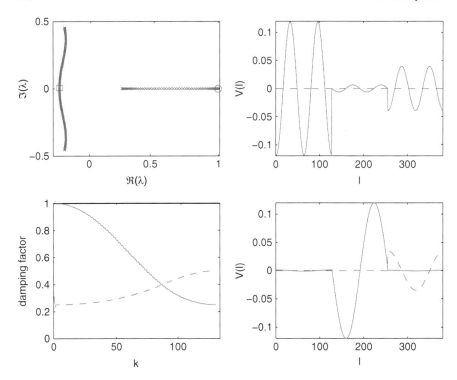

Fig. 1 *Top left*: Spectrum of the lattice Boltzmann time-stepper/Richardson smoother for homogeneous diffusion. *Bottom left*: Richardson amplification factors for each of the eigenmodes, as a function of wave number k. The *solid line* corresponds to the amplifications factors of the real positive eigenvalues; the *dashed line* corresponds to the amplifications factors of the complex eigenvalues with negative real part. *Right*: a typical eigenmode corresponding to a real positive eigenvalue (*top*) and a complex eigenvalue (*bottom*) of the time-stepper. The *solid line* contains the real part; the *dashed line* is the imaginary part. The corresponding eigenvalues are depicted by a *square*, resp. *circle* in the *top left* panel. Eigenmodes are displayed in moment representation: the *leftmost* part of each eigenvector corresponds to density, the *middle* part to momentum and the *rightmost* part to energy

4.1 Lifting and Restriction

We consider operators that define a mapping between the distribution functions $f_i(x,t)$ and the density $\rho(x,t)$. A *restriction operator*, mapping the distribution $f(x,t) = (f_i(x,t))_i$ to the density $\rho(x,t)$,

$$\mathcal{R} : f(x,t) \mapsto \rho(x,t) = \mathcal{R}(f(x,t)), \tag{11}$$

can easily be obtained using (3). The converse *lifting operator*, reconstructing distributions from the density,

A Multilevel Algorithm to Compute States of Lattice Boltzmann Models 159

$$\mathcal{L} : \rho(x,t) \mapsto f(x,t) = \mathcal{L}(\rho(x,t)), \tag{12}$$

is considerably more involved.

One could simply initialize the distributions as

$$f_i(x,t) = w_i \rho(x,t), \qquad \sum_i w_i = 1, \tag{13}$$

where $w_i = w_i^{eq}$ is an obvious choice, since this corresponds to the local diffusive equilibrium. Via a Chapman–Enskog expansion, it can be shown that this corresponds to a zeroth order approximation of the slaving relations [23]. Clearly, initializing also the higher order terms in the Chapman–Enskog expansion yields a more accurate reconstruction.

A numerical procedure to obtain a second order approximation to the slaving relations, using only the time-stepper (2) is the constrained runs scheme [7, 8]. Here, during a preparatory step, a number of lattice Boltzmann time-steps is performed, after each of which the density is reset. This scheme can be shown to converge linearly, with a convergence factor $|1 - \omega|$ [23]; hence, it is also stable whenever $\omega \in [0, 2]$. Higher order variants, which are more accurate but may become unstable, have been investigated in [24].

4.2 Coarse-Level Correction

Once the fine-scale state has been transferred to the coarse level, we need to define the coarse linear system that needs to be solved. For this, we start from the approximate coarse model (4), and construct a corresponding time-stepper, which we write in compact form as

$$\rho^{n+1} = \Phi(\rho^n),$$

where $\rho = \rho(\mathbf{x})$.

The fixed-point equation at this level can then equivalently be written as

$$(I - J_{\Phi}(\bar{\rho})) \, \Delta\rho = -R(\bar{\rho}), \tag{14}$$

which we write in a shorthand notation

$$A_c \mathbf{X} = B. \tag{15}$$

Here, we use $\mathbf{X} = \Delta\rho$, and $J_{\Phi}(\bar{\rho}) = (\partial\Phi/\partial\rho)(\bar{\rho})$ and $R(\bar{\rho}) = \bar{\rho} - \Phi(\bar{\rho})$ denote the Jacobian of Φ and the residual of the coarse fixed-point equation, respectively, evaluated at a current guess $\bar{\rho}$.

4.3 The Multilevel Algorithm

With all building blocks in place, we are now ready to formulate the complete multilevel algorithm. One iteration, starting from $\mathbf{x}^{(m)}$, consists of the following steps:

1. *Presmoothing:* Perform ν_1 Richardson iterations,

$$\bar{\mathbf{x}}^{(m)} = S^{\nu_1}(\mathbf{x}^{(m)}, b). \tag{16}$$

2. *Coarse-level correction:*

   - Compute the defect: $d^{(m)} = b - A\bar{\mathbf{x}}^{(m)}$.
   - Restrict defect: $D^{(m)} = \mathcal{R}(d^{(m)})$.
   - Coarse-level solve: $A_c \mathbf{V}^{(m)} = D^{(m)}$.
   - Lift correction: $v^{(m)} = \mathcal{L}(\mathbf{V}^{(m)})$.
   - Update fine-scale solution: $\hat{\mathbf{x}}^{(m)} = \bar{x}^{(m)} + v^{(m)}$.

3. *Postsmoothing:* Perform ν_2 Richardson iterations,

$$\mathbf{x}^{(m+1)} = S^{\nu_2}(\hat{\mathbf{x}}^{(m)}, b). \tag{17}$$

Remark 3 (Comparison with multigrid). The algorithmic structure resembles that of a classical multigrid method to find steady states of partial differential equations [1, 22]. For the PDE case, the amplification factors of a coarse-level Richardson iteration are similar to the solid line in Fig. 1 (bottom left). Multigrid combines the good smoothing properties for high wavenumber modes with a coarse grid correction. One first performs a number of smoothing steps, after which one transfers the residuals to the coarser grid (*restriction*) and solves for the error. This restriction is possible, because a smooth residual can be accurately represented on a coarser mesh. The resulting error is then transferred back to the finer grid (*prolongation*), and added to the current guess. Here, exactly the same procedure is followed; only the prolongation and restriction have been replaced by equation-free lifting and restriction. Classical multigrid can be used for the coarse level correction.

Remark 4 (Full approximation scheme). Note that, as in multigrid, we transfer the *residual* to the coarse level, and solve the error equation there. Instead of using Newton's method with a multilevel algorithm to solve the linear systems in each iteration, the nonlinear equation (6) can also be solved using a nonlinear multilevel algorithm directly, in the spirit of a nonlinear multilevel method (the full approximation scheme) [1, 22].

5 Convergence of the Two-Level Cycle

We now proceed to show how smoothing and coarse-level correction work together to yield fast convergence. To this end, we adapt the exposition that was presented in [1] for the classical multigrid method to our setting. While the iterative method

A Multilevel Algorithm to Compute States of Lattice Boltzmann Models 161

will rely on the link between the fine-scale lattice Boltzmann and the coarse PDE level, the aim is to find a solution for the fine-scale lattice Boltzmann model. We will therefore investigate numerically how the error (the difference of the current approximation with respect to the exact lattice Boltzmann solution) evolves throughout the iterative procedure.

5.1 An Intuitive Algebraic Picture

We start with an illustrative numerical example. We consider the pure diffusion 3-speed lattice Boltzmann problem, i.e., (1) with $q = 1$, $\beta = 0$ and $R_i \equiv 0$ using Dirichlet boundary conditions. The domain is $[0, 1]$, and the model parameters are chosen to be $D = 1$, $\omega = 1.1$ and $\delta x = 1/128$. The time step then follows from (4). We again look for a fixed point of the time-stepper as the solution of the linear system (10). To this end, we perform two iterations of the multilevel algorithm using $\nu_1 = 2$ presmoothing and $\nu_2 = 2$ postsmoothing steps, starting from an initial guess $\mathbf{f}$ that has the moment representation

$$\rho(\mathbf{x}) = \varphi(\mathbf{x}) = \xi(\mathbf{x}) = \sin(3\mathbf{x}) + \sin(45\mathbf{x}).$$

The evolution of the error throughout the algorithm is depicted in Fig. 2. The figure shows the effect of presmoothing, coarse-level correction and postsmoothing during both iterations. We see that during the presmoothing step the error in ϕ and ξ decreases rapidly, as well as the error in the high wavenumber modes of ρ. However, the low wavenumber modes of ρ remain virtually unaffected by the presmoothing. In the next step, we solve for the density error at the coarse-level correction, and lift the resulting error the fine-scale representation using the zeroth-order term of the Chapman–Enskog expansion. We remark that during this lifting step, the error in ξ also decreases substantially, as a consequence of the form of the slaving relations (5). Finally, in the postsmoothing phase, the error in the higher order moments is again decreased. Note that the coarse-level equation does not correspond exactly to the behaviour of the density of the lattice Boltzmann model, and hence the resulting density error does not vanish. This is due to the fact that, at the coarse scale, only an approximate model for the density evolution is used. In particular, the diffusion coefficient at the coarse level might be different from the diffusion coefficient of the density in the full fine-scale model. However, the lattice Boltzmann model converges to the limiting density equation for $\omega \to 1$, so the density error that remains after the coarse-level solve will be larger if ω is further away from 1. This is illustrated in Fig. 3, where the experiment is repeated with $\omega = 1.25$.

We now proceed to providing an intuitive, graphical explanation of this behaviour, see Fig. 4. The exposition is very similar in spirit to the intuitive picture that was given in [1, Chap. 5]. In panel (a), we show the total error e_f in the current guess, and decompose this error according to the subspaces that correspond to slaved and non-slaved eigenmodes (the dashed coordinate system). We can also make a second

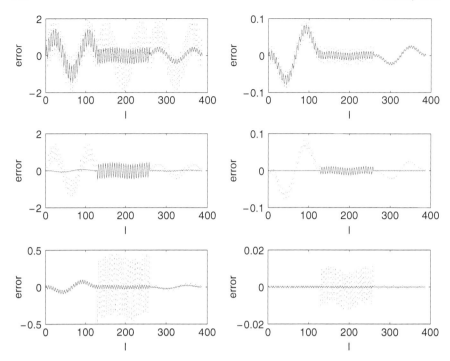

Fig. 2 Evolution of error throughout the multilevel iteration for a pure diffusion lattice Boltzmann model with $\omega = 1.1$. (The other parameters are in the text.) *Left*: first iteration. *Right*: second iteration. *Dashed*, resp. *solid*, *lines* represent the error before, resp. after the presmoothing phase (*top*), the coarse-level correction (*middle*) and the postsmoothing phase (*bottom*). Eigenmodes are displayed in moment representation: the *leftmost* part corresponds to density, the *middle* part to momentum and the *rightmost* part to energy

decomposition that corresponds to the transition between the coarse and fine-scale levels of description. A first subspace contains the fine-scale modes that can be represented given only the density, i.e. the range of the lifting operator Range($\mathcal{L}$). Because the lifting operator can only reconstruct the fine-scale state approximately, the subspace of the slaved modes and Range($\mathcal{L}$) do not coincide exactly. The complementary space contains all remaining fine-scale modes, i.e. the modes that map to a zero density, the null space of the restriction operator Null($\mathcal{R}$). Also this space does not correspond exactly to the slaved modes. We will now look into the behaviour of the multilevel algorithm in these decompositions. The first step in the algorithm is presmoothing, depicted in panel (b). Here, the error in non-slaved modes decrease quickly, while the error in the slaved modes is virtually unaltered. (Note from the experiment before that this is only approximately true, and only for the low wavenumber components.) Next, we transfer the error to the coarse level and perform a coarse-level solve, see panel (c). In this phase, the part of the error that is in Range($\mathcal{L}$) is significantly reduced. Due to the fact that the density equation only *approximates* the behaviour of the density of the lattice Boltzmann model, this

A Multilevel Algorithm to Compute States of Lattice Boltzmann Models

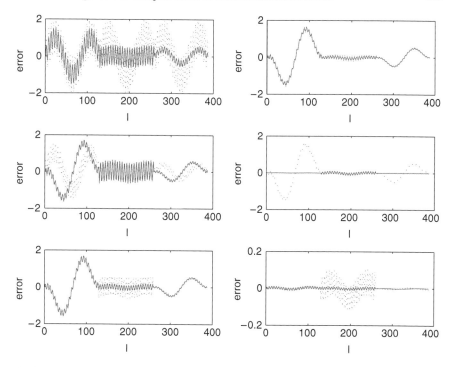

Fig. 3 Evolution of error throughout the multilevel iteration for a pure diffusion lattice Boltzmann model with $\omega = 1.25$. (The other parameters are in the text.) *Left*: first iteration. *Right*: second iteration. *Dashed*, resp. *solid*, *lines* represent the error before, resp. after the presmoothing phase (*top*), the coarse-level correction (*middle*) and the postsmoothing phase (*bottom*). Eigenmodes are displayed in moment representation: the *leftmost* part corresponds to density, the *middle* part to momentum and the *rightmost* part to energy

part of the error is not put exactly back to zero. We also observe that, in this step, the error in the non-slaved modes has increased again. This is due to the lifting operator, which reconstructs the fine-scale state based on the density alone, reintroducing artifacts. In a final postsmoothing step, see panel (d), these non-slaved modes are again removed by a number of additional smoothing steps.

5.2 Numerical Convergence Tests

Let us now proceed to numerically illustrate convergence. We again consider the pure diffusion 3-speed lattice Boltzmann problem, i.e. (1) with $q = 1$, $\beta = 0$ and $R_i \equiv 0$ using Dirichlet boundary conditions. The domain is $[0, 1]$, and the model parameters are chosen to be $D = 1$, $\delta x = 1/128$, and $\omega = 1.1$, resp. $\omega = 1.25$. The time step then follows from (4). We again look for a fixed point of the time-stepper as the solution of the linear system (10) using the multilevel algorithm with $\nu_1 = 2$

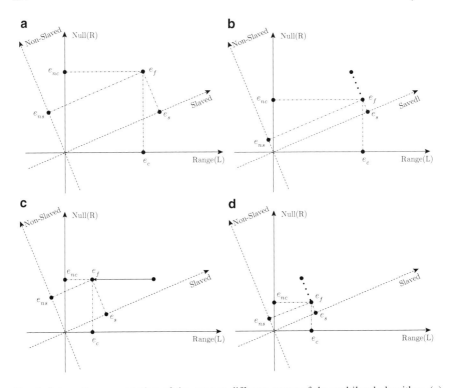

Fig. 4 Schematic representation of the error at different stages of the multilevel algorithm: (**a**) initial error; (**b**) the effect of presmoothing; (**c**) the effect of coarse-level correction; and (**d**) the effect of postsmoothing. Shown is how the error decomposes in terms of (1) the range of the lifting operator and the nullspace of restriction; and (2) slaved and non-slaved eigenmodes of the lattice Boltzmann time-stepper

presmoothing and $v_2 = 2$ postsmoothing steps, starting from a random initial guess. Figure 5 shows the two-norm of the error as a function of the iteration number. We see linear convergence, with a convergence rate that depends on ω. As follows from the algebraic picture above, this can be explained by noting that the coarse-level correction reduces more error in the density when ω is closer to 1. We verified that the convergence rates are mesh-independent.

6 Conclusions

We presented a multilevel algorithm that accelerates the iterative computation of steady states of lattice Boltzmann models. The algorithm exploits the link between a fine-scale description and a coarse-scale limiting equation. As in multigrid, the error is decomposed into fine-scale and coarse components, which are each handled

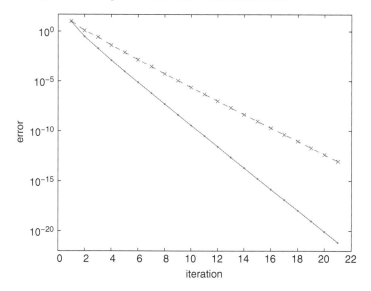

Fig. 5 Two-norm of the error of the multilevel iteration as a function of the iteration number for a fixed point computation of a lattice Boltzmann time-stepper with $\omega = 1.1$ (*solid*) and $\omega = 1.25$ (*dashed*), starting from a random initial guess

independently. In multigrid, the fine and coarse scales correspond to high and low wavenumber modes. In our algorithm, the error is decomposed into a coarse-scale component, which is defined by a reduced description in terms of density alone, and the remaining fine-scale components. The fine-scale component of the error is quickly damped by straightforward time integration; the coarse component, on the other hand, is reduced via the solution of an approximate coarse-scale equation.

The algorithm relies on the definition of appropriate lifting and restriction operators to transfer the error from the coarse to the fine level and vice versa. Such operators have been developed in the equation-free framework [11, 12] for a variety of fine-scale systems. In cases where fine-scale steady states exist (such as the kinetic formulations in this paper), our algorithm can provide an efficient means to compute these fixed points. We note that, on the coarse level, one could also use a multigrid algorithm for the solution of the coarse level system, resulting in a method with overall linear complexity, also in higher spatial dimension.

We emphasize that the method computes solutions for the full fine-scale equations (6). Hence, no modeling errors are made, and no additional dissipation is introduced. The modeling approximations that are made in the coarse model only influence the convergence speed of the method. Indeed, when the coarse model only provides a poor approximation of the density evolution of the fine-scale model, the coarse level correction step will be less effective, resulting in slower convergence.

In future work, we will extend this work to more general discretization of kinetic equations, and perform a more rigourous two-grid analysis.

Acknowledgements This work was supported by the Research Foundation – Flanders through Research Projects G.0130.03 and G.0170.08 and by the Interuniversity Attraction Poles Programme of the Belgian Science Policy Office through grant IUAP/V/22 (GS). The scientific responsibility rests with its authors.

References

1. Briggs, W., Henson, V.E., McCormick, S.: A multigrid tutorial. SIAM: Philadelphia, PA, (2000)
2. Chalub, F.A.C.C., Markowich, P.A., Perthame, B., Schmeiser, C.: Kinetic models for chemotaxis and their drift-diffusion limits. Monatshefte für Mathematik **142(1–2)** (2004) 123–141
3. Chopard, B., Dupuis, A., Masselot, A., Luthi, P.: Cellular automata and lattice Boltzmann techniques: An approach to model and simulate complex systems. Advances in Complex Systems **5(2)** (2002) 103–246
4. Coscia, V., Delitala, M., Frasca, P.: On the mathematical theory of vehicular traffic flow II: Discrete velocity kinetic models. International Journal of Non-Linear Mechanics **42(3)** (2007) 411–421
5. Dawson, S., Chen, S., Doolen, G.: Lattice Boltzmann computations for reaction-diffusion equations. The Journal of Chemical Physics **98(2)** (1993) 1514–1523
6. Gear, C.W., Kevrekidis, I.G.: Projective methods for stiff differential equations: Problems with gaps in their eigenvalue spectrum. SIAM Journal on Scientific Computing **24(4)** (2003) 1091–1106
7. Gear, C.W., Kevrekidis, I.G.: Constraint-defined manifolds: a legacy code approach to low-dimensional computation. Journal on Scientific Computing **25(1)** (2005) 17–28
8. Gear, C.W., Kaper, T.J., Kevrekidis, I.G., Zagaris, A.: Projecting to a slow manifold: Singularly perturbed systems and legacy codes. IAM Journal on Applied Dynamical Systems **4(3)** (2005) 711–732
9. He, X., Luo, L.S.: Lattice Boltzmann model for the incompressible Navier–Stokes equation. Journal of Statistical Physics **88(3)** (1997) 927–944
10. Junk, M., Klar, A.: Discretizations for the incompressible Navier–Stokes equations based on the lattice Boltzmann method. SIAM Journal on Scientific Computing **22(1)** (2000) 1–19
11. Kevrekidis, I.G., Samaey, G.: Equation-free multiscale computation: Algorithms and applications. Annual Review on Physical Chemistry **60** (2009) 321–344
12. Kevrekidis, I.G., Gear, C.W., Hyman, J.M., Kevrekidis, P.G., Runborg, O., Theodoropoulos, C.: Equation-free, coarse-grained multiscale computation: Enabling microscopic simulators to perform system-level tasks. Communications in Mathematical Sciences **1(4)** (2003) 715–762
13. Li, C., Ebert, U., Brok, W.J.M., Hundsdorfer, W.: Spatial coupling of particle and fluid models for streamers: Where nonlocality matters. Journal of Physics D: Applied Physics **41(3)** (2008) 032005
14. Luo, L.S.: Unified theory of lattice Boltzmann models for nonideal gases. Physical Review Letters **81(8)** (1998) 1618–1621
15. Othmer, H.G., Hillen, T.: The diffusion limit of transport equations II: Chemotaxis equations. SIAM Journal on Applied Mathematics **62(4)** (2002) 1222–1250
16. Qian, Y., Orszag, S.: Scalings in diffusion-driven reaction $A + B \rightarrow C$: Numerical simulations by lattice BGK models. Journal of Statistical Physics **81(1–2)** (1995) 237–253
17. Qiao, L., Erban, R., Kelley, C.T., Kevrekidis, I.G.: Spatially distributed stochastic systems: Equation-free and equation-assisted preconditioned computations. The Journal of Chemical Physics **125** (2006) 204108
18. Samaey, G., Vanroose, W.: An analysis of equivalent operator preconditioning for equation-free Newton–Krylov computations. SIAM Journal on Numerical Analysis **48(2)** (2010) 633–658

19. Samaey, G., Vanroose, W., Roose, D., Kevrekidis, I.G.: Newton–Krylov solvers for the equation-free computation of coarse traveling waves. Computer Methods in Applied Mechanics and Engineering **197(43–44)** (2008) 3480–3491
20. Succi, S.: The lattice Boltzmann equation for fluid dynamics and beyond. Oxford University Press, London (2001)
21. Theodoropoulos, C., Qian, Y.H., Kevrekidis, I.G.: Coarse stability and bifurcation analysis using time-steppers: A reaction-diffusion example. Proceedings of the National Academy of Science **97** (2000) 9840–9845
22. Trottenberg, U., Oosterlee, C., Schuller, A.: Multigrid. Academic Press, NY (2001)
23. Van Leemput, P., Vanroose, W., Roose, D.: Mesoscale analysis of the equation-free constrained runs initialization scheme. Multiscale Modeling and Simulation **6(4)** (2007) 1234–1255
24. Vandekerckhove, C., Kevrekidis, I.G., Roose, D.: An efficient Newton–Krylov implementation of the constrained runs scheme for initializing on a slow manifold. Journal on Scientific Computing **39(2)** (2009) 167–188
25. Vanroose, W., Samaey, G., Van Leemput, P.: Coarse-grained analysis of a lattice Boltzmann model for planar streamer fronts. Technical Report TW479, Department of Computer Science, K.U.Leuven (2007)

Time Step Expansions and the Invariant Manifold Approach to Lattice Boltzmann Models

David J. Packwood, Jeremy Levesley, and Alexander N. Gorban

Abstract The classical method for deriving the macroscopic dynamics of a lattice Boltzmann system is to use a combination of different approximations and expansions. Usually a Chapman–Enskog analysis is performed, either on the continuous Boltzmann system, or its discrete velocity counterpart. Separately a discrete time approximation is introduced to the discrete velocity Boltzmann system, to achieve a practically useful approximation to the continuous system, for use in computation. Thereafter, with some additional arguments, the dynamics of the Chapman–Enskog expansion are linked to the discrete time system to produce the dynamics of the completely discrete scheme.

In this paper we put forward a different route to the macroscopic dynamics. We begin with the system discrete in both velocity space and time. We hypothesize that the alternating steps of advection and relaxation, common to all lattice Boltzmann schemes, give rise to a *slow invariant manifold*. We perform a time step expansion of the discrete time dynamics using the invariance of the manifold. Finally we calculate the dynamics arising from this system.

By choosing the fully discrete scheme as a starting point we avoid mixing approximations and arrive at a general form of the microscopic dynamics up to the second order in the time step. We calculate the macroscopic dynamics of two commonly used lattice schemes up to the first order, and hence find the precise form of the deviation from the Navier–Stokes equations in the dissipative term, arising from the discretization of velocity space.

Finally we perform a short wave perturbation on the dynamics of these example systems, to find the necessary conditions for their stability.

D.J. Packwood (✉), J. Levesley, and A.N. Gorban
Department of Mathematics, University of Leicester, Leicester LE2 2PQ, UK
e-mail: dp123@le.ac.uk, ag153@le.ac.uk, jl1@le.ac.uk

A.N. Gorban and D. Roose (eds.), *Coping with Complexity: Model Reduction and Data Analysis*, Lecture Notes in Computational Science and Engineering 75, DOI 10.1007/978-3-642-14941-2_9, © Springer-Verlag Berlin Heidelberg 2011

1 Introduction

The Boltzmann Equation is a key tool within statistical mechanics, used to describe the time evolution of gases, and with some extensions, other fluids also. In this work we are concerned with calculating the dynamics of an ideal gas. This is achieved by calculating the statistical behaviour of single particles, that is the distribution of their positions in phase space. In particular by fixing a time point and integrating across velocity space, it is possible to calculate the macroscopic quantities of a fluid across space. Performing such an integration allows us to take a 'snapshot' of the dynamics at any time point. If we are concerned, however, with discovering the rates at which the macroscopic variables change, we need to apply additional techniques and assumptions to the Boltzmann Equation. A common choice of technique to derive these macroscopic dynamics is the 'Chapman–Enskog' procedure [1, 2]. This method involves calculating the dynamics of the distribution of the particles at different orders, within the Boltzmann Equation following a perturbation by a small parameter, the Knudsen number. Under such a perturbation the convective dynamics of the fluid will appear at the zero order, and the dissipative dynamics at the first order. The final result is that following such a treatment, the Navier–Stokes equations will be revealed. At the second-order and beyond additional terms give rise to Burnett and super-Burnet type equations respectively.

Despite the Boltzmann Equation recovering the Navier–Stokes equations, to first order in the Knudsen number at least, a practical investigation into the macroscopic dynamics of the Boltzmann system is not necessarily complete. In order to solve the Boltzmann equation numerically, a discretization of both time and velocity space is necessary. To be clear, in many cases for lattice-Boltzmann methods, a discretization of space is not necessary, as a good choice of velocity set and time step size can result in the Boltzmann equation being solved on a discrete lattice subgroup of space points, hence suffering no extra error in this regard.

Within the discrete velocity, continuous time scheme the method of choice to evaluate the macroscopic dynamics of the system has remained to be the Chapman–Enskog procedure. This is presented in a number of different ways [3, 4], however the crux of the approach remains the expansion in the small parameter the Knudsen number. The necessary discretization of the velocity performed and described in some detail in Sect. 3, already gives rise to an additional error from the Navier–Stokes equations. Due to the approximation to the Maxwell distribution, the Navier–Stokes equations are only recovered exactly as the Mach number tends to zero. Where the Mach number is large, additional viscous (dissipative) terms appear [5].

In order to move from continuous time to the lattice system, an Euler step is used to approximate the time derivative. Collisions should happen exactly once per time step, therefore the rate at which collisions occur is given by the time step itself. Furthermore, if the time step is small, in the same scale as the Knudsen number inherited from the continuous time system, the asymptotics of the order by order expansion may be compromised. Bearing in mind the complexity of combining the necessary number of terms from the Euler approximation, along with the existing expansions from the continuous system and taking into account the additional small parameter, we present, in this work, a possibly simpler route to the macroscopic dynamics.

Our starting point is, in fact, not the continuous Boltzmann equation, but the discrete time system itself, in this sense we work in parallel to the historic lattice gas automata idea. We choose that the time step is small, and it is this parameter which we use for our asymptotic analysis. By choosing a discrete scheme such that the zero order dynamics give the Euler equations, we show in Sect. 3 that we retrieve the same computational system as the discrete time Boltzmann anyway. We pursue the dispersive dynamics as the higher order dynamics, in the time step, of the difference scheme that we have chosen. Such a perspective is motivated by, for example Goodman and Lax [6] where it was shown that a particular difference scheme applied to the partial differential equation,

$$\frac{\partial}{\partial t}u(x,t) + u(x,t)\frac{\partial}{\partial x}u(x,t) = 0, \tag{1}$$

recovers at the second order in the space difference parameter Δ, the KdV equation,

$$\frac{\partial}{\partial t}u(x,t) + u(x,t)\frac{\partial}{\partial x}u(x,t) + \frac{1}{6}\Delta^2 u(x,t)\frac{\partial^3}{\partial x^3}u(x,t) = 0 \tag{2}$$

In tandem with the discrete time step asymptotics we hypothesize the existence of a slow Invariant Manifold [7], we calculate the general form of this manifold and use it to find an expression for the macroscopic dynamics of general discrete time systems, with both discrete and continuous velocities. As we have stated our choice of the small parameter to be used in the asymptotics is the discrete time step itself ϵ. The dynamics of the quasi-equilibrium approximation (the Maxwell distribution in the continuum) define the zero order dynamics, higher order dynamics are given by the correction to the equilibrium of the same order [8]. We match the dynamics of the distribution function at microscopic and macroscopic [9] levels to find an expression for the first order non-equilibrium component of the distribution function. Together the zero and first order components of the distribution are sufficient to calculate the macroscopic dynamics up to the first (dissipative) order.

As well as deriving a general expression of the macroscopic dynamics, we additionally provide two examples of both discrete and continuous velocity systems. We find that despite the qualitative difference between continuous and discrete time systems, in discrete time we can still recover the Navier–Stokes equations in a continuous velocity system. In the discrete velocity system, however, the discretizations examined display, as expected, additional errors in the dissipative part. The precise form of these errors is subject to the discrete velocity set chosen and given for the two common examples we use.

Finally we test, by a short wave perturbation, the stability of the dynamics of the discrete velocity system.

2 Background and Notation

We begin with a short summary of the background to this work which will introduce some of the requisite ideas and notation. This brief discussion will be sufficient for this work, for many more details regarding the background and theory to Boltzmann systems a large number of works exist by several authors [2–4, 10].

The Boltzmann Equation is concerned with the time evolution of what might be described physically as the density of particles. This density function is denoted f and is a function of space, velocity space and time $f \equiv f(x, v, t)$. It is given as follows;

$$\frac{\partial}{\partial t} f + v \cdot \nabla_x f = Q^c(f). \tag{3}$$

Already we need a certain amount of details to formalize what we have written. We begin with the space variable x, this is a smooth k-dimensional manifold, for example Euclidean space, or a torus, in finite dimensions. The velocity variable v ranges over the same space. In the continuous time system above t is a single real variable, however we will dispense with this very soon. The final notation to mention in (3) is the collision integral Q^c, a differentiable transformation of the population function. More detailed properties which we require of this function will be given later.

In fact we are not at all concerned with the continuous time Boltzmann evolution, but the corresponding discrete time system:

$$f(x + \epsilon v, v, t + \epsilon) = f(x, t) + Q(f(x, t)). \tag{4}$$

This is the discrete time Boltzmann system which is at the core of this work. In contrast with the continuous time system we have introduced the time step ϵ, which will play a key role in our analysis. The time step should be small and it restricts admissible values of time to the subgroup $\epsilon Z \in R$.

The left two terms in (4) are collectively termed advection or free flight, if the collision integral is omitted, the exact evolution of the population function at a point may be written as $f(x, v, t + \epsilon) = f(x - \epsilon v, v, t)$. The physical interpretation is that particles move freely under their own momentum with no interaction between themselves. For our analysis we can use the fact that advection is smooth and we will be able to take a Taylor series expansion to any finite order that we need. The rightmost term is again called the collision integral and is denoted slightly differently to notate that it may differ somewhat in form from its continuous counterpart.

Rewriting the right hand side of (4) leads us to a perspective which will be key to our approach. Rewriting the collision part as $F(f) = f + Q(f)$, allows us to present (4) differently:

$$f(x, v, t + \epsilon) = F(f(x - \epsilon v, v, t)). \tag{5}$$

Time Step Expansions and the Invariant Manifold Approach 173

In this way, the discrete time Boltzmann system can then be thought of as a superposition of the advection and collision operations. Advection and collision steps happen in turn, this is qualitatively different from the continuous time system where both operations might be said to be happening simultaneously at all times.

The evolution of the particle density function f can be said to describe the microscopic evolution of the system. To recover the macroscopic system we need to sum across the velocity space. Here then is the key distinction between the continuous ($v \in \mathbb{R}^k$) and discrete ($v \in \{v^\alpha\}\alpha = 0\ldots n$) velocity systems. In the discrete velocity case, instead of considering the density function to be a function of velocity, we define different density functions f^α, one per velocity vector and we can denote then by f the ordered set of them $f \equiv f^\alpha(x,t)$, $\alpha = 0, \ldots, n$. For example the density of the fluid is calculated in the continuous and discrete schemes respectively as

$$\rho(x,t) = \int_{\mathbb{R}^k} f(x,v,t)dv, \ \rho(x,t) = \sum_{\alpha=0}^{n} f^\alpha(x,t) \tag{6}$$

In fact which macroscopic variables M we use are not important to our initial analysis. The properties that are important for the macroscopic moments are firstly that the operator m, which recovers the macroscopic moments from the density function f ($m(f) = M$) is linear. The second property is that these moments are invariant under the collision operation, that is $m(f) = m(F(f))$.

Defining the macroscopic variables allows us to discuss another property of the collision operation F. The quasi-equilibrium is the unique vector $f = f_M^{eq}$ such that $F(f_M^{eq}) = f_M^{eq}$ and $m(f_M^{eq}) = M$. Existence and uniqueness of this quasi-equilibrium will be assumed in this analysis. There is exactly one equilibrium point per value of M and together they form the quasi-equilibrium manifold through the space of the density function.

3 The Discretization of Velocity Space

While we calculate the dynamics of the continuous velocity system later, they serve only for the purpose of calculating the error incurred by an approximation. For practical computations we would use a discrete velocity system. What we have control of is which discrete approximation to use. Two different approaches will lead us to the same result.

Firstly we consider an actual discretization of the Maxwell–Boltzmann distribution in D dimensions, that is the known quasi-equilibrium in continuous velocity space.

$$f^{eq} = \rho(2\pi T)^{-\frac{D}{2}} \exp\left(\frac{-(v-u)^2}{2T}\right). \tag{7}$$

The $2 + D$ thermodynamic moments here are the density ρ, momentum u and temperature T. This distribution is multiplied with low order polynomials of the velocity and integrated to retrieve the macroscopic moments.

$$\int_{\mathbb{R}^D} f^{\text{eq}} dv = \rho,$$
$$\int_{\mathbb{R}^D} v f^{\text{eq}} dv = \rho u, \tag{8}$$
$$\int_{\mathbb{R}^D} v^2 f^{\text{eq}} dv = \rho D T + \rho u^2.$$

We can view the velocity set as a quadrature approximation to these integrals [11]. With such a perspective the choice of which quadrature to use may seem obvious, integrals of Gaussian type (as are the moment integrals) can be integrated exactly using a Gauss-Hermite quadrature. Unfortunately several stumbling blocks prevent us from performing this exact integration.

The first problem is the necessary change of variable in the equilibrium distribution. For a Hermitian quadrature of any order, the nodes should be distributed symmetrically about the centre of the Gaussian (in this case u), and the coordinate of integration should be normalized. Effectively it is necessary to apply a change of variable to the moment integrals so that the exponential term is of the simple form $\exp(-v^2)$. Unfortunately, to do this in practice would require us to know both u and T before performing the integration. Of course we may still be able to solve such a system, perhaps by some iterative method (moving the quadrature nodes) up to an arbitrary degree of accuracy, but only if we can evaluate the density function anywhere we choose. In practice of course we can only store a finite number of evaluations of the density function, and these points of evaluation are chosen pre-emptively to be the same across all lattice sites and time steps (in order that advection is an exact operation). Because of these factors we should not expect to be able to integrate exactly a density function of Gaussian type with a general momentum and temperature.

The popular method to partially rectify this problem is to assume that u is close to zero. If we choose to work in an athermal system (where T is some constant) then we can expand the Maxwell distribution about the point $u = 0$ up to the second order giving us

$$f^{\text{eq}} = \rho(2\Pi T)^{\frac{-D}{2}} \exp\left(\frac{-v^2}{2T}\right)\left(1 + \frac{vu}{T} - \frac{(T - v^2)u^2}{2T^2}\right) \tag{9}$$

The second order expansion is taken to get sufficient terms that the temperature moment integral is calculated correctly. With a constant T, and u no longer affecting the midpoint of the distribution, the same nodes and weights can be used for all space points and time steps. Assuming the proper transformation of variable these nodes and weights are the standard hermitian ones.

The alternative method of defining the quasi-equilibrium is by solution of a linear system. Since the moments are calculated by linear combinations of the discrete quasiequilibria, these equilibria can be found by the inversion of a matrix of components of the discrete velocities, to illustrate this we give an extremely simple example. We consider an athermal one dimensional system with three discrete velocities $\{-1, 0, 1\}$ and fixed temperature $1/3$. We create a matrix where each line represents an elementwise power of the velocities (up to the second order) and the corresponding vector of moments we desire. Together this creates the following linear system.

$$\begin{bmatrix} 1 & 1 & 1 \\ -1 & 0 & 1 \\ 1 & 0 & 1 \end{bmatrix} \cdot f^{eq} = \begin{bmatrix} \rho \\ \rho u \\ \rho/3 + \rho u^2 \end{bmatrix} \tag{10}$$

Solution of the above system yields a vector of equilibria

$$f^{eq} = \left\{ \frac{1}{6} \left(\rho - 3\rho u + 3\rho u^2 \right), \frac{2}{3} \left(\rho - \frac{3}{2}\rho u^2 \right), \frac{1}{6} \left(\rho + 3\rho u + 3\rho u^2 \right) \right\} \tag{11}$$

Inspection reveals this to be exactly the same, with weights pre-included, as the discretized Taylor approximation to the Boltzmann (9) where the same velocity set is applied. We note that in this example we have exactly the necessary amount of velocities (that is one per moment) and hence have a square matrix. As in the case of any linear system, were we to have fewer velocities it would become likely that the system would become unsolvable. For each additional discrete velocity that we add above the necessary amount, we can impose an additional linear constraint. Often the choice would be to zero higher order moments of the equilibrium distribution, that is enforcing that the sum of the equilibria multiplied with polynomials of the discrete velocity components, with order higher than two, should be zero. Another option is to create additional, non-hydrodynamic moments in order to suppress instabilities [12].

4 Invariant Manifolds for Discrete Time Boltzmann Systems

We will, by finding a general form for an invariant manifold, calculate general microscopic dynamics for discrete time Boltzmann systems. To do this we need to make some assumptions regarding the stability of collisions and the smoothness of the distribution function.

For the next subsections we introduce the notation that fields of distribution functions of moments will be written in gothic font, hence the field of macroscopic variables, for example, will be denoted $\mathfrak{M}$.

We assume that collisions are stable, and for any admissible initial state f iterations $F^p(f)$ converge to unique equilibrium point f^{eq} exponentially fast and uniformly:

$$\| F^p(f) - f^{eq}_{m(f)} \| < C \exp(-\lambda p) \| f - f^{eq}_{m(f)} \|, \tag{12}$$

where the Lyapunov exponent $\lambda > 0$ and pre-factor $C > 0$ are the same for all admissible f. In the limit $\epsilon = 0$ there is no free flight, the field of macroscopic variables $\mathfrak{M}$ does not change, and the field of distributions $\mathfrak{f}$ converges to the local equilibrium field $\mathfrak{f}_M^{eq}$ by repeated application of the collision operation. Since each collision occurs instantaneously, the superposed collisions become a projection onto the local equilibrium in zero time.

In order to discuss small $\epsilon > 0$ we need to evaluate the change of macroscopic variables in free flight during time step ϵ. To find the qth term of the non-equilibrium density function of a discrete velocity system, we make two assumptions:

- The first q derivatives of f_M along vector fields v^α exist and are bounded.
- The differentials $(D_M^{(q)} f_M^{eq})$, are uniformly bounded for all M.

Our expression for the manifold will be of the form of an aysmptotic expansion in the small parameter ϵ,

$$f^{\text{inv}} = f^{(0)} + \epsilon f^{(1)} + \epsilon f^{(2)} + o(\epsilon). \tag{13}$$

Our first goal is to find a prescription for this $f^{(1)}$ term. The zero order term of this is simply given by the quasi-equilibrium distributions, that is $f^{(0)} \equiv f^{eq}$. For the first order term we will take expansions of the distribution function in terms of time and of moments and equate them. That is for each order in epsilon we can take the effect of a complete LBM step (advection and collision) and match the effect on the distribution function to that of taking the Taylor approximation of the manifold through the moments up to the same order. In other words we match the dynamics of the microscopic and macroscopic scales on an order by order basis.

4.1 The Invariance Equation

The procedure we use can also be described in terms of the invariant manifold hypothesis. Coupled steps of advection and collision form a chain of states of the population function belonging to the manifold. Since the number of discrete velocities used is normally larger than the number of macroscopic moments, there are an infinite number of possible population distributions which can give rise to the same configuration of moments, however only one of these distributions exists on the manifold. We use a Taylor approximation to the manifold and match it with a single coupled step to find the components, at different orders of the time step, of the distribution function.

If we consider $\mathfrak{f}_{\mathfrak{M}}^{\text{inv}}$ to be the field of population distributions on the manifold with corresponding field of macroscopic moments $\mathfrak{M}$ then in a continuous time system this invariance property can be defined as

$$D_t \mathfrak{f}_{\mathfrak{M}}^{\text{inv}} = D_{\mathfrak{M}} \mathfrak{f}_{\mathfrak{M}}^{\text{inv}} \cdot D_t \mathfrak{M}. \tag{14}$$

Here the derivative $D_{\mathfrak{M}}$ indicates the derivative through the field of macroscopic moments $\mathfrak{M}$, which the field of distributions on the manifold $\mathfrak{f}_{\mathfrak{M}}^{\text{inv}}$ is parameterized

Time Step Expansions and the Invariant Manifold Approach 177

by. Altogether the rate of change of the population function is equal to the rate of change of the moments multiplied by the change of the populations with respect to the moments. The discrete time analogy of this is given by,

$$\left(f_{\mathfrak{M}}^{\text{inv}}\right)' = f_{\mathfrak{M}'}^{\text{inv}}. \tag{15}$$

where the prime notates the next time step, therefore the left hand side of this equation is given by (5).

From now on we dispense with the gothic notation for fields of variables and simply use a standard font to describe distribution functions or macroscopic variables.

4.2 The Expansion of the Distribution Function Following a Step in the LBM Chain

A little abuse of notation will make the same calculations in this section both brief and meaningful. No distinction will be made here between continuous and discrete velocity systems, however implicitly there is one. As previously $f \equiv f(x, v, t)$ for the continuous case, and $f \equiv f^\alpha(x, t)$, $\alpha = 0, \ldots, n$ in the discrete case. For the purposes of our analysis, in both the continuous and discrete velocity systems the space coordinate x is continuous. Due to this, even though we may choose to only evaluate f at a discrete number of lattice sites, the space derivative $D_x f$ remains well defined, even for the discrete velocity case. The notation is similar for moment derivatives, where $D_M f_M$ refers to the change in moments of the function f evaluated at the moments given by the subscript to f. When the moment operator m is applied this refers to the integral in the continuous case and the summation in the discrete case as given in (6).

With these ideas in mind the procedure below is valid for both cases. The first ingredient for the time step expansion is the Taylor series of the advection operation up to the required order in ϵ. For the first order we have

$$f(x - \epsilon v) = f - \epsilon v \cdot D_x f + o(\epsilon). \tag{16}$$

Combining this with (13) we have to the first order,

$$f(x - \epsilon v) = f^{(0)} - \epsilon v \cdot D_x f^{(0)} + \epsilon f^{(1)} + o(\epsilon). \tag{17}$$

Applying a collision operation gives the complete, composite discrete time step,

$$\left(f_M^{\text{inv}}\right)' = F\left(f^{(0)} - \epsilon v \cdot D_x f^{(0)} + \epsilon f^{(1)} + o(\epsilon)\right). \tag{18}$$

The second ingredient is to use a linearised version of the collision operation, this is sufficient to get the first order populations correctly. Here the linearisation is made about the equilibrium corresponding to the populations to be collided,

$$f \mapsto f^{eq}_{m(f)} + \left(D_f F\right)_{f^{eq}_{m(f)}} \left(f - f^{eq}_{m(f)}\right). \tag{19}$$

Due to the linearity we can move the error term in (18) outside the collision altogether. The linearisation is then made about the equilibrium defined by the moments of the first order advected populations

$$M'_1 = m\left(f^{(0)}_M - \epsilon v \cdot D_x f^{(0)}_M\right) = M + m\left(-\epsilon v \cdot D_x f^{(0)}_M\right), \tag{20}$$

Finally then for the first order approximation to the next step through the time step expansion we have,

$$\left(f^{inv}_M\right)' = f^{eq}_{M'_1} + \left(D_f F\right)_{f^{eq}_{M'_1}} \left(f^{(0)}_M + \epsilon f^{(1)}_M - \epsilon v \cdot D_x f^{(0)}_M - f^{eq}_{M'_1}\right) + o(\epsilon). \tag{21}$$

4.3 The Expansion of the Invariance Equation Following a Time Step

With the expansion of the left hand of (15) complete we consider the right hand side. Here we find the Taylor expansion of the invariant manifold up to the linear term so,

$$f_{M'} = f_M + (D_M f_M) \cdot m\left(-\epsilon v \cdot D_x f_M\right) + o(\epsilon). \tag{22}$$

Substituting (13) into (22) we have

$$f_{M'} = f^{(0)}_M + \epsilon f^{(1)}_M + (D_M f^{(0)}_M) \cdot m\left(-\epsilon v \cdot D_x f^{(0)}_M\right) + o(\epsilon). \tag{23}$$

We can now equate (21) and (23) for a first order approximation to (15),

$$f^{eq}_{M'_1} + \left(D_f F\right)_{f^{eq}_{M'_1}} \left(f^{(0)}_M + \epsilon f^{(1)}_M - \epsilon v \cdot D_x f^{(0)}_M - f^{eq}_{M'_1}\right)$$
$$= f^{(0)}_M + \epsilon f^{(1)}_M + (D_M f^{(0)}_M) \cdot m\left(-\epsilon v \cdot D_x f^{(0)}_M\right). \tag{24}$$

Of course in a similar style to (22),

$$f^{eq}_{M'_1} = f^{(0)}_M + (D_M f^{(0)}_M) \cdot m\left(-\epsilon v \cdot D_x f^{(0)}_M\right). \tag{25}$$

Time Step Expansions and the Invariant Manifold Approach 179

Substituting back into (24) we have

$$
\left(D_f F\right)_{f^{eq}_{M'_1}} \left(\epsilon f_M^{(1)} - \epsilon v \cdot D_x f_M^{(0)} - (D_M f_M^{(0)}) \cdot m \left(-\epsilon v \cdot D_x f_M^{(0)}\right)\right) = \epsilon f_M^{(1)}
$$

(26)

This equation forms the prototype to find $f_M^{(1)}$ for different possible collision operations, it implicitly gives the first order approximation to the invariance equation (15). It depends on the choice of the velocity set, the quasiequilibrium and the collision integral.

4.4 Example First Order Invariant Manifolds

We consider two possible examples of collisions. The first example is the simple Ehrenfest step [13],

$$
F(f) = f^{eq}_{m(f)}
$$

(27)

we immediately have,

$$
\left(D_f F\right)^{eq}_{f_{m(f)}} (f) = 0.
$$

(28)

Substituting back into (26),

$$
f_M^{(1)} = 0.
$$

(29)

This of course expected since using Ehrenfests steps for the collisions we should expect to return at every time step to the quasi-equilibrium manifold.

The second example of a collision operator is the BGK collision [14],

$$
F(f) = f + \omega \left(f^{eq}_{m(f)} - f\right).
$$

(30)

Differentiating we have

$$
\left(D_f F\right)_{f^{eq}_{m(f)}} (f) = (1 - \omega) \cdot f
$$

(31)

Substituting this into (26),

$$
(1 - \omega) \cdot \left(\epsilon f_M^{(1)} - \epsilon v \cdot D_x f_M^{(0)} - (D_M f_M^{(0)}) \cdot m \left(-\epsilon v \cdot D_x f_M^{(0)}\right)\right) = \epsilon f_M^{(1)}.
$$

(32)

We can multiply out (32) and solve for, $f_M^{(1)}$.

$$
\frac{\omega}{1 - \omega} f_M^{(1)} = \left(-v \cdot D_x f_M^{(0)}\right) - (D_M f_M^{(0)}) \cdot m \left(-v \cdot D_x f_M^{(0)}\right)
$$

(33)

Figure 1 graphically demonstrates the $f_M^{(1)}$ for the BGK collision type. In particular for this example there is a critical parameter value at $\omega = 1$. For $\omega = 1$ we recover

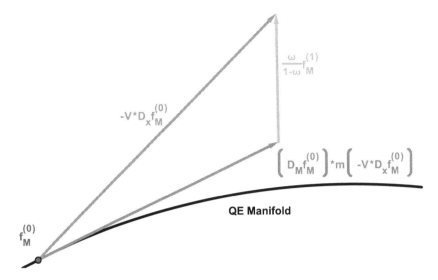

Fig. 1 Graphical representation of the $f_M^{(1)}$ for the BGK collision. Adding the $f_M^{(1)}$ term to the quasiequilibrium manifold gives the invariant manifold to first order in ϵ. In particular the collision parameter ω is critical, for $\omega > 1$ the direction of the $f_M^{(1)}$ term is inverted and consequently the invariant manifold is below (in the sense of this illustration) the quasiequilibrium. Therefore at each step the advection operation crosses the quasiequilibrium and the collision returns below it

the Ehrenfest step, for $\omega > 1$ we have the normal BGK over-relaxation where both of the coupled steps of advection and collision cross the quasiequilibrium manifold, one in each direction.

4.5 Second Order Manifolds and an Example

The next goal is to find an equation analagous to (26) for the second order term of the invariant manifold. During the next section we use a linear collision operator, in this case the linearised collision we use produces the exact same result as the original collision. We restart the procedure using second order expansions where appropriate, the first of these is the Taylor expansion of the advected populations.

$$f(x - \epsilon v) = f - \epsilon v \cdot D_x f + \frac{\epsilon^2}{2} v \cdot D_x (v \cdot D_x f) + o(\epsilon^2) \qquad (34)$$

The second order population expansion is also used.

$$f^{\text{inv}} = f^{(0)} + \epsilon f^{(1)} + \epsilon^2 f^{(2)} + o(\epsilon^2). \qquad (35)$$

Altogether the second order expansion of the advected populations is,

$$f(x - \epsilon v) = f^{(0)} - \epsilon v \cdot D_x f^{(0)} + \frac{\epsilon^2}{2} v \cdot D_x \left(v \cdot D_x f^{(0)} \right)$$
$$+ \epsilon f^{(1)} - \epsilon^2 v \cdot D_x f^{(1)} + \epsilon^2 f^{(2)} + o(\epsilon^2) \tag{36}$$

We use the linearized collision integral in replacement of the original collision operation,

$$(f_M^{inv})' = f_{M_2'}^{eq} + (D_f F)_{f_{M_2'}^{eq}} \left(f_M^{(0)} - \epsilon v \cdot D_x f_M^{(0)} + \frac{\epsilon^2}{2} v \cdot D_x \left(v \cdot D_x f_M^{(0)} \right) \right.$$
$$\left. + \epsilon f_M^{(1)} - \epsilon^2 v \cdot D_x f_M^{(1)} + \epsilon^2 f_M^{(2)} - f_{M_2'}^{eq} \right) + o(\epsilon^2). \tag{37}$$

where M_2' are the moments of the post advection populations to *second* order,

$$M_2' = m \left(f_M^{(0)} - \epsilon v \cdot D_x f_M^{(0)} + \frac{\epsilon^2}{2} v \cdot D_x \left(v \cdot D_x f_M^{(0)} \right) - \epsilon^2 v \cdot D_x f_M^{(1)} \right), \tag{38}$$

For the right hand side of (15) we use a second order approximation to the invariant manifold,

$$f_{M'}^{inv} = f_M^{inv}(\mathbf{x}, v_\alpha) + \Delta M_2 \cdot D_M f_M^{inv} + \frac{1}{2} \Delta M_2 \cdot D_M (\Delta M_2 \cdot D_M f_M^{inv}) + o((\Delta M_2)^2). \tag{39}$$

where for berevity ΔM_2 is the difference between the moments of the post and pre advection populations to second order,

$$\Delta M_2 = m \left(-\epsilon v \cdot D_x f_M^{(0)} + \frac{\epsilon^2}{2} v \cdot D_x \left(v \cdot D_x f_M^{(0)} \right) - \epsilon^2 v \cdot D_x f_M^{(1)} \right) = M_2' - M. \tag{40}$$

Substituting (35) and (40) into (39) we have

$$f_{M'}^{inv} = f_M^{(0)} + \epsilon f_M^{(1)} + \epsilon^2 f_M^{(2)} + m \left(-\epsilon v \cdot D_x f_M^{(0)} + \frac{\epsilon^2}{2} v \cdot D_x \left(v \cdot D_x f_M^{(0)} \right) \right.$$
$$\left. - \epsilon^2 v \cdot D_x f_M^{(1)} \right) \cdot D_M f_M^{(0)}$$
$$+ \frac{1}{2} m \left(-\epsilon v \cdot D_x f_M^{(0)} \right) \cdot D_M \left(m \left(-\epsilon v \cdot D_x f_M^{(0)} \right) \cdot D_M f_M^{(0)} \right)$$
$$+ \epsilon m \left(-\epsilon v \cdot D_x f_M^{(0)} \right) \cdot D_M f_M^{(1)} + o(\epsilon^2). \tag{41}$$

With the expansions of both sides complete we can equate (37) and (41),

$$
f^{\text{eq}}_{M'_2} + (D_f F)_{f^{\text{eq}}_{M'_2}} \left(f^{(0)}_M - \epsilon v \cdot D_x f^{(0)}_M + \frac{\epsilon^2}{2} v \cdot D_x \left(v \cdot D_x f^{(0)}_M \right) + \epsilon f^{(1)}_M \right.
$$

$$
\left. - \epsilon^2 v \cdot D_x f^{(1)}_M + \epsilon^2 f^{(2)}_M - f^{\text{eq}}_{M'_2} \right) = f^{(0)}_M + \epsilon f^{(1)}_M + \epsilon^2 f^{(2)}_M
$$

$$
+ m \left(-\epsilon v \cdot D_x f^{(0)}_M + \frac{\epsilon^2}{2} v \cdot D_x \left(v \cdot D_x f^{(0)}_M \right) - \epsilon^2 v \cdot D_x f^{(1)}_M \right) \cdot D_M f^{(0)}_M
$$

$$
+ \frac{1}{2} m \left(-\epsilon v \cdot D_x f^{(0)}_M \right) \cdot D_M \left(m \left(-\epsilon v \cdot D_x f^{(0)}_M \right) \cdot D_M f^{(0)}_M \right)
$$

$$
+ \epsilon m \left(-\epsilon v \cdot D_x f^{(0)}_M \right) \cdot D_M f^{(1)}_M \tag{42}
$$

Analogously to the first order case we note that,

$$
f^{\text{eq}}_{M'_2} = f^{(0)}_M + m \left(-\epsilon v \cdot D_x f^{(0)}_M + \frac{\epsilon^2}{2} v \cdot D_x \left(v \cdot D_x f^{(0)}_M \right) \right.
$$

$$
\left. - \epsilon^2 v \cdot D_x f^{(1)}_M \right) \cdot D_M f^{(0)}_M
$$

$$
+ m \left(-\epsilon v \cdot D_x f^{(0)}_M \right) \cdot D_M \left(m \left(-\epsilon v \cdot D_x f^{(0)}_M \right) \cdot D_M f^{(0)}_M \right). \tag{43}
$$

Substituting this back into (42) we have the final prototype for $f^{(2)}_M$ which this time is given implicitly,

$$
(D_f F)_{f^{\text{eq}}_{M'_2}} \left(-\epsilon v \cdot D_x f^{(0)}_M + \frac{\epsilon^2}{2} v \cdot D_x \left(v \cdot D_x f^{(0)}_M \right) + \epsilon f^{(1)}_M - \epsilon^2 v \cdot D_x f^{(1)}_M + \epsilon^2 f^{(2)}_M \right.
$$

$$
- m \left(-\epsilon v \cdot D_x f^{(0)}_M + \frac{\epsilon^2}{2} v \cdot D_x \left(v \cdot D_x f^{(0)}_M \right) - \epsilon^2 v \cdot D_x f^{(1)}_M \right) \cdot D_M f^{(0)}_M
$$

$$
\left. - \frac{1}{2} m \left(-\epsilon v \cdot D_x f^{(0)}_M \right) \cdot D_M \left(m \left(-\epsilon v \cdot D_x f^{(0)}_M \right) \cdot D_M f^{(0)}_M \right) \right)
$$

$$
= \epsilon f^{(1)}_M + \epsilon^2 f^{(2)}_M + \epsilon m \left(-\epsilon v \cdot D_x f^{(0)}_M \right) \cdot D_M f^{(1)}_M \tag{44}
$$

We return to the BGK collision for a specific example of an $f^{(2)}_M$ term. Equation (44) becomes,

$$
(1 - \omega) \cdot \left(-\epsilon v \cdot D_x f^{(0)}_M + \frac{\epsilon^2}{2} v \cdot D_x \left(v \cdot D_x f^{(0)}_M \right) + \epsilon f^{(1)}_M - \epsilon^2 v \cdot D_x f^{(1)}_M + \epsilon^2 f^{(2)}_M \right.
$$

$$
- m \left(-\epsilon v \cdot D_x f^{(0)}_M + \frac{\epsilon^2}{2} v \cdot D_x \left(v \cdot D_x f^{(0)}_M \right) - \epsilon^2 v \cdot D_x f^{(1)}_M \right) \cdot D_M f^{(0)}_M
$$

Time Step Expansions and the Invariant Manifold Approach

$$-\frac{1}{2}m\left(-\epsilon v \cdot D_x f_M^{(0)}\right) \cdot D_M \left(m\left(-\epsilon v \cdot D_x f_M^{(0)}\right) \cdot D_M f_M^{(0)}\right)$$

$$= \epsilon f_M^{(1)} + \epsilon^2 f_M^{(2)} + \epsilon m \left(-\epsilon v \cdot D_x f_M^{(0)}\right) \cdot D_M f_M^{(1)} \quad (45)$$

Rearranging and equating terms with ϵ order 2 gives us,

$$
\begin{aligned}
\frac{\omega}{1-\omega} f_M^{(2)} =\ & \frac{1}{2}v \cdot D_x \left(v \cdot D_x f_M^{(0)}\right) - v \cdot D_x f_M^{(1)} \\
& - m\left(\frac{1}{2}v \cdot D_x \left(v \cdot D_x f_M^{(0)}\right) - v \cdot D_x f_M^{(1)}\right) \cdot D_M f_M^{(0)} \\
& - \frac{1}{2}m\left(-v \cdot D_x f_M^{(0)}\right) \cdot D_M \left(m\left(-v \cdot D_x f_M^{(0)}\right) \cdot D_M f_M^{(0)}\right) \\
& - \frac{1}{1-\omega}m\left(-v \cdot D_x f_M^{(0)}\right) \cdot D_M f_M^{(1)}
\end{aligned}
\quad (46)
$$

We will not use these populations in the examples of macroscopic dynamics which we calculate in the next section. For the dissipative dynamics the first order populations are sufficient. We expect that this second order part should give rise to macroscopic dynamics equating to the Burnett equations in a continuous velocity system, with some additional error terms if a discrete velocity set is used.

5 Macroscopic Equations

In this section we are concerned with deriving equations for the macroscopic dynamics arising from several different example lattices. We expect that the lattice parameter ϵ should partly govern these dynamics and that the 1st order macroscopic dynamics should be governed by the 1st order population functions.

In order to find these dynamics we project the microscopic flow (advection) up to the required order, following one time step, onto the invariant manifold up to the same order [7, 15].

We can immediately perform a Taylor expansion in time on the macroscopic dynamics,

$$M' = M + \epsilon \frac{\partial}{\partial t} M + o(\epsilon) \quad (47)$$

We expect that the final model should be given in terms of a time derivative of the moments, we write this in a power series in terms of ϵ,

$$\frac{\partial}{\partial t} M = \Psi^{(0)} + \epsilon \Psi^{(1)} + o(\epsilon) \quad (48)$$

Combining these two we have

$$M' = M + \epsilon \Psi^{(0)} + o(\epsilon) \tag{49}$$

Equating (20) and (49) we have

$$\Psi^{(0)} = m\left(-v \cdot D_x f_M^{(0)}\right) \tag{50}$$

The corresponding second order approximation of the moments in time is

$$M' = M + \epsilon(\Psi^{(0)} + \epsilon \Psi^{(1)}) + \frac{\epsilon^2}{2}\frac{\partial}{\partial t}\Psi^{(0)} + o(\epsilon^2) \tag{51}$$

Equating terms on the second order of ϵ we have,

$$m\left(\frac{1}{2}v \cdot D_x\left(v \cdot D_x f_M^{(0)}\right)\right) + m\left(-v \cdot D_x f_M^{(1)}\right) = \Psi^{(1)}(t) + \frac{1}{2}\frac{\partial}{\partial t}\Psi^{(0)}(t) \tag{52}$$

or

$$\Psi^{(1)}(t) = m\left(\frac{1}{2}v \cdot D_x\left(v \cdot D_x f_M^{(0)}\right)\right) + m\left(-v \cdot D_x f_M^{(1)}\right) - \frac{1}{2}\frac{\partial}{\partial t}\Psi^{(0)}(t) \tag{53}$$

Although in this notation the evolution of the macroscopic moments of the continuous and discrete velocity systems can be written in the same way, in practical examples we have no reason to suspect that they should be necessarily equivalent. In the next two sections we calculate some example, equivalent, discrete and continuous systems, in order to compare them.

6 Discrete Velocity Examples

We will now demonstrate the exact first order dynamics of a popular choice of lattice scheme in one and two dimensions. The athermal schemes we consider are typically described in shorthand by the dimension within which they operate and the number of velocities used to form the lattice in the form DnQm, where m and n are integers representing the number of dimensions and velocities respectively. The general quasi-equilibrium for these systems, including the two examples we use can be written in a general form

$$f_M^{\alpha,(0)} = W_\alpha \rho \left(1 + \frac{v^\alpha \cdot u}{c_s^2} + \frac{(v^\alpha \cdot u)^2}{2c_s^4} - \frac{u^2}{2c_s^2}\right) \tag{54}$$

Time Step Expansions and the Invariant Manifold Approach 185

This equilibrium defines an athermal system where the temperature is fixed. To complete the definition of the discrete system requires only the selection of a velocity set and some accompanying weights W_α.

6.1 An Athermal Three Velocity Lattice (D1Q3)

Our 1-D example lattice is one of the most common, the athermal 1-D lattice with three velocities. In this example the velocity vectors are $\{-1, 0, 1\}$ and the speed of sound $c_s = 1/\sqrt{3}$ hence the equilibrium populations are derived from the general formula for athermal quasi-equilibria in any dimension where the additional parameters the weights W_α are $\{\frac{1}{6}, \frac{2}{3}, \frac{1}{6}\}$.

For this case the populations are,

$$\frac{1}{6}\left\{\rho(1 - 3u + 3u^2), 2\rho\left(1 - \frac{3u^2}{2}\right), \rho(1 + 3u + 3u^2)\right\}. \tag{55}$$

For this lattice with unit distances we note that $v^4 = v^2$, $v^1 = v^3$ etc. We calculate the two components of $\Psi^{(0)}$ using the formulas for the moments. We have for the density derivative,

$$\Psi_1^{(0)} = -\sum_\alpha v^\alpha \frac{\partial}{\partial x} f_M^{\alpha,(0)} = -\frac{\partial}{\partial x} \sum_\alpha W_\alpha v^\alpha f_M^{\alpha,(0)} = -\frac{\partial}{\partial x} \rho u, \tag{56}$$

and for the momentum derivative

$$\Psi_2^{(0)} = -\sum_\alpha W_\alpha (v^\alpha)^2 \frac{\partial}{\partial x} f_M^{\alpha,(0)} = -\frac{\partial}{\partial x} \sum_\alpha W_\alpha (v^\alpha)^2 f_M^{\alpha,(0)}$$

$$= -\frac{\partial}{\partial x}\left(\frac{\rho}{3} + \rho u^2\right). \tag{57}$$

Now we examine the individual moments of the first order part in the case of the one dimensional lattice, as before we begin with the density,

$$\Psi_1^{(1)} = \frac{1}{2}\sum_\alpha (v^\alpha)^2 \frac{\partial^2}{\partial x^2} f_M^{\alpha,(0)} - \sum_\alpha v^\alpha \frac{\partial}{\partial x} f_M^{\alpha,(1)} - \frac{1}{2}\frac{\partial}{\partial t}\Psi_1^{(0)}. \tag{58}$$

The second term here is the space derivative of the momentum of the $f_M^{(1)}$ which equals zero due to all moments of non equilibrium components being zero and the first term can be calculated immediately from the quasi-equilibrium. Therefore,

$$\Psi_1^{(1)} = \frac{1}{2}\frac{\partial^2}{\partial x^2}\left(\frac{\rho}{3} + \rho u^2\right) - \frac{1}{2}\frac{\partial}{\partial t}\Psi_1^{(0)}. \tag{59}$$

The time derivative of $\Psi_1^{(0)}$ can be calculated by the chain rule,

$$\frac{\partial}{\partial t}\Psi_1^{(0)} = \frac{\partial \Psi_1^{(0)}}{\partial \rho}\frac{\partial \rho}{\partial t} + \frac{\partial \Psi_1^{(0)}}{\partial \rho u}\frac{\partial \rho u}{\partial t} = -\frac{\partial}{\partial x}\Psi_2^{(0)} = \frac{\partial^2}{\partial x^2}\left(\frac{\rho}{3} + \rho u^2\right) \tag{60}$$

Substituting this back in we have,

$$\Psi_1^{(1)} = \frac{1}{2}\frac{\partial^2}{\partial x^2}\left(\frac{\rho}{3} + \frac{(\rho u)^2}{\rho}\right) - \frac{1}{2}\frac{\partial^2}{\partial x^2}\left(\frac{\rho}{3} + \rho u^2\right) = 0. \tag{61}$$

For the momentum moment we have

$$\Psi_2^{(1)} = \frac{1}{2}\sum_\alpha (v^\alpha)^3 \frac{\partial^2}{\partial x^2} f_M^{\alpha,(0)} - \sum_\alpha (v^\alpha)^2 \frac{\partial}{\partial x} f_M^{\alpha,(1)} - \frac{1}{2}\frac{\partial}{\partial t}\Psi_2^{(0)}. \tag{62}$$

Recalling that $v^3 = v^1$ we can simplify the first term so,

$$\frac{1}{2}\sum_\alpha (v^\alpha)^3 \frac{\partial^2}{\partial x^2} f_M^{\alpha,(0)} = \frac{1}{2}\sum_\alpha v^\alpha \frac{\partial^2}{\partial x^2} f_M^{\alpha,(0)} = \frac{1}{2}\frac{\partial^2}{\partial x^2}\rho u. \tag{63}$$

For the second term we need to calculate the $f_M^{\alpha,(1)}$ terms. To do this we need to specify a collision type, we use the BGK collision described above (33).

$$\frac{\omega}{1-\omega}f_M^{(1)} = \frac{\partial}{\partial x}\left(2u\frac{\partial}{\partial x}\rho + (-2 - 3u^2)\frac{\partial}{\partial x}\rho u + 6u\frac{\partial}{\partial x}\rho u^2\right.,$$
$$-u\frac{\partial}{\partial x}\rho + \left(1 + \frac{3}{2}u^2\right)\frac{\partial}{\partial x}\rho u - 3u\frac{\partial}{\partial x}\rho u^2,$$
$$\left.2u\frac{\partial}{\partial x}\rho + (-2 - 3u^2)\frac{\partial}{\partial x}\rho u + 6u\frac{\partial}{\partial x}\rho u^2\right). \tag{64}$$

This gives then

$$\sum_\alpha (v^\alpha)^2 \frac{\partial}{\partial x} f_M^{\alpha,(1)} = \frac{1-\omega}{\omega}\frac{\partial}{\partial x}\left(\frac{2}{3}u\frac{\partial}{\partial x}\rho + \left(-\frac{2}{3} - u^2\right)\frac{\partial}{\partial x}\rho u + 2u\frac{\partial}{\partial x}\rho u^2\right). \tag{65}$$

Again using the chain rule,

$$\frac{\partial}{\partial t}\Psi_2^{(0)} = \frac{\partial \Psi_2^{(0)}}{\partial \rho}\frac{\partial \rho}{\partial t} + \frac{\partial \Psi_2^{(0)}}{\partial \rho u}\frac{\partial \rho u}{\partial t} = -\frac{\partial}{\partial x}\left(\frac{1}{3} - u^2\right)\Psi_1^{(0)} - \frac{\partial}{\partial x}2u\Psi_2^{(0)}$$
$$= \frac{\partial}{\partial x}\left(\frac{2}{3}u\frac{\partial}{\partial x}\rho + \left(\frac{1}{3} - u^2\right)\frac{\partial}{\partial x}\rho u + 2u\frac{\partial}{\partial x}\rho u^2\right) \tag{66}$$

Time Step Expansions and the Invariant Manifold Approach

Substituting this all back in we have,

$$
\begin{aligned}
\Psi_2^{(1)} &= \left(\frac{\omega-1}{\omega}-\frac{1}{2}\right)\frac{\partial}{\partial x}\left(\frac{2}{3}u\frac{\partial}{\partial x}\rho+\left(-\frac{2}{3}-u^2\right)\frac{\partial}{\partial x}\rho u+2u\frac{\partial}{\partial x}\rho u^2\right)\\
&= \frac{\omega-2}{2\omega}\frac{\partial}{\partial x}\left(\frac{2}{3}u\frac{\partial}{\partial x}\rho+\left(-\frac{2}{3}-u^2\right)\frac{\partial}{\partial x}\rho u+2u\frac{\partial}{\partial x}\rho u^2\right)\\
&= \frac{\omega-2}{2\omega}\frac{\partial}{\partial x}\left(u^3\frac{\partial}{\partial x}\rho+\rho\left(3u^2-\frac{2}{3}\right)\frac{\partial}{\partial x}u\right).
\end{aligned}
\tag{67}
$$

The moment gradients are then to first order in ϵ,

$$
\begin{aligned}
\frac{\partial}{\partial t}\rho &= -\frac{\partial}{\partial x}\rho u\\
\frac{\partial}{\partial t}\rho u &= -\frac{\partial}{\partial x}\left(\frac{1}{3}\rho+\rho u^2\right)-\epsilon\frac{2-\omega}{2\omega}\frac{\partial}{\partial x}\left(u^3\frac{\partial}{\partial x}\rho+\rho\left(3u^2-\frac{2}{3}\right)\frac{\partial}{\partial x}u\right)
\end{aligned}
\tag{68}
$$

6.2 An Athermal Nine Velocity Model (D2Q9)

The 2-D example we consider is a popular 2d lattice consisting of nine different velocities. If we identify v_1 as the horizontal component of a vector and v_2 the vertical component then the set of velocities is

$$
v_\alpha = \left\{\begin{pmatrix}0\\0\end{pmatrix},\begin{pmatrix}1\\0\end{pmatrix},\begin{pmatrix}0\\1\end{pmatrix},\begin{pmatrix}-1\\0\end{pmatrix},\begin{pmatrix}0\\-1\end{pmatrix},\right.
$$
$$
\left.\begin{pmatrix}1\\1\end{pmatrix},\begin{pmatrix}-1\\1\end{pmatrix},\begin{pmatrix}-1\\-1\end{pmatrix},\begin{pmatrix}1\\-1\end{pmatrix}\right\}
\tag{69}
$$

The equilibrium is then given by the polynomial formula (54) with corresponding weights

$$
W_\alpha = \left\{\frac{4}{9},\frac{1}{9},\frac{1}{9},\frac{1}{9},\frac{1}{9},\frac{1}{36},\frac{1}{36},\frac{1}{36},\frac{1}{36}\right\}
\tag{70}
$$

As before we calculate the components of $\Psi^{(0)}$ using the formulas for the moments although this time we have two momentum density momentums for the two dimensions. We have for the density derivative,

$$
\begin{aligned}
\Psi_1^{(0)} &= \sum_\alpha\left(-v^\alpha\cdot D_x f_M^{\alpha,(0)}\right)\\
&= -\frac{\partial}{\partial x_1}\sum_\alpha v_1^\alpha f_M^{\alpha,(0)}-\frac{\partial}{\partial x_2}\sum_\alpha v_2^\alpha f_M^{\alpha,(0)}\\
&= -\frac{\partial}{\partial x_1}\rho u_1-\frac{\partial}{\partial x_2}\rho u_2,
\end{aligned}
\tag{71}
$$

for the first momentum derivative

$$
\begin{aligned}
\Psi_2^{(0)} &= \sum_\alpha v_1^\alpha \left(-v^\alpha \cdot D_x f_M^{\alpha,(0)} \right) \\
&= -\frac{\partial}{\partial x_1} \sum_\alpha (v_1^\alpha)^2 f_M^{\alpha,(0)} - \frac{\partial}{\partial x_2} \sum_\alpha v_1^\alpha v_2^\alpha f_M^{\alpha,(0)} \\
&= -\frac{\partial}{\partial x_1} \left(\frac{1}{3}\rho + \rho u_1^2 \right) - \frac{\partial}{\partial x_2} \rho u_1 u_2,
\end{aligned} \tag{72}
$$

and for the second momentum derivative

$$
\begin{aligned}
\Psi_3^{(0)} &= \sum_\alpha v_2^\alpha \left(-v^\alpha \cdot D_x f_M^{\alpha,(0)} \right) \\
&= -\frac{\partial}{\partial x_1} \sum_\alpha v_1^\alpha v_2^\alpha f_M^{\alpha,(0)} - \frac{\partial}{\partial x_2} \sum_\alpha (v_2^\alpha)^2 f_M^{\alpha,(0)} \\
&= -\frac{\partial}{\partial x_1} \rho u_1 u_2 - \frac{\partial}{\partial x_2} \left(\frac{1}{3}\rho + \rho u_2^2 \right).
\end{aligned} \tag{73}
$$

The first order density moment is given by,

$$
\Psi_1^{(1)} = \frac{1}{2} \sum_\alpha \left(v \cdot D_x \left(v \cdot D_x f_M^{(0)} \right) \right) + \sum_\alpha \left(-v \cdot D_x f_M^{(1)} \right) - \frac{1}{2} \frac{\partial}{\partial t} \Psi_1^{(0)}. \tag{74}
$$

Again we observe that the second term is the space gradient multiplied with the momentum densities of the first order populations and hence is zero, for the first term we have

$$
\begin{aligned}
\sum_\alpha & \left(v \cdot D_x \left(v \cdot D_x f_M^{(0)} \right) \right) \\
&= \sum_\alpha \left(\frac{\partial^2}{\partial x_1^2} (v_1^\alpha)^2 f_M^{\alpha,(0)} + 2 \frac{\partial^2}{\partial x_1 \partial x_2} v_1^\alpha v_2^\alpha f_M^{\alpha,(0)} + \frac{\partial^2}{\partial x_2^2} (v_2^\alpha)^2 f_M^{\alpha,(0)} \right) \\
&= \frac{\partial^2}{\partial x_1^2} \left(\frac{1}{3}\rho + \rho u_1^2 \right) + 2 \frac{\partial^2}{\partial x_1 \partial x_2} \rho u_1 u_2 + \frac{\partial^2}{\partial x_2^2} \left(\frac{1}{3}\rho + \rho u_2^2 \right), \tag{75}
\end{aligned}
$$

and for the third term

$$
\begin{aligned}
\frac{\partial}{\partial t} \Psi_1^{(0)} &= \frac{\partial \Psi_1^{(0)}}{\partial \rho} \frac{\partial \rho}{\partial t} + \frac{\partial \Psi_1^{(0)}}{\partial \rho u_1} \frac{\partial \rho u_1}{\partial t} + \frac{\partial \Psi_1^{(0)}}{\partial \rho u_2} \frac{\partial \rho u_2}{\partial t} = -\frac{\partial}{\partial x_1} \Psi_2^{(0)} - \frac{\partial}{\partial x_2} \Psi_3^{(0)} \\
&= -\frac{\partial}{\partial x_1} \left(-\frac{\partial}{\partial x_1} \left(\frac{1}{3}\rho + \rho u_1^2 \right) - \frac{\partial}{\partial x_2} \rho u_1 u_2 \right) \\
&\quad - \frac{\partial}{\partial x_2} \left(-\frac{\partial}{\partial x_1} \rho u_1 u_2 - \frac{\partial}{\partial x_2} \left(\frac{1}{3}\rho + \rho u_2^2 \right) \right). \tag{76}
\end{aligned}
$$

Time Step Expansions and the Invariant Manifold Approach

hence subtracting these we have $\Psi_1^{(1)} = 0$.

For the first second order momentum density we have

$$\Psi_2^{(1)} = \frac{1}{2}\sum_\alpha v_1^\alpha \left(v \cdot D_x \left(v \cdot D_x f_M^{(0)}\right)\right) + \sum_\alpha v_1^\alpha \left(-v \cdot D_x f_M^{(1)}\right) - \frac{1}{2}\frac{\partial}{\partial t}\Psi_2^{(0)}. \quad (77)$$

Examining each term in turn more closely we have for the first term

$$\sum_\alpha v_1^\alpha \left(v \cdot D_x \left(v \cdot D_x f_M^{(0)}\right)\right)$$

$$= \sum_\alpha v_1^\alpha \left(\frac{\partial^2}{\partial x_1^2}(v_1^\alpha)^2 f_M^{\alpha,(0)} + 2\frac{\partial^2}{\partial x_1 \partial x_2}v_1^\alpha v_2^\alpha f_M^{\alpha,(0)} + \frac{\partial^2}{\partial x_2^2}(v_2^\alpha)^2 f_M^{\alpha,(0)}\right)$$

$$= \frac{\partial}{\partial x_1}\left(\rho\frac{\partial}{\partial x_1}u_1 + u_1\frac{\partial}{\partial x_1}\rho + \frac{1}{3}\rho\frac{\partial}{\partial x_2}u_2 + \frac{1}{3}u_2\frac{\partial}{\partial x_2}\rho\right)$$

$$+ \frac{\partial}{\partial x_2}\left(\frac{1}{3}\rho\frac{\partial}{\partial x_1}u_2 + \frac{1}{3}u_2\frac{\partial}{\partial x_1}\rho + \frac{1}{3}\rho\frac{\partial}{\partial x_2}u_1 + \frac{1}{3}u_1\frac{\partial}{\partial x_2}\rho\right). \quad (78)$$

The first order populations are given in Appendix, these give us for the second term

$$\sum_\alpha v_1^\alpha \left(-v \cdot D_x f_M^{(1)}\right) = -\sum_\alpha v_1^\alpha \left(\frac{\partial}{\partial x_1}v_1^\alpha f_M^{\alpha,(1)} + \frac{\partial}{\partial x_2}v_2^\alpha f_M^{(1)}\right)$$

$$= \frac{\omega - 1}{\omega}\left(\frac{\partial}{\partial x_1}\left(u_1^3\frac{\partial}{\partial x_1}\rho + u_1^2 u_2\frac{\partial}{\partial x_2}\rho + \left(3\rho u_1^2 - \frac{2}{3}\rho\right)\frac{\partial}{\partial x_1}u_1\right.\right.$$

$$\left.+ 2\rho u_1 u_2\frac{\partial}{\partial x_2}u_1 + \rho u_1^2\frac{\partial}{\partial x_2}u_2\right) \quad (79)$$

$$+ \frac{\partial}{\partial x_2}\left(u_1^2 u_2\frac{\partial}{\partial x_1}\rho + u_1 u_2^2\frac{\partial}{\partial x_2}\rho + 2\rho u_1 u_2\frac{\partial}{\partial x_1}u_1 + 2\rho u_1 u_2\frac{\partial}{\partial x_2}u_2\right.$$

$$\left.\left.+ \left(\rho u_2^2 - \frac{1}{3}\rho\right)\frac{\partial}{\partial x_2}u_1 + \left(\rho u_1^2 - \frac{1}{3}\rho\right)\frac{\partial}{\partial x_1}u_2\right)\right).$$

and for the third term

$$\frac{\partial}{\partial t}\Psi_2^{(0)} = \frac{\partial\Psi_2^{(0)}}{\partial\rho}\frac{\partial\rho}{\partial t} + \frac{\partial\Psi_2^{(0)}}{\partial\rho u_1}\frac{\partial\rho u_1}{\partial t} + \frac{\partial\Psi_2^{(0)}}{\partial\rho u_2}\frac{\partial\rho u_2}{\partial t}$$

$$= \left(\frac{\partial}{\partial x_1}\left(-\frac{1}{3}u^2 + u_1^2\right) + \frac{\partial}{\partial x_2}u_1 u_2\right)\Psi_1^{(0)}$$

$$+ \left(-2\frac{\partial}{\partial x_1}u_1 - \frac{\partial}{\partial x_2}u_2\right)\Psi_2^{(0)} - \frac{\partial}{\partial x_2}u_1\Psi_3^{(0)}$$

$$= \frac{\partial}{\partial x_1}\left((u_1 + u_1^3)\frac{\partial}{\partial x_1}\rho + \left(\frac{1}{3}u_2 + u_1^2 u_2\right)\frac{\partial}{\partial x_2}\rho\right.$$
$$+ \left(\frac{1}{3}\rho + 3\rho u_1^2\right)\frac{\partial}{\partial x_1}u_1 + \left(\frac{1}{3}\rho + \rho u_1^2\right)\frac{\partial}{\partial x_2}u_2 + 2\rho u_1 u_2 \frac{\partial}{\partial x_2}u_1\right)$$
$$+ \frac{\partial}{\partial x_2}\left(\left(\frac{1}{3}u_2 + u_1^2 u_2\right)\frac{\partial}{\partial x_1}\rho + \left(\frac{1}{3}u_1 + u_1 u_2^2\right)\frac{\partial}{\partial x_2}\rho\right.$$
$$+ 2\rho u_1 u_2 \frac{\partial}{\partial x_1}u_1 + \rho u_2^2 \frac{\partial}{\partial x_2}u_1 + \rho u_1^2 \frac{\partial}{\partial x_1}u_2 + 2\rho u_1 u_2 \frac{\partial}{\partial x_2}u_2\right). \quad (80)$$

Combining all three terms we have,

$$\Psi_2^{(1)} = \left(\frac{\omega - 1}{\omega} - \frac{1}{2}\right)\left(\frac{\partial}{\partial x_1}\left(u_1^3 \frac{\partial}{\partial x_1}\rho + u_1^2 u_2 \frac{\partial}{\partial x_2}\rho + \left(3\rho u_1^2 - \frac{2}{3}\rho\right)\frac{\partial}{\partial x_1}u_1\right.\right.$$
$$+ 2\rho u_1 u_2 \frac{\partial}{\partial x_2}u_1 + \rho u_1^2 \frac{\partial}{\partial x_2}u_2\right)$$
$$+ \frac{\partial}{\partial x_2}\left(u_1^2 u_2 \frac{\partial}{\partial x_1}\rho + u_1 u_2^2 \frac{\partial}{\partial x_2}\rho + 2\rho u_1 u_2 \frac{\partial}{\partial x_1}u_1 + 2\rho u_1 u_2 \frac{\partial}{\partial x_2}u_2\right.$$
$$\left.\left. + \left(\rho u_2^2 - \frac{1}{3}\rho\right)\frac{\partial}{\partial x_2}u_1 + \left(\rho u_1^2 - \frac{1}{3}\rho\right)\frac{\partial}{\partial x_1}u_2\right)\right)$$

$$(81)$$

The final macroscopic equations for this particular lattice and quasiequilibrium then are to first order

$$\frac{\partial}{\partial t}\rho = -D_x\rho u$$
$$\frac{\partial}{\partial t}\rho u_1 = -\frac{\partial}{\partial x_1}\left(\frac{\rho}{3} + \rho u_1^2\right) - \frac{\partial}{\partial x_2}\rho u_1 u_2$$
$$- \epsilon \frac{2 - \omega}{2\omega}\left(\frac{\partial}{\partial x_1}\left(u_1^3 \frac{\partial}{\partial x_1}\rho + u_1^2 u_2 \frac{\partial}{\partial x_2}\rho + \left(3\rho u_1^2 - \frac{2}{3}\rho\right)\frac{\partial}{\partial x_1}u_1\right.\right.$$
$$+ 2\rho u_1 u_2 \frac{\partial}{\partial x_2}u_1 + \rho u_1^2 \frac{\partial}{\partial x_2}u_2\right)$$
$$+ \frac{\partial}{\partial x_2}\left(u_1^2 u_2 \frac{\partial}{\partial x_1}\rho + u_1 u_2^2 \frac{\partial}{\partial x_2}\rho + 2\rho u_1 u_2 \frac{\partial}{\partial x_1}u_1 + 2\rho u_1 u_2 \frac{\partial}{\partial x_2}u_2\right.$$
$$\left.\left. + \left(\rho u_2^2 - \frac{1}{3}\rho\right)\frac{\partial}{\partial x_2}u_1 + \left(\rho u_1^2 - \frac{1}{3}\rho\right)\frac{\partial}{\partial x_1}u_2\right)\right)$$

$$(82)$$

The second momentum density is available easily through symmetry.

Time Step Expansions and the Invariant Manifold Approach 191

7 Continuous Velocity Examples

We are now concerned with calculating, via the invariant manifold populations, the macroscopic moments approximated by the LBM chain in a continuous velocity system. We select two examples, chosen to match the previous discrete velocity schemes.

7.1 The Athermal 1-D Model

The first continuous velocity model we will examine is one chosen to match the zero order dynamics of the discrete model studied in Sect. 6.1, the one dimensional system with the three discrete velocities $\{-1, 0, 1\}$. The continuous population function acting as the quasi-equilibrium is a specific case of the Maxwell distribution where the temperature is fixed, in this case to $1/3$.

$$f^{(0)} = \rho \sqrt{\frac{3}{2\pi}} \exp\left(-\frac{3}{2}(v-u)^2\right) \tag{83}$$

With such a system the macroscopic variables are calculated as integrals rather than the sums in the discrete case.

$$\int_{-\infty}^{\infty} f^{(0)} dv = \rho$$
$$\int_{-\infty}^{\infty} v f^{(0)} dv = \rho u \tag{84}$$
$$\int_{-\infty}^{\infty} v^2 f^{(0)} dv = \frac{1}{3}\rho + \rho u^2$$

Clearly this matches the moments retrieved in the discrete velocity case. Due to this we can, analogously to the discrete case, immediately write down the zero order macroscopic dynamics following (50).

$$\Psi_0^{(1)} = -\int_{\infty}^{\infty} v \left(\frac{\partial}{\partial x} f^{(0)}\right) dv = -\frac{\partial}{\partial x} \int_{\infty}^{\infty} v f^{(0)} dv = -\frac{\partial}{\partial x} \rho u \tag{85}$$

$$\Psi_0^{(2)} = -\int_{\infty}^{\infty} v^2 \left(\frac{\partial}{\partial x} f^{(0)}\right) dv = -\frac{\partial}{\partial x} \int_{\infty}^{\infty} v^2 f^{(0)} dv = -\frac{\partial}{\partial x} \left(\frac{1}{3}\rho + \rho u^2\right) \tag{86}$$

In order to calculate the first order moments we expect that we shall require the first order continuous populations. These are also derived exactly as in the discrete case with the replacement of the sum, by the integral, in the calculation of the moments. Since we replicate the discrete case the collision we select is again the BGK collision and we derive the first order populations from (33).

$$\frac{\omega}{1-\omega} f^{(1)} = \rho \sqrt{\frac{3}{2\pi}} \exp\left(-\frac{3}{2}(v-u)^2\right) \cdot \left(1 - 3v^2 + 6vu - 3u^2\right) \cdot \left(\frac{\partial}{\partial x} u\right) \quad (87)$$

Again we calculate the first order moments from the template given by (53)

$$\Psi_1^{(1)} = \frac{1}{2} \int_\infty^\infty v^2 \left(\frac{\partial^2}{\partial x^2} f^{(0)}\right) dv - \int_\infty^\infty v \left(\frac{\partial}{\partial x} f^{(1)}\right) dv - \frac{1}{2}\frac{\partial}{\partial t}\Psi_1^{(0)} \quad (88)$$

Exactly as the discrete case the second term here is the space derivative of the momentum of the $f^{(1)}$ which equals zero due to all moments of non equilibrium components being zero and the first term can be calculated immediately from the quasi-equilibrium therefore,

$$\Psi_1^{(1)} = \frac{1}{2}\frac{\partial^2}{\partial x^2}\left(\frac{\rho}{3} + \rho u^2\right) - \frac{1}{2}\frac{\partial}{\partial t}\Psi_1^{(0)}. \quad (89)$$

Again the time derivative of $\Psi^{(0)}$ can be calculated by the chain rule,

$$\frac{\partial}{\partial t}\Psi_1^{(0)} = \frac{\partial \Psi_1^{(0)}}{\partial \rho}\frac{\partial \rho}{\partial t} + \frac{\partial \Psi_1^{(0)}}{\partial \rho u}\frac{\partial \rho u}{\partial t} = -\frac{\partial}{\partial x}\Psi_2^{(0)} = \frac{\partial^2}{\partial x^2}\left(\frac{\rho}{3} + \rho u^2\right) \quad (90)$$

Substituting we have,

$$\Psi_1^{(1)} = \frac{1}{2}\frac{\partial^2}{\partial x^2}\left(\frac{\rho}{3} + \rho u^2\right) - \frac{1}{2}\frac{\partial^2}{\partial x^2}\left(\frac{\rho}{3} + \rho u^2\right) = 0. \quad (91)$$

For the continuous velocity momentum moment we have

$$\Psi_2^{(1)} = \frac{1}{2} \int_\infty^\infty v^3 \left(\frac{\partial^2}{\partial x^2} f^{(0)}\right) dv - \int_\infty^\infty v^2 \left(\frac{\partial}{\partial x} f^{(1)}\right) dv - \frac{1}{2}\frac{\partial}{\partial t}\Psi_2^{(0)} \quad (92)$$

Rearranging and performing the first two integrals gives us

$$\Psi_2^{(1)} = \frac{1}{2}\frac{\partial^2}{\partial x^2}\left(\rho u + \rho u^3\right) - \frac{\partial}{\partial x}\left(-\frac{2}{3}\rho\left(\frac{\partial}{\partial x}u\right)\right) - \frac{1}{2}\frac{\partial}{\partial t}\Psi_2^{(0)} \quad (93)$$

Again using the chain rule, this term is exactly as in the discrete case,

$$\frac{\partial}{\partial t}\Psi_2^{(0)} = \frac{\partial \Psi_2^{(0)}}{\partial \rho}\frac{\partial \rho}{\partial t} + \frac{\partial \Psi_2^{(0)}}{\partial \rho u}\frac{\partial \rho u}{\partial t} = -\frac{\partial}{\partial x}\left(\frac{1}{3} - u^2\right)\Psi_1^{(0)} - \frac{\partial}{\partial x}2u\Psi_2^{(0)}$$

$$= \frac{\partial}{\partial x}\left(\frac{2}{3}u\frac{\partial}{\partial x}\rho + \left(\frac{1}{3} - u^2\right)\frac{\partial}{\partial x}\rho u + 2u\frac{\partial}{\partial x}\rho u^2\right) \quad (94)$$

Time Step Expansions and the Invariant Manifold Approach 193

Substituting this all back in we have,

$$\Psi_2^{(1)} = \left(\frac{\omega-1}{\omega} - \frac{1}{2}\right)\frac{\partial}{\partial x}\left(-\frac{2}{3}\rho\frac{\partial}{\partial x}u\right) = \frac{\omega-2}{2\omega}\frac{\partial}{\partial x}\left(-\frac{2}{3}\rho\frac{\partial}{\partial x}u\right). \qquad (95)$$

The moment gradients are then, for the continuous velocity system, to first order in ϵ,

$$\begin{aligned}
\frac{\partial}{\partial t}\rho &= -\frac{\partial}{\partial x}\rho u + o(\epsilon) \\
\frac{\partial}{\partial t}\rho u &= -\frac{\partial}{\partial x}\left(\frac{1}{3}\rho + \rho u^2\right) - \epsilon\frac{2-\omega}{2\omega}\frac{\partial}{\partial x}\left(-\frac{2}{3}\rho\frac{\partial}{\partial x}u\right) + o(\epsilon)
\end{aligned} \qquad (96)$$

We immediately observe that several of the dissipative terms that appeared in the discrete velocity system do not occur when we use continuous velocities.

7.2 The Athermal 2D Model

The next continuous velocity model we examine is the widely used athermal 2d model. Again we use a specific choice of the Maxwellian distribution which matches the zero order moments given by the discrete velocity set.

$$f^{(0)} = \rho\frac{3}{2\pi}\exp\left(-\frac{3}{2}\left((v_1-u_1)^2 + (v_2-u_2)^2\right)\right) \qquad (97)$$

Again macroscopic variables are calculated by integrals over velocity space

$$\begin{aligned}
\int_{\mathbb{R}^2} f^{(0)}dv &= \rho \\
\int_{\mathbb{R}^2} v_1 f^{(0)}dv &= \rho u_1 \\
\int_{\mathbb{R}^2} v_2 f^{(0)}dv &= \rho u_2 \\
\int_{\mathbb{R}^2} (v_1^2+v_2^2)f^{(0)}dv &= \frac{2}{3}\rho + \rho(u_1^2+u_2^2)
\end{aligned} \qquad (98)$$

Again we calculate the zero order moments,

$$\begin{aligned}
\Psi_1^{(0)} &= \int_{\mathbb{R}^2} -v\cdot D_x f_M^{\alpha,(0)}dv \\
&= -\frac{\partial}{\partial x_1}\int_{\mathbb{R}^2} v_1 f_M^{\alpha,(0)}dv - \frac{\partial}{\partial x_2}\int_{\mathbb{R}^2} v_2 f_M^{\alpha,(0)}dv \\
&= -\frac{\partial}{\partial x_1}\rho u_1 - \frac{\partial}{\partial x_2}\rho u_2
\end{aligned} \qquad (99)$$

and for the first momentum derivative

$$
\begin{aligned}
\Psi_2^{(0)} &= \int_{\mathbb{R}^2} v_1 \left(-v \cdot D_x f_M^{\alpha,(0)} \right) dv \\
&= -\frac{\partial}{\partial x_1} \int_{\mathbb{R}^2} v_1^2 f_M^{\alpha,(0)} dv - \frac{\partial}{\partial x_2} \int_{\mathbb{R}^2} v_1 v_2 f_M^{\alpha,(0)} dv \\
&= -\frac{\partial}{\partial x_1} \left(\frac{1}{3}\rho + \rho u_1^2 \right) - \frac{\partial}{\partial x_2} \rho u_1 u_2
\end{aligned}
\tag{100}
$$

for the second momentum derivative

$$
\begin{aligned}
\Psi_3^{(0)} &= \int_{\mathbb{R}^2} v_2 \left(-v \cdot D_x f_M^{\alpha,(0)} \right) dv \\
&= -\frac{\partial}{\partial x_1} \int_{\mathbb{R}^2} v_1 v_2 f_M^{\alpha,(0)} dv - \frac{\partial}{\partial x_2} \int_{\mathbb{R}^2} v_2^2 f_M^{\alpha,(0)} dv \\
&= -\frac{\partial}{\partial x_1} \rho u_1 u_2 - \frac{\partial}{\partial x_2} \left(\frac{1}{3}\rho + \rho u_2^2 \right)
\end{aligned}
\tag{101}
$$

We again calculate the first order populations following (33).

$$
\begin{aligned}
\frac{\omega}{1-\omega} f^{(1)} &= \\
\rho \frac{3}{2\pi} & \exp\left(-\frac{3}{2} \left((v_1 - u_1)^2 + (v_2 - u_2)^2 \right) \right) \\
& \cdot \Bigg(\left(1 - 3v_1^2 + 6v_1 u_1 - 3u_1^2 \right) \frac{\partial}{\partial x_1} u_1 \\
& + (-3v_1 v_2 + 3v_1 u_2 + 3v_2 u_1 - 3u_1 u_2) \frac{\partial}{\partial x_2} u_1 \\
& + (-3v_1 v_2 + 3v_1 u_2 + 3v_2 u_1 - 3u_1 u_2) \frac{\partial}{\partial x_1} u_2 \\
& + \left(1 - 3v_2^2 + 6v_2 u_2 - 3u_2^2 \right) \frac{\partial}{\partial x_2} u_2 \Bigg)
\end{aligned}
\tag{102}
$$

The first order density moment is given by,

$$
\Psi_1^{(1)} = \frac{1}{2} \int_{\mathbb{R}^2} v \cdot D_x \left(v \cdot D_x f^{(0)} \right) dv - \int_{\mathbb{R}^2} v \cdot D_x f_M^{(1)} dv - \frac{1}{2} \frac{\partial}{\partial t} \Psi_1^{(0)}
\tag{103}
$$

Performing the integrals of the first two terms we note that the second term is again zero therefore

$$
\Psi_1^{(1)} = \frac{1}{2} \frac{\partial^2}{\partial x_1^2} \left(\frac{1}{3}\rho + \rho u_1^2 \right) + \frac{\partial^2}{\partial x_1 x_2} \rho u_1 u_2 + \frac{1}{2} \frac{\partial^2}{\partial x_2^2} \left(\frac{1}{3}\rho + \rho u_2^2 \right) - \frac{1}{2} \frac{\partial}{\partial t} \Psi_1^{(0)}
\tag{104}
$$

Time Step Expansions and the Invariant Manifold Approach 195

and exactly as in the discrete velocity system we have for the third term

$$\frac{\partial}{\partial t}\Psi_1^{(0)} = \frac{\partial\Psi_1^{(0)}}{\partial\rho}\frac{\partial\rho}{\partial t} + \frac{\partial\Psi_1^{(0)}}{\partial\rho u_1}\frac{\partial\rho u_1}{\partial t} + \frac{\partial\Psi_1^{(0)}}{\partial\rho u_2}\frac{\partial\rho u_2}{\partial t} = -\frac{\partial}{\partial x_1}\Psi_2^{(0)} - \frac{\partial}{\partial x_2}\Psi_3^{(0)}$$

$$= -\frac{\partial}{\partial x_1}\left(-\frac{\partial}{\partial x_1}\left(\frac{1}{3}\rho + \rho u_1^2\right) - \frac{\partial}{\partial x_2}\rho u_1 u_2\right)$$

$$-\frac{\partial}{\partial x_2}\left(-\frac{\partial}{\partial x_1}\rho u_1 u_2 - \frac{\partial}{\partial x_2}\left(\frac{1}{3}\rho + \rho u_2^2\right)\right)$$

$$(105)$$

hence subtracting these we have $\Psi_1^{(1)} = 0$.

For the first second order momentum density we have

$$\Psi_2^{(1)} = \frac{1}{2}\int_{\mathbb{R}^2} v_1\left(v \cdot D_x\left(v \cdot D_x f^{(0)}\right)\right) dv + \int_{\mathbb{R}^2} v_1\left(-v \cdot D_x f^{(1)}\right) dv - \frac{1}{2}\frac{\partial}{\partial t}\Psi_2^{(0)}$$

$$(106)$$

Again performing the integrations from the first two terms we have

$$\Psi_2^{(1)} = \frac{1}{2}\left(\frac{\partial^2}{\partial x_1^2}\left(\rho u_1^3 + \rho u_1\right) + \frac{\partial^2}{\partial x_1\partial x_2}\left(\frac{1}{3}\rho u_2 + \rho u_1^2 u_2\right)\right.$$

$$\left. + \frac{\partial^2}{\partial x_2}\left(\frac{1}{3}\rho u_1 + \rho u_1 u_2^2\right)\right)$$

$$+ \frac{\omega - 1}{2\omega}\left(\frac{\partial}{\partial x_1}\left(-\frac{2}{3}\rho\frac{\partial}{\partial x_1}u_1\right)\right.$$

$$\left. - \frac{\partial}{\partial x_2}\left(-\frac{1}{3}\rho\left(\frac{\partial}{\partial x_1}u_2 + \frac{\partial}{\partial x_2}u_1\right)\right)\right) - \frac{1}{2}\frac{\partial}{\partial t}\Psi_2^{(0)}$$

$$(107)$$

and for the third term

$$\frac{\partial}{\partial t}\Psi_2^{(0)} = \frac{\partial\Psi_2^{(0)}}{\partial\rho}\frac{\partial\rho}{\partial t} + \frac{\partial\Psi_2^{(0)}}{\partial\rho u_1}\frac{\partial\rho u_1}{\partial t} + \frac{\partial\Psi_2^{(0)}}{\partial\rho u_2}\frac{\partial\rho u_2}{\partial t}$$

$$= \left(\frac{\partial}{\partial x_1}\left(-\frac{1}{3}u^2 + u_1^2\right) + \frac{\partial}{\partial x_2}u_1 u_2\right)\Psi_1^{(0)}$$

$$+ \left(-2\frac{\partial}{\partial x_1}u_1 - \frac{\partial}{\partial x_2}u_2\right)\Psi_2^{(0)} - \frac{\partial}{\partial x_2}u_1\Psi_3^{(0)}$$

$$= \frac{\partial}{\partial x_1}\left((u_1 + u_1^3)\frac{\partial}{\partial x_1}\rho + \left(\frac{1}{3}u_2 + u_1^2 u_2\right)\frac{\partial}{\partial x_2}\rho\right.$$

$$\left. + \left(\frac{1}{3}\rho + 3\rho u_1^2\right)\frac{\partial}{\partial x_1}u_1 + \left(\frac{1}{3}\rho + \rho u_1^2\right)\frac{\partial}{\partial x_2}u_2 + 2\rho u_1 u_2\frac{\partial}{\partial x_2}u_1\right)$$

$$+ \frac{\partial}{\partial x_2} \left(\left(\frac{1}{3} u_2 + u_1^2 u_2 \right) \frac{\partial}{\partial x_1} \rho + \left(\frac{1}{3} u_1 + u_1 u_2^2 \right) \frac{\partial}{\partial x_2} \rho + 2 \rho u_1 u_2 \frac{\partial}{\partial x_1} u_1 \right.$$

$$\left. + \rho u_2^2 \frac{\partial}{\partial x_2} u_1 + \rho u_1^2 \frac{\partial}{\partial x_1} u_2 + 2 \rho u_1 u_2 \frac{\partial}{\partial x_2} u_2 \right) \tag{108}$$

Combining all three terms we have

$$\Psi_2^{(1)} = \left(\frac{\omega - 1}{\omega} - \frac{1}{2} \right) \left(\frac{\partial}{\partial x_1} \left(-\frac{2}{3} \rho \frac{\partial}{\partial x_1} u_1 \right) \right.$$

$$\left. - \frac{\partial}{\partial x_2} \left(-\frac{1}{3} \rho \left(\frac{\partial}{\partial x_1} u_2 + \frac{\partial}{\partial x_2} u_1 \right) \right) \right) \tag{109}$$

The final macroscopic equations for this particular lattice and quasiequilibrium then are

$$\frac{\partial}{\partial t} \rho = -D_x \rho u$$

$$\frac{\partial}{\partial t} \rho u_1 = -\frac{\partial}{\partial x_1} \left(\frac{\rho}{3} + \rho u_1^2 \right) - \frac{\partial}{\partial x_2} \rho u_1 u_2$$

$$- \epsilon \frac{2 - \omega}{2\omega} \left(\frac{\partial}{\partial x_1} \left(-\frac{2}{3} \rho \frac{\partial}{\partial x_1} u_1 \right) - \frac{\partial}{\partial x_2} \left(-\frac{1}{3} \rho \left(\frac{\partial}{\partial x_1} u_2 + \frac{\partial}{\partial x_2} u_1 \right) \right) \right) \tag{110}$$

and again the second momentum density can be found by reflection. Once again many of the dissipative terms vanish in the continuous velocity system.

8 Macroscopic Stability

In the previous sections we have demonstrated the discrete velocity systems studied do not recover the exact macroscopic dissipative dynamics of the continuous system. We are now concerned with the stability of the discrete dynamics under a short wave perturbation. In each example we are concerned with the stability of the linear part of the dynamics (as calculated above) only.

8.1 The Athermal 1-D Model

We consider perturbations by a Fourier mode around a constant flow, that is we write

$$\rho = \rho_0 + A e^{i(\lambda t + \kappa x)}$$

$$u = u_0 + B e^{i(\lambda t + \kappa x)} \tag{111}$$

Time Step Expansions and the Invariant Manifold Approach 197

We combine this with a composite coefficient for the first order part

$$v = \epsilon \frac{2 - \omega}{2\omega} \tag{112}$$

Substituting these into the macroscopic equations and with some rearrangement for the u term we have

$$A\lambda = -\rho_0 B\kappa - u_0 A\kappa$$
$$B\lambda = -\frac{1}{3\rho_0} A\kappa - u_0 B\kappa - v \left(\frac{u_0^3}{\rho_0} Ai\kappa^2 + \left(3u_0^2 - \frac{2}{3} \right) Bi\kappa^2 \right) \tag{113}$$

We take eigenvalues of the matrix

$$\begin{pmatrix} -u_0\kappa & -\rho_0\kappa \\ -\frac{1}{3\rho_0}\kappa - v\frac{u_0^3}{\rho_0}i\kappa^2 & -u_0\kappa - v\left(3u_0^2 - \frac{2}{3}\right)i\kappa^2 \end{pmatrix} \tag{114}$$

which give us two values for λ

$$\lambda = \kappa \left(-u_0 - \frac{3}{2}vu_0^2 i\kappa + \frac{1}{3}vi\kappa \right.$$
$$\left. \pm \sqrt{vu_0^3 i\kappa - \frac{9}{4}v^2 u_0^4 \kappa^2 + v^2 u_0^2 \kappa^2 - \frac{1}{9}v^2\kappa^2 + \frac{1}{3}} \right) \tag{115}$$

In order for the manifold to remain bounded in time we investigate parameters which give $\Im(\lambda) \geq 0$. We begin by checking asymptotics of two parameters, for large κ we have

$$\lambda = v\kappa^2 \left(-\frac{3}{2}u_0^2 i + \frac{1}{3}i \pm \sqrt{-\left(\frac{3}{2}u_0^2 + \frac{1}{3} \right)} \right)$$
$$= 0, v\kappa^2 \left(-3u_0^2 + \frac{2}{3} \right) i \tag{116}$$

and for large u_0

$$\lambda = \kappa \left(-\frac{3}{2}vu_0^2 i\kappa \pm \sqrt{-\frac{9}{4}v^2 u_0^4 \kappa^2} \right)$$
$$= 0, -3vu_0^2 i\kappa^2 \tag{117}$$

We can see from this that for non-zero κ the first condition that should be satisfied for stability is $u_0^2 < 2/9$, for large u_0 it is necessary for κ to equal 0. Additionally, stability is absolutely contingent on the composite coefficient v being positive, this is the dual condition that time steps are positive and that relaxation parameter of the

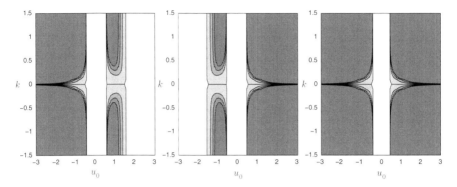

Fig. 2 The first two figures show stability for each of the two eigenvalues in the D1Q3 system with $v = 1$, the third figure plots the minimum of the two. Contours are plotted at $\Im(\lambda) = (-0.3, -0.2, -0.1, 0)$, the *white* region and it's boundary indicate the stable region and, the other *colours*, the decay from stability

collision ω is in the interval $0 \leq \omega \leq 2$ (repeated steps of the collision integral in isolation go towards the quasiequilibrium). In the case that either v is negative or that u_0 is outside the given region, the magnitude of the Fourier perturbation will grow exponentially causing a rapid divergence from the constant flow.

We can confirm these results numerically by plotting the contours of the two eigenvalues equal to zero. In fact in Fig. 2 we additionally plot contours below zero to show the decay from stability.

8.2 The Athermal 2-D Model

We extend the stability analysis from the one dimensional case with a perturbation in the additional space direction. The perturbed system is given by,

$$\begin{aligned}
\rho &= \rho_0 + Ae^{i(\lambda t + \kappa_1 x_1 + \kappa_2 x_2)} \\
u_1 &= u_{10} + B_1 e^{i(\lambda t + \kappa_1 x_1 + \kappa_2 x_2)} \\
u_2 &= u_{20} + B_2 e^{i(\lambda t + \kappa_1 x_1 + \kappa_2 x_2)}
\end{aligned} \tag{118}$$

In this case we investigate the short wave asymptotics as $|\kappa_1|, |\kappa_2| \to \infty$. The eigenvalues of the system under such conditions are

$$\begin{aligned}
\lambda_{1,2} &= \left(\frac{1}{3} - \frac{3}{2}u_{10}^2\right) i v \kappa_1^2 + \left(\frac{1}{3} - \frac{3}{2}u_{20}^2\right) i v \kappa_2^2 - 3i v u_{10} u_{20} \kappa_1 \kappa_2 \\
&\pm \sqrt{-\left(\left(\frac{1}{3} - \frac{3}{2}u_{10}^2\right) v \kappa_1^2 + \left(\frac{1}{3} - \frac{3}{2}u_{20}^2\right) v \kappa_2^2 - 3v u_{10} u_{20} \kappa_1 \kappa_2\right)}, \\
\lambda_3 &= \left(\frac{1}{3} - u_{10}^2\right) i v \kappa_1^2 + \left(\frac{1}{3} - u_{20}^2\right) i v \kappa_2^2 - 2i v u_{10} u_{20} \kappa_1 \kappa_2.
\end{aligned} \tag{119}$$

In the 1-D examples all terms were in even powers of κ whereas in this case there are cross terms in the product $\kappa_1\kappa_2$. Because of this it is necessary to consider the different permutations of signs for these terms. Since the condition that the third eigenvalue imposes is weaker than that of the first two, which are equivalent, it is sufficient to find the region of stability using just one of these. Again assuming that the coefficient v is positive, the region is given by parameters satisfying the two conditions.

$$\left(\frac{1}{3} - \frac{3}{2}u_{10}^2\right) + \left(\frac{1}{3} - \frac{3}{2}u_{20}^2\right) - 3u_{10}u_{20} \geq 0$$
$$\left(\frac{1}{3} - \frac{3}{2}u_{10}^2\right) + \left(\frac{1}{3} - \frac{3}{2}u_{20}^2\right) + 3u_{10}u_{20} \geq 0$$
(120)

The plot of the region generated by these inequalities is given in Fig. 6. Similarly to the one dimensional example, in the event that v is negative or the constant flow speed moves outside this region, the magnitude of the Fourier perturbation will increase exponentially in time.

Again for specific parameters the stability can be calculated numerically. In the first case examine the case where $\kappa_2, u_{20} = 0$. Figure 3 shows the stability plot for the three eigenvalues and their minimum, in this case we see that while the eigenvalues are different from their counterparts in the 1-D system, the stability region is exactly the same. In Fig. 4 we vary κ_2 and u_{20} to see what affect this has on the stability region. For a more complete picture we plot u_{10} against κ_2 and again plot the stability region. In Fig. 5 we vary κ_1 and u_{20} across the different plots.

9 Conclusion

In the analysis of the continuous Boltzmann equation, the Chapman–Enskog procedure is known to reproduce the Navier–Stokes equations [2–4]. This is achieved by a perturbation by a small parameter, the Knudsen number. At the zero and first

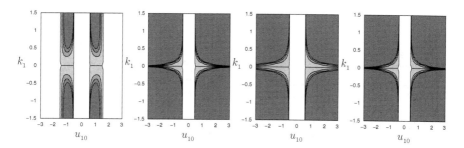

Fig. 3 The first three figures show stability for each of the three eigenvalues in the athermal D2Q9 system with parameters $v = 1, u_2 = 0, \kappa_2 = 0$, the fourth figure plots the minimum of them. In each case the contours are plotted at $\lambda = (-0.3, -0.2, -0.1, 0)$ therefore the *white* region and it's boundary describe the stable area

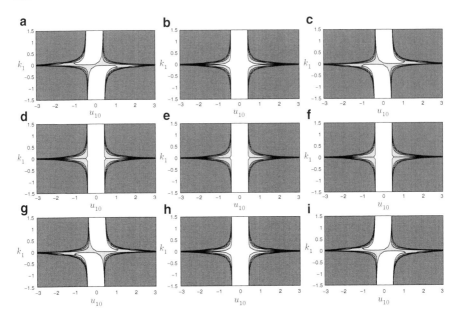

Fig. 4 Stability regions for the athermal D2Q9 system. Contours are plotted at $\Im(\lambda) = (-0.3, -0.2, -0.1, 0)$, the *white* region and it's boundary indicate the stable region and, the other *colours*, the decay from stability. The parameters are $v = 1$ and additionally (**a**) $\kappa_2 = -0.1, u_{20} = -0.5$; (**b**) $\kappa_2 = -0.1, u_{20} = 0$; (**c**) $\kappa_2 = -0.1, u_{20} = 0.5$; (**d**) $\kappa_2 = 0, u_{20} = -0.5$; (**e**) $\kappa_2 = 0, u_{20} = 0$; (**f**) $\kappa_2 = 0, u_{20} = 0.5$; (**g**) $\kappa_2 = 0.1, u_{20} = -0.5$; (**h**) $\kappa_2 = 0.1, u_{20} = 0$; (**i**) $\kappa_2 = 0.1, u_{20} = 0.5$

orders of this parameter, respectively, the convective and diffusive dynamics appear. At higher orders which were not discussed here the Burnett equations arise.

The discrete Boltzmann schemes studied here are defined by the requirement that the Euler equations are recovered at the zero order. In common with the continuous scheme, dissipative terms arise at the first order, however in the discrete case there appear additional viscous terms. In parallel with Goodman and Lax [6] we view the additional dissipative part of the fluid as a direct consequence of the discrete scheme used. In this work we have used the idea of invariant manifolds [7] to calculate the macroscopic dynamics arising from discrete time Boltzmann schemes. This technique is based on an expansion in a different small parameter, the time step ϵ. Dynamics at the zero and first orders again correspond to the conservative and dissipative parts of a fluid respectively. Although in this work we calculate these dynamics up to the first order only, the methodology can be extended to calculate higher order systems.

To compute a solution to the Boltzmann system it is also necessary to discretize velocity space [16]. We have presented two alternative modes of thought to reason why this should be, which produce equivalent systems. We have calculated the exact macroscopic dynamics up to first order of two common discrete velocity schemes, and their continuous counterparts. Although the dynamics of these two schemes

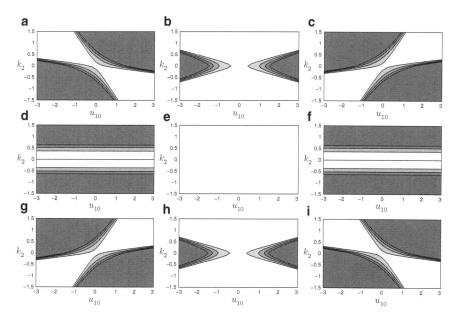

Fig. 5 Stability regions for the athermal D2Q9 system. Contours are plotted at $\Im(\lambda) = (-0.3, -0.2, -0.1, 0)$, the *white* region and it's boundary indicate the stable region and, the other *colours*, the decay from stability. The parameters are $\nu = 1$ and additionally (**a**) $\kappa_1 = -0.1, u_{20} = -0.5$; (**b**) $\kappa_1 = -0.1, u_{20} = 0$; (**c**) $\kappa_1 = -0.1, u_{20} = 0.5$; (**d**) $\kappa_1 = 0, u_{20} = -0.5$; (**e**) $\kappa_1 = 0, u_{20} = 0$; (**f**) $\kappa_1 = 0, u_{20} = 0.5$; (**g**) $\kappa_1 = 0.1, u_{20} = -0.5$; (**h**) $\kappa_1 = 0.1, u_{20} = 0$; (**i**) $\kappa_1 = 0.1, u_{20} = 0.5$

match at the zero order, in the discrete velocity case additional erroneous terms arise at the first order. Such errors might be expected due to the way the quasi-equilibria in the discrete case are defined. If we view the discrete velocity summation as a quadrature approximation to the continuous velocity integral, then we should expect an error of integration. At the zero order we find no such error. This is due to an equilibrium being constructed specifically that the zero order moments are calculated exactly. This equilibrium consists of merely the first three terms of the Taylor expansion of the continuous equilibrium about the zero momentum position. It should, perhaps then, be no surprise that the dissipative dynamics in the discrete system approach those of the continuous system only in the limit of momentum going to zero.

Finally we perform a stability analysis of the linear part of the macroscopic dynamics of the discrete velocity schemes under a short wave perturbation. In common with other authors using similar Fourier techniques [17, 18], and with our own earlier assumptions, we find that two lattice parameters are critical for stability. These are the time step ϵ which must be positive, and the relaxation parameter ω, which must be chosen for non-zero flow speed in the interval $(-1, 1)$. We also analytically and graphically give the permissible range of macroscopic quantities for stability. For the athermal systems study the density ρ can be any value, whereas

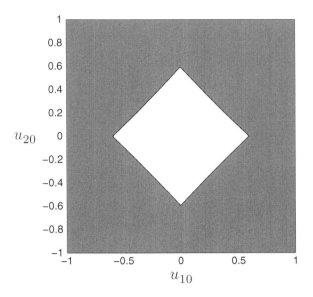

Fig. 6 Stability regions for the athermal D2Q9 system. Contours are plotted at $\Im(\lambda) = (-0.3, -0.2, -0.1, 0)$, the *white* region and it's boundary indicate the stable region and, the other *colours*, the decay from stability. The parameters are $\nu = 1$ and $|\kappa_1|, |\kappa_2| \to \infty$

the momentum u should be within an area centered around the zero point. The exact shape of this region is determined by the choice of velocity discretization.

Appendix: First Order Populations for the D2Q9 Lattice with Standard Polynomial Quasi-Equilibria

$$\frac{\omega}{1-\omega} f^{1\ldots5,(1)} = \frac{4}{9}\left(-u_1\frac{\partial}{\partial x_1}\rho - u_2\frac{\partial}{\partial x_2}\rho + \left(1+\frac{3}{2}u^2\right)\frac{\partial}{\partial x_1}\rho u_1 \right.$$
$$+ \left(1+\frac{3}{2}u^2\right)\frac{\partial}{\partial x_2}\rho u_2 - 3u_1\frac{\partial}{\partial x_1}\rho u_1^2 - 3u_2\frac{\partial}{\partial x_2}\rho u_2^2$$
$$\left. -3u_1\frac{\partial}{\partial x_2}\rho u_1 u_2 - 3u_2\frac{\partial}{\partial x_1}\rho u_1 u_2\right),$$
$$\frac{1}{9}\left(2u_1\frac{\partial}{\partial x_1}\rho - u_2\frac{\partial}{\partial x_2}\rho + \left(-2-3u_1^2+\frac{3}{2}u_2^2\right)\frac{\partial}{\partial x_1}\rho u_1\right.$$
$$+ \left(1-3u_1^2+\frac{3}{2}u_2^2\right)\frac{\partial}{\partial x_2}\rho u_2 + 6u_1\frac{\partial}{\partial x_1}\rho u_1^2$$
$$+ \frac{3}{2}\frac{\partial}{\partial x_1}\rho u_2^2 - 3u_2\frac{\partial}{\partial x_2}\rho u_2^2 - 3u_2\frac{\partial}{\partial x_1}\rho u_1 u_2$$
$$\left. + (3+6u_1)\frac{\partial}{\partial x_2}\rho u_1 u_2\right),$$

$$\frac{1}{9}\left(-u_1\frac{\partial}{\partial x_1}\rho + 2u_2\frac{\partial}{\partial x_2}\rho + \left(1 + \frac{3}{2}u_1^2 - 3u_2^2\right)\frac{\partial}{\partial x_1}\rho u_1\right.$$
$$+ \left(-2 + \frac{3}{2}u_1^2 - 3u_2^2\right)\frac{\partial}{\partial x_2}\rho u_2 + 6u_2\frac{\partial}{\partial x_2}\rho u_2^2$$
$$\frac{3}{2}\frac{\partial}{\partial x_2}\rho u_1^2 - 3u_1\frac{\partial}{\partial x_1}\rho u_1^2 - 3u_1\frac{\partial}{\partial x_2}\rho u_1 u_2$$
$$\left.+ (3 + 6u_2)\frac{\partial}{\partial x_1}\rho u_1 u_2\right),$$
$$\frac{1}{9}\left(2u_1\frac{\partial}{\partial x_1}\rho - u_2\frac{\partial}{\partial x_2}\rho + \left(-2 - 3u_1^2 + \frac{3}{2}u_2^2\right)\frac{\partial}{\partial x_1}\rho u_1\right.$$
$$+ \left(1 - 3u_1^2 + \frac{3}{2}u_2^2\right)\frac{\partial}{\partial x_2}\rho u_2 + 6u_1\frac{\partial}{\partial x_1}\rho u_1^2$$
$$- \frac{3}{2}\frac{\partial}{\partial x_1}\rho u_2^2 - 3u_2\frac{\partial}{\partial x_2}\rho u_2^2 - 3u_2\frac{\partial}{\partial x_1}\rho u_1 u_2$$
$$\left.+ (-3 + 6u_1)\frac{\partial}{\partial x_2}\rho u_1 u_2\right),$$
$$\frac{1}{9}\left(-u_1\frac{\partial}{\partial x_1}\rho + 2u_2\frac{\partial}{\partial x_2}\rho + \left(1 + \frac{3}{2}u_1^2 - 3u_2^2\right)\frac{\partial}{\partial x_1}\rho u_1\right.$$
$$+ \left(-2 + \frac{3}{2}u_1^2 - 3u_2^2\right)\frac{\partial}{\partial x_2}\rho u_2 + 6u_2\frac{\partial}{\partial x_2}\rho u_2^2 \tag{121}$$
$$- \frac{3}{2}\frac{\partial}{\partial x_2}\rho u_1^2 - 3u_1\frac{\partial}{\partial x_1}\rho u_1^2 - 3u_1\frac{\partial}{\partial x_2}\rho u_1 u_2$$
$$\left.+ (-3 + 6u_2)\frac{\partial}{\partial x_1 5}\rho u_1 u_2\right).$$

$$\frac{\omega}{1-\omega}f^{6\ldots 8,(1)} = \frac{1}{36}\left(2u_1\frac{\partial}{\partial x_1}\rho + 2u_2\frac{\partial}{\partial x_2}\rho + 3u_1\frac{\partial}{\partial x_2}\rho + 3u_2\frac{\partial}{\partial x_1}\rho\right.$$
$$+ \left(-2 - 3u_1^2 - 9u_1 u_2 - 3u_2^2\right)\frac{\partial}{\partial x_1}\rho u_1$$
$$+ \left(-2 - 3u_1^2 - 9u_1 u_2 - 3u_2^2\right)\frac{\partial}{\partial x_2}\rho u_2$$
$$- 3\frac{\partial}{\partial x_1}\rho u_2 - 3\frac{\partial}{\partial x_2}\rho u_1 - 3\frac{\partial}{\partial x_1}\rho u_2^2 - 3\frac{\partial}{\partial x_2}\rho u_1^2$$
$$+ (6u_1 + 9u_2)\frac{\partial}{\partial x_1}\rho u_1^2 + (9u_1 + 6u_2)\frac{\partial}{\partial x_2}\rho u_2^2$$
$$+ (-6 + 9u_1 + 6u_2)\frac{\partial}{\partial x_1}\rho u_1 u_2$$
$$\left.+ (-6 + 9u_1 + 6u_2)\frac{\partial}{\partial x_2}\rho u_1 u_2\right),$$

$$\frac{1}{36}\left(2u_1\frac{\partial}{\partial x_1}\rho + 2u_2\frac{\partial}{\partial x_2}\rho - 3u_1\frac{\partial}{\partial x_2}\rho - 3u_2\frac{\partial}{\partial x_1}\rho\right.$$
$$+\left(-2 - 3u_1^2 + 9u_1u_2 - 3u_2^2\right)\frac{\partial}{\partial x_1}\rho u_1$$
$$+\left(-2 - 3u_1^2 + 9u_1u_2 - 3u_2^2\right)\frac{\partial}{\partial x_2}\rho u_2$$
$$+3\frac{\partial}{\partial x_1}\rho u_2 + 3\frac{\partial}{\partial x_2}\rho u_1 + 3\frac{\partial}{\partial x_1}\rho u_2^2 - 3\frac{\partial}{\partial x_2}\rho u_1^2$$
$$+\left(6u_1 - 9u_2\right)\frac{\partial}{\partial x_1}\rho u_1^2 + \left(-9u_1 + 6u_2\right)\frac{\partial}{\partial x_2}\rho u_2^2$$
$$+\left(-6 - 9u_1 + 6u_2\right)\frac{\partial}{\partial x_1}\rho u_1 u_2$$
$$+\left.\left(6 - 9u_1 + 6u_2\right)\frac{\partial}{\partial x_2}\rho u_1 u_2\right),$$

$$\frac{1}{36}\left(2u_1\frac{\partial}{\partial x_1}\rho + 2u_2\frac{\partial}{\partial x_2}\rho + 3u_1\frac{\partial}{\partial x_2}\rho + 3u_2\frac{\partial}{\partial x_1}\rho\right.$$
$$+\left(-2 - 3u_1^2 - 9u_1u_2 - 3u_2^2\right)\frac{\partial}{\partial x_1}\rho u_1$$
$$+\left(-2 - 3u_1^2 - 9u_1u_2 - 3u_2^2\right)\frac{\partial}{\partial x_2}\rho u_2 \qquad (122)$$
$$-3\frac{\partial}{\partial x_1}\rho u_2 - 3\frac{\partial}{\partial x_2}\rho u_1 + 3\frac{\partial}{\partial x_1}\rho u_2^2 + 3\frac{\partial}{\partial x_2}\rho u_1^2$$
$$+\left(6u_1 + 9u_2\right)\frac{\partial}{\partial x_1}\rho u_1^2 + \left(9u_1 + 6u_2\right)\frac{\partial}{\partial x_2}\rho u_2^2$$
$$+\left(6 + 9u_1 + 6u_2\right)\frac{\partial}{\partial x_1}\rho u_1 u_2$$
$$+\left.\left(6 + 9u_1 + 6u_2\right)\frac{\partial}{\partial x_2}\rho u_1 u_2\right),$$

$$\frac{\omega}{1-\omega}f^{9,(1)} = \frac{1}{36}\left(2u_1\frac{\partial}{\partial x_1}\rho + 2u_2\frac{\partial}{\partial x_2}\rho - 3u_1\frac{\partial}{\partial x_2}\rho - 3u_2\frac{\partial}{\partial x_1}\rho\right.$$
$$+\left(-2 - 3u_1^2 + 9u_1u_2 - 3u_2^2\right)\frac{\partial}{\partial x_1}\rho u_1$$
$$+\left(-2 - 3u_1^2 + 9u_1u_2 - 3u_2^2\right)\frac{\partial}{\partial x_2}\rho u_2$$
$$+3\frac{\partial}{\partial x_1}\rho u_2 + 3\frac{\partial}{\partial x_2}\rho u_1 - 3\frac{\partial}{\partial x_1}\rho u_2^2 + 3\frac{\partial}{\partial x_2}\rho u_1^2$$

$$+ (6u_1 - 9u_2) \frac{\partial}{\partial x_1} \rho u_1^2 + (-9u_1 + 6u_2) \frac{\partial}{\partial x_2} \rho u_2^2$$

$$+ (6 - 9u_1 + 6u_2) \frac{\partial}{\partial x_1} \rho u_1 u_2 \qquad (123)$$

$$+ \left. (-6 - 9u_1 + 6u_2) \frac{\partial}{\partial x_2} \rho u_1 u_2 \right)$$

References

1. Chapman, S., Cowling, T. G.: The mathematical theory of non-uniform gases. Cambridge University Press, London (1939)
2. Succi, S.: The lattice Boltzmann equation for fluid dynamics and beyond. Oxford University Press, New York (2001)
3. Karlin, I. V., Ansumali, S., Frouzakus, C. E., Chikatamarla, S. S.: Elements of the lattice Boltzmann method I: Linear advection equation. Commun. Comput. Phys. **1(1)** (2006) 1–45
4. He, X., Luo, L. S.: Lattice Boltzmann model for the incompressible Navier–Stokes equation. J. Stat. Phys. **88(3–4)** (1997) 927–944
5. Dellar, P. J.: Bulk and shear viscosities in lattice Boltzmann equations. Phys. Rev. E **64(3)** (2001) 031203
6. Goodman, J., Lax, P. D.: On dispersive difference schemes I. Commun. Pur. Appl. Math. **151** (1988) 591–613
7. Gorban, A. N., Karlin, I. V.: Invariant Manifolds for Physical and Chemical Kinetics. Springer, Berlin (2005)
8. Gorban, A. N., Karlin, I. V., Ilg, P., Ottinger, H. C.: Corrections and enhancements of quasi-equilibrium states. J. Non-Newtonian Fluid. Mech. **96** (2001) 203–219
9. Karlin, I. V., Tatarinova, L. L., Gorban, A. N., Ottinger, H. C.: Irreversibility in the short memory approximation. Physica A **327** (2003) 399–424
10. Yeomans, J. M.: Mesoscale simulations: Lattice Boltzmann and particle algorithms. Physica A **369** (2006) 159–184
11. He, X., Luo., L. S.: Theory of the lattice Boltzmann method: From the Boltzmann equation to the lattice Boltzmann equation. Phys Rev E **56(6)** (1997) 6811–6817
12. Dellar, P. J.: Non-hydrodynamic modes and general equations of state in lattice. Physica A **362** (2006) 132–238
13. Ehrenfest, P., Ehrenfest, T.: The conceptual foundations of the statistical approach in mechanics. Dover Publications, New York (1990)
14. Bhatnagar, P. L., Gross, E. P., Krook, M.: A model for collision processes in gases. I. Small amplitude processes in charged and neutral one-component systems. Phys. Rev. **94(3)** (1954) 511–525
15. Gorban, A. N.: Basic types of coarse-graining. In: Model Reduction and Coarse-Graining Approaches for Multiscale Phenomena. Gorban, A. N., Kazantzis, N., Kevrekidis, I. G., Öttinger, H.-C., Theodoropoulos, C. (eds.), Springer, Berlin (2006) 117–176; arXiv:cond-mat/0602024v2
16. Ansumali, S., Chikatamarla, S. S., Frouzakis, C. E., Karlin, I. V., Kevrekidis, I. G.: Lattice Boltzmann methods and kinetic theory. In: Model Reduction and Coarse-Graining Approaches for Multiscale Phenomena. Gorban, A. N., Kazantzis, N. , Kevrekidis, I. G., Öttinger, H.-C., Theodoropoulos, C. (eds.), Springer, Berlin (2006) 403–422
17. Sterling, J. D., Chen, S.: Stability analysis of lattice Boltzmann methods. J. Comput. Phys **123** (1996) 196–206
18. Servan-Camas, B., Tsai, F. T. C.: Non-negativity and stability analyses of lattice Boltzmann method for advection-diffusion equation. J. Comput. Phys. **228** (2009) 236–256

The Shock Wave Problem Revisited: The Navier–Stokes Equations and Brenner's Two Velocity Hydrodynamics

Francisco J. Uribe

Abstract The problem of plane shock waves in dilute gases is considered. Its importance relies upon the fact that is an example in which the Navier–Stokes equations have been shown to be susceptible of improvement. Several alternatives to these equations are briefly mentioned in order to give an idea of the importance that the problem of extending the Navier–Stokes equations has these days. We focus our attention on the recent proposal of extension of the Navier–Stokes equations given by Brenner. Our analysis considers the stationary case for the shock wave problem and we compare it with a recent non-stationary analysis.

1 Introduction

The plane shock wave problem provides a tough test for competing hydrodynamic theories that try to improve on the Navier–Stokes description of fluids. Being an essentially non-linear problem, the linearization of the hydrodynamic equations destroys it, although useful information can be obtained by such linearization at each critical (equilibrium) point. There are several proposed theoretical approaches in the literature of dilute gases and it seems that the shock wave problem can be seen as a good test problem for discerning between them. For a list of some theoretical treatments see Table 1. Among them, the Navier–Stokes equations provide a good description for fluids that is worth analyzing so that the different good points they have can perhaps be considered as a must for other hydrodynamics theories that try to improve on them. An important point to notice is that in the problem considered here the boundary conditions at solid boundaries are not needed; this eliminates an additional issue which is hard to deal with, especially in rarefied gases.

F.J. Uribe

Departamento de Física, Universidad Autónoma Metropolitana – Iztapalapa, Av San Rafael Atlixco # 186, cp. 14090, México, Distrito Federal, México
e-mail: paco@xanum.uam.mx

A.N. Gorban and D. Roose (eds.), *Coping with Complexity: Model Reduction and Data Analysis*, Lecture Notes in Computational Science and Engineering 75, DOI 10.1007/978-3-642-14941-2_10, © Springer-Verlag Berlin Heidelberg 2011

208   F.J. Uribe

Table 1 Different theoretical descriptions and computational methods used to do research about the limitations and extensions of Navier–Stokes equations

Computational methods	Moments method	Chapman–Enskog method	Boltzmann equation	Others
Direct simulation Monte Carlo (DSMC) [1,2]	Grad's 13 moments [3–5]	Euler equations [6]	Original [7,8]	Brenner's theory [9–14]
Cellular Automata [15]	Regularized 13 moments [16,17]	Navier–Stokes equations [6,18]	Computational methods [19,20]	Principle of the invariant manifold [21–23]
Lattice Boltzmann gas [24–26]	Weiss' 21 moments [27]	Burnett equations [6,28–35]	Collisional models [36–40]	Holian conjecture [41]
Smooth particles methods [42]	Müller and Ruggeri [43]	Super Burnett equations [30]	Grad's ansatz [44]	Mott–Smith anzats [45]
Molecular dynamics [32,46,47]	Eu's moment method [48,49]	Variants and regularizations [50–56]	Mathematical aspects [57–59]	Maxwellian iteration [57,60–62]

As seen from Table 1 the number of theories that have been proposed to deal with the problem is a big one, and a throughout discussion of all of them will not be attempted here, although we will comment briefly on some of them except for Brenner's theory that is considered in some detail.

Let's start with the conservation equations [6],

$$\frac{\partial \rho}{\partial t} + \nabla \cdot (\rho \mathbf{u}) = 0, \tag{1}$$

$$\rho \left(\frac{\partial \mathbf{u}}{\partial t} + (\mathbf{u} \cdot \nabla)\mathbf{u} \right) = -\nabla \cdot \mathbf{P}, \tag{2}$$

$$\frac{\partial T}{\partial t} + \mathbf{u} \cdot \nabla T = -\frac{2}{3 k_B n} \{\mathbf{P} : \nabla \mathbf{u} + \nabla \cdot \mathbf{q}\}, \tag{3}$$

where t represents time, $\rho(\mathbf{r}, t)$ is the mass density with $\mathbf{r} = (x, y, z)$ representing position, $\mathbf{u}(\mathbf{r}, t)$ is the hydrodynamic velocity, $\mathbf{P}(\mathbf{r}, t)$ is the pressure tensor, $T(\mathbf{r}, t)$ is the temperature, k_B is the Boltzmann constant, $n(\mathbf{r}, t)$ the number density, the symbol ":" represents the full contraction of the tensors, and $\mathbf{q}(\mathbf{r}, t)$ is the heat flux.

It is convenient to express the conservation equations in the so-called conservation form,

$$\frac{\partial \rho}{\partial t} + \nabla \cdot (\rho \mathbf{u}) = 0,$$

The Shock Wave Problem Revisited

$$\frac{\partial \rho \mathbf{u}}{\partial t} + \nabla \cdot (\rho \mathbf{u}\mathbf{u}) + \nabla \cdot \mathbf{P} = 0,$$

$$\frac{\partial}{\partial t}\left(\rho\left(\frac{3k_BT}{2m} + \frac{\|\mathbf{u}\|^2}{2}\right)\right) + \nabla \cdot \left(\rho\left(\frac{3k_BT}{2m} + \frac{\|\mathbf{u}\|^2}{2}\right)\mathbf{u}\right) + \nabla\cdot(\mathbf{P}\cdot\mathbf{u})$$
$$+\nabla\cdot\mathbf{q} = 0. \tag{4}$$

The previous equations represent the point of departure also for the phenomenological approach; they can be applied for liquids or dense gases by changing $\frac{3k_BT}{2m}$ by the specific energy per unit mass (e). In Brenner's two fluid hydrodynamics he proposes that two velocity vectors should be used, one of them is the hydrodynamic velocity $\mathbf{u}$ that appears in the previous equations and will be denoted by $\mathbf{u}_m$, Brenner calls this velocity the barycentric velocity [11], when considering Brenner's formulation. The other will be explained when we consider the constitutive equations in Sect. 2. As a result the equation for energy conservation is modified, see Sect. 3.

For a stationary plane shock wave moving along the x direction $(\mathbf{i})$ all the relevant quantities depend only on x and the hydrodynamic velocity takes the form $\mathbf{u} = u(x)\mathbf{i}$.

The conservation equations can then be integrated to give;

$$\rho(x)u(x) = C_1,$$
$$\rho(x)u^2(x) + \mathbf{P}_{xx} = C_2,$$
$$\left(\frac{3}{2}\frac{k_BT(x)}{m} + \frac{u^2(x)}{2}\right)\rho(x)u(x) + \mathbf{P}_{xx}u(x) + \mathbf{q}_x = C_3, \tag{5}$$

where the C_i are constants,[1] $\mathbf{P}_{xx}$ is the xx component of the pressure tensor, and $\mathbf{q}_x$ is the x component of the heat flux. For a shock wave one is interested in solutions that tend to a definite limit for $x = \pm\infty$. Denoting the corresponding limits for ρ, u, T, and the pressure p by subscripts 0 and 1 (these two limits are also referred to as up-flow and down-flow). Assuming that the dissipative effects are zero for $x \to \pm\infty$ (the heat flux is zero and the pressure tensor contains only the hydrostatic pressure), the conservation equations in integrated form of the conservation equations given by (5) lead to the Rankine–Hugoniot jump conditions:

$$\rho_0 u_0 = \rho_1 u_1,$$
$$\rho_0 u_0^2 + p_0 = \rho_1 u_1^2 + p_1,$$
$$\left(\frac{3}{2}\frac{k_BT_0}{m} + \frac{u_0^2}{2}\right)\rho_0 u_0 + p_0 u_0 = \left(\frac{3}{2}\frac{k_BT_1}{m} + \frac{u_1^2}{2}\right)\rho_1 u_1 + p_1 u_1. \tag{6}$$

---

[1] These constants can be evaluated at either part of the shock that correspond to thermodynamic equilibrium points where there are no gradients. Denoting the equilibrium values at one end of the shock by the subscript 0, we obtain $C_1 = \rho_0 u_0$, $C_2 = \rho_0 u_0^2$, and $C_3 = \left(\frac{3}{2}\frac{k_BT_0}{m}\right)\rho_0 u_0 + p_0$, where p_0 is the pressure.

So far we have only been concerned with the conservation equations that hold in general, the details of solving the shock wave problem depend now on the type of theory used. The Chapman–Enskog method gives expressions for the fluxes $(\mathbf{P}, \mathbf{q})$ in terms of the so-called hydrodynamic variables $(\rho, p, \mathbf{u}, T)$ and their derivatives, by making an expansion in terms of the Knudsen number. The results are the different levels of description given in the third column of Table 1. Moment methods treat the fluxes as additional variables, and the Boltzmann equation is used to derive partial differential equations (PDE) that are satisfied by them; the actual PDE obtained depend on the assumed closure for the distribution function, that is, the number of moments considered, usually the expansion is performed around a local Maxwellian, but in stationary states far from equilibrium other choices can be made [63]. Actually, the Mott–Smith ansatz considers a combination of two Maxwellians corresponding to the values at up-flow and down-flow, and use certain functions of the velocity and the Boltzmann equation to derive differential equations (for what can be called the moments) to solve the shock wave problem. Comparisons with the DSMC method are available as well as a critique of this approach and some moments methods [2, 64]. Several moments methods are mentioned in the second column of Table 1 but the brief description given here is simplified and cannot do full justice to them; the interested reader may take a look at the references that we provide at the end. Computational methods are now ubiquitous in the physical sciences and provide an extremely important tool for validating the theory, especially when relevant information is missing or is partially known, for example: the interaction potential of a specific system. Experiments provide the ultimate way of validating a theory, and for the shock wave problem the interplay of theory, computational methods, and experiments is rather complete. For rarefied gases the DSMC method, as implemented by Bird, seems to be the method of choice, although faster algorithms have been developed in recent years, as the information preservation method [65]. We do not have space to comment, even briefly, on the work summarized in Table 1 but we would like to end this discussion by mentioning the Principle of the Invariant Manifold that has received a considerable amount of attention in recent years. As far as the authors know, the shock wave problem has not been studied using the principle and it seems worth to take a look at this.

There are two main approaches to study hydrodynamics; one is the phenomenological approach and the other one is based on statistical mechanics and in particular in the Boltzmann equation. Actually, the Navier–Stokes equations were first derived using the phenomenological approach which applies for any fluid (not only a dilute gas); actual computations can be done provided that the transport coefficients (shear viscosity and thermal conductivity) are known. The thermodynamic theory that is at the heart of such derivation is called Linear Irreversible Thermodynamics (LIT) [66], it is well accepted[2] but as far as we know a consensus about a thermodynamic theory extending the scope of LIT has not been achieved yet. In contrast,

---

[2] This statement has been challenged, so to speak, by Brenner [13] since in his opinion he is doing LIT. Here we find controversies since as pointed out by him one of the referees suggested the term extended LIT implying that the term may not have a consensus among researches.

the kinetic approach provide expressions for the transport coefficients in terms of the molecular interatomic potential but its applicability is restricted to gases if the Boltzmann equation is used. Modifications to the Navier–Stokes equations based on the kinetic approach have not been entirely successful as one may infer from the several theories proposed; it seems that there is not agreement in which is the "best" theory. Recently the phenomenological two fluid velocity theory proposed by Brenner has received quite a deal of attention [9–14] as well as criticism [67]. We decided to study this theory because we thought that the phenomenological approach may bring new light on the problem of extending the Navier–Stokes equations of hydrodynamics, and secondly because we noticed incorrect statements with respect to one instance of Brenner's theory in the shock wave problem. In particular, Greenshields and Reese [14] did not found solutions for the non stationary shock wave problem in one case, and here we provide evidence of the existence of an stationary solution. Finally, we would like to mention that a two temperature phenomenological theory capable of dealing with shocks has been proposed recently by Hoover and Hoover [68].

The structure of this work is as follows. After the introduction we consider the Navier–Stokes equations, Sect. 3 deals with Brenner's extension to them, and we end the paper with some concluding remarks.

2 The Navier–Stokes Equations

For the Navier–Stokes equations we have the following constitutive equations,[3]

$$\mathbf{P} = p\mathbf{I} - \mu\left(\nabla\mathbf{u} + (\nabla\mathbf{u})^{T} - \frac{2}{3}\nabla\cdot\mathbf{u}\mathbf{I}\right), \quad \mathbf{q} = -k\nabla T, \tag{7}$$

where μ is the shear viscosity, k the thermal conductivity, the superscript T denotes the transpose of the matrix, and $\mathbf{I}$ is the identity matrix. In the case of dilute gases both transport coefficients depend only on the temperature. Brenner argued that the velocity that appears in (7) is different from the one that appears in the conservation equations, he proposed then that in the previous equation one should use $\mathbf{u} \to \mathbf{u}_v$, where $\mathbf{u}_v$ is called volume velocity.[4] One of the main points of Brenner's theory is that the volume velocity is different from the barycentric velocity defined below (4), $\mathbf{u}_m \neq \mathbf{u}_v$.

---

[3] The superscript T that appears in (7) denotes the transpose and should not be confused with the temperature for which we use the symbol T in the same equation.

[4] In his first papers Brenner used the terminology given here [11], in most recent works he changed it and introduced the work velocity [13]. We decided to keep the terminology used in his earlier works.

For the case of a stationary plane shock wave we are led to:

$$\mathbf{P}_{xx} = p - \frac{4}{3}\mu(T(x))\frac{du}{dx}, \quad \mathbf{q}_x = -k(T(x))\frac{dT}{dx}. \tag{8}$$

Substitution of the constitutive relations, given by (8), into the integrated form of the conservation equations for momentum and energy – see (5) – leads to the following two differential equations for u and T,

$$C_1 u(x) + C_1 \frac{k_B T(x)}{mu(x)} - \frac{4}{3}\mu(T(x))\frac{du}{dx} = C_2,$$

$$\left(\frac{3}{2}\frac{k_B T(x)}{m} + \frac{1}{2}u(x)^2\right)C_1 + (C_2 - C_1 u(x))u(x) - k(T(x))\frac{dT}{dx} = C_3, \tag{9}$$

where we used mass conservation to express ρ in terms of u – see (5) – the equation of state $p = nk_B T$, and the relation $\mathbf{P}_{xx} = C_2 - C_1 u(x)$ (that follows from the integrated form of the momentum conservation equation) in the integrated form of the energy conservation equation. Solving (9) for the derivatives gives:

$$\frac{d}{dx}u(x) = \frac{3}{4}\frac{C_1(u(x))^2 m + C_1 k_B T(x) - C_2 mu(x)}{\mu(T(x))mu(x)},$$

$$\frac{d}{dx}T(x) = \frac{1}{2}\frac{3C_1 k_B T(x) - C_1(u(x))^2 m + 2C_2 mu(x) - 2C_3 m}{k(T(x))m}. \tag{10}$$

From the previous equation we obtain the following equation for the orbits expressing T as a function of u:

$$\frac{dT}{du} = \frac{\mu(T)}{k(T)}\frac{2u(T)\left(3C_1 k_B T - C_1(u(T))^2 m + 2C_2 mu(T) - 2C_3 m\right)}{3\left(C_1(u(T))^2 m + C_1 k_B T - C_2 mu(T)\right)}. \tag{11}$$

To first order in the Sonine expansion the thermal conductivity are related by [6]:

$$k = \frac{5}{2}c_v\mu, \tag{12}$$

where c_v is the specific heat of the gas at constant volume, that is:

$$c_v = \frac{3k_B}{2m}, \tag{13}$$

The Shock Wave Problem Revisited

for a monatomic gas. Substitution of (12) into (11) gives:

$$\frac{dT}{du} = \frac{4m}{15k_B} \frac{2u(T)\left(3C_1 k_B T - C_1(u(T))^2 m + 2C_2 mu(T) - 2C_3 m\right)}{3\left(C_1(u(T))^2 m + C_1 k_B T - C_2 mu(T)\right)}.$$

(14)

While (14) contains the interesting result that if the constants $C_i, (i = 1, 2, 3)$ are the same, then the orbits are also the same independently of the form of the interaction potential, provided the mass of the particle is the same; it is convenient to express the equation of the orbits using the following reduced variables,

$$u^\star(s) \equiv \frac{u(x)}{u_0}, \quad \rho^\star(s) \equiv \frac{\rho(x)}{\rho_0}, \quad \tau(s) \equiv \frac{k_B T(x)}{mu_0^2}, \quad P_{xx}^\star(s) \equiv \frac{\mathbf{P}_{xx}(x)}{\rho_0 u_0^2},$$
$$q^\star(s) \equiv \frac{\mathbf{q}_x(x)}{\rho_0 u_0^3},$$

(15)

where $s = s(x)$ is a reduced distance whose definition will be given below, and u_0 is the velocity at up-flow (the part of the shock with lower temperature).

Then, the integrated form of the conservation equations, (5), are given by

$$\rho^\star(s)u^\star(s) = 1,$$
$$u^\star(s) + P_{xx}^\star(s) = 1 + \tau_0,$$
$$\frac{3}{2}\tau(s) + \frac{u^\star(s)^2}{2} + P_{xx}^\star(s)u^\star(s) + q^\star(s) = \frac{5}{2}\tau_0 + \frac{1}{2},$$

(16)

where $\tau_0 \equiv \dfrac{k_B T_0}{mu_0^2}$ is related to the Mach number (M), defined as $M \equiv u_0/u_S$ where u_S is the sound velocity at up-flow, by

$$M = \sqrt{\frac{3}{5\tau_0}}.$$

(17)

The reduced form of the constitutive relations for the Navier–Stokes equations, (8), can be written as,

$$P_{xx}^\star(s) = \tau(s)/u^\star(s) - \frac{4}{3}\mu^\star(\tau(s))\frac{du^\star}{ds}\frac{\mu_0}{\rho_0 u_0}\frac{ds}{dx},$$
$$q^\star(s) = -k^\star(\tau(s))\frac{d\tau}{ds}\frac{mk_0}{\rho_0 k_B u_0}\frac{ds}{dx},$$

(18)

where the shear viscosity and the thermal conductivity are reduced by their values at up-flow (μ_0 and k_0, respectively). Substitution of the reduced constitutive equations

214   F.J. Uribe

into the reduced integrated form of the conservation of momentum and energy – see (16) – leads finally to:

$$\frac{du^\star}{ds} = -\frac{3}{4} \frac{\rho_0 u_0 \left(-u^\star(s)^2 - \tau(s) + u^\star(s) + \tau_0 u^\star(s)\right)}{\mu_0 \mu^\star(\tau(s)) \dfrac{ds}{dx} u^\star(s)},$$

$$\frac{d\tau}{ds} = \frac{1}{2} \frac{\rho_0 u_0 k_B \left(3\tau(s) - (u^\star(s))^2 + 2u^\star(s) + 2u^\star(s)\tau_0 - 5\tau_0 - 1\right)}{m k_0 k^\star(\tau(s)) \dfrac{ds}{dx}}$$

$$= \frac{2}{15} \frac{\rho_0 u_0 \left(3\tau(s) - (u^\star(s))^2 + 2u^\star(s) + 2u^\star(s)\tau_0 - 5\tau_0 - 1\right)}{\mu_0 \mu^\star(\tau(s)) \dfrac{ds}{dx}},$$

$$\tag{19}$$

where the last relation follows from (12). Regarding τ as a function $u^\star$ we obtain that the equation for the orbits take the form:

$$\frac{d\tau}{du^\star} = -\frac{8}{45} \frac{\left(3\tau - u^{\star 2} + 2u^\star + 2u^\star \tau_0 - 5\tau_0 - 1\right) u^\star}{\left(-u^{\star 2} - \tau + u^\star + \tau_0 u^\star\right)}. \tag{20}$$

Equation (20) shows that for a given Mach number the orbits in reduced variables are locally the same for any functional dependence of the transport coefficients on the temperature and on the specific reduced distance chosen. We think this is an interesting result. It was suggested by previous computations in which we noted that the orbits predicted by using the Holian conjecture [41] were the same as the Navier–Stokes equations [34] for the rigid sphere model; the equivalence for the orbits was later understood on an analytic basis [69]. The result given here follows the steps from that work [69]. In fact, the equation for the "universal" orbit given by (20) can be found in the work by Gilbarg and Paolucci [18] (in terms of the reduced variables used by them). Apparently the result has been overlooked in the recent literature, but with the DSMC data available now it acquires a new significance, as we will see in the final remarks.

For the soft sphere model; $\mu = \tau^\Gamma$, with $\Gamma \in [0.5, 1.0]$, and taking $\dfrac{ds}{dx} = (3/4)(\rho_0 u_0)/\mu_0$ (19) take the form:

$$\frac{du^\star}{ds} = -\frac{\left(-u^\star(s)^2 - \tau(s) + u^\star(s) + \tau_0 u^\star(s)\right)}{\tau(s)^\Gamma u^\star(s)},$$

$$\frac{d\tau}{ds} = \frac{8}{45} \frac{\left(3\tau(s) - (u^\star(s))^2 + 2u^\star(s) + 2u^\star(s)\tau_0 - 5\tau_0 - 1\right)}{\tau^\star(s)^\Gamma}, \tag{21}$$

The Shock Wave Problem Revisited   215

that for the rigid sphere model ($\Gamma = 1/2$) reduced to the ones used by Holian et al. [41].

We end this section by giving the Rankine–Hugoniot jump conditions in the reduced variables given by (15): at up-flow the reduced velocity is equal to one ($u_0^\star = 1$) and the reduced temperature is given by τ_0, the corresponding reduced variables at down-flow (the equilibrium point with higher temperature) are then given by

$$u_1^\star = \frac{5}{4}\tau_0 + \frac{1}{4}, \quad \tau_1 = \frac{7}{8}\tau_0 + \frac{3}{16} - \frac{5}{16}\tau_0^2. \tag{22}$$

3 Brenner's Theory

The two velocity hydrodynamic theory developed by Brenner is certainly controversial, this makes the subject very difficult to describe properly without extensive discussions. Therefore, we will concentrate on its predictions for the shock wave problem and in particular with the problem as considered by Greenshields and Reese [14]. For the reader interested in pursuing the theory in more detail the following references provide a good set for the extensive literature available [9–14, 67]. As mentioned before there are two velocities; the barycentric one denoted by $\mathbf{u}_m$ which is the one that appears in the conservation equations, see (23) and (24) below. The only difference with respect to the standard conservation equations being that in the equation for energy conservation instead of the heat flux one has to replace it by the diffusive flux density ($\mathbf{j}_e$), this vector includes Fourier's law and an additional term (see (31) coming from the difference between $\mathbf{u}_m$ and $\mathbf{u}_v$), the last velocity is the one second velocity introduced by Brenner and appears in the constitutive equations as was mentioned before. The diffusive volume flux density; $\mathbf{j}_v \equiv \mathbf{u}_v - \mathbf{u}_m a$, is the relevant vector to consider and a constitutive relation for it will be used, see (27). Alternatively, for gases, one may in principle try to use information coming from solutions of the Boltzmann equation to derive, if possible, the corresponding constitutive relation in much as the same way as is done with the Navier–Stokes–Fourier constitutive relations. These are the main ingredients necessary to get a well posed problem for the shock wave problem for the two velocity hydrodynamics. We now consider the details of the formulation and refer the interested reader to the original sources [9–14] for more information in case he needs it.

We follow the description given by Greenshields and Reese [14] with respect to Brenner's formulation, and in particular their notation as far as possible. The conservation equations – (1)–(3) – take in the formulation by Greenshields and Reese the form,[5]

---

[5] The relations $\mathbf{u} \to \mathbf{u}_m$, $\mathbf{P} \to \mathbf{P}$ and $\mathbf{q} \to \mathbf{j}_e$ are used. In addition Greenshields and Reese used the momentum density defined by $\mathbf{m} \equiv \rho \mathbf{u}_m$ and their expression for momentum conservation is written in terms of it.

$$\frac{\partial \rho}{\partial t} + \nabla \cdot (\rho \mathbf{u}_m) = 0, \qquad (23)$$

$$\frac{\partial \rho \mathbf{u}_m}{\partial t} + \nabla \cdot (\rho \mathbf{u}_m \mathbf{u}_m) + \nabla \cdot \mathbf{P} = 0$$

$$\frac{\partial E}{\partial t} + \nabla \cdot \left(\mathbf{u}_m E \right) + \nabla \cdot \mathbf{j}_e + \nabla \cdot \left(\mathbf{P} \cdot \mathbf{u}_m \right) = 0. \qquad (24)$$

where $E = \rho(e + \|\mathbf{u}_m\|^2)$ is the total energy density, $e = \dfrac{3 k_B T}{2m}$ the specific internal energy (for a monatomic gas), and $\mathbf{j}_e$ is called by Greenshields and Reese the diffusive flux density of internal energy. The pressure tensor $\mathbf{P}$ is equal to $p\mathbf{I} + \mathbf{T}$ with p the pressure, $\mathbf{I}$ the 3×3 identity matrix, and $\mathbf{T} = \mathbf{P} - p\mathbf{I}$ the viscous stress tensor.

The form of the conservation equations given by (23) and (24) is rather convenient since for a stationary plane shock wave ($\mathbf{u}_m(\mathbf{r}, t) = u(x)\mathbf{i}$) they can easily be integrated to give:

$$\rho(x)u(x) = C_1,$$

$$\rho(x)u^2(x) + \mathbf{P}_{xx} = C_2,$$

$$\left(\frac{3}{2} \frac{k_B T(x)}{m} + \frac{u^2(x)}{2} \right) \rho(x)u(x) + \mathbf{P}_{xx}u(x) + \mathbf{j}_{ex} = C_3. \qquad (25)$$

Notice that up to now, the only difference between the previous integrated form of the conservation equations – see (5) – is the term $\mathbf{j}_{ex}$ that appears in the energy equation instead of $\mathbf{q}_x$, this is one of the differences in Brenner's formulation. The explicit form of the pressure tensor in Brenner's formulation comes from an argument by him in the sense that the velocity ($\mathbf{u}_v$) that appears in the constitutive relation for it should be different from the one appearing in the conservation equations,[6] the two velocities are related by

$$\mathbf{u}_v = \mathbf{u}_m + \mathbf{j}_v, \qquad (26)$$

$\mathbf{j}_v$ is called the diffusive volume flux density, and the following relation has been used:

$$\mathbf{j}_v = \alpha_v \frac{1}{\rho} \nabla \rho, \qquad (27)$$

with $\mathbf{j}_v$ the so-called volume diffusivity and the exact form α_v has been referred by Greenshields and Reeds as an open problem. A proposal by Brenner is that it should be the same as the volume diffusivity $\alpha = k/\rho c_p$ with c_p the specific heat at constant pressure, this identification ($\alpha_v = \alpha$) will be used later on. Using (26)

---

[6] This idea is new and critiques can of course be found as well as arguments supporting it; our purpose here is not to give a detailed argumentation about its validity but to see the consequences.

The Shock Wave Problem Revisited

and (27) we have that:

$$\mathbf{u}_v = \mathbf{u}_m + \frac{\alpha_v}{\rho}\nabla\rho. \tag{28}$$

The idea is now to take for granted the form of the pressure tensor given in the Navier–Stokes regime – see (7) – with $\mathbf{u}$ equal to $\mathbf{u}_v$. Using (28) then it leads to the following constitutive relation for the pressure tensor:

$$\mathbf{P} = p\mathbf{I} - \mu\left(\nabla\mathbf{u}_m + (\nabla\mathbf{u}_m)^T - \frac{2}{3}\nabla\cdot\mathbf{u}_m\mathbf{I}\right) - 2\mu\overline{\nabla\left\{\frac{\alpha_v}{\rho}\nabla\rho\right\}}, \tag{29}$$

where the overline with the circle is an standard notation representing the corresponding symmetric traceless tensor, that is:

$$\overline{\nabla\left\{\frac{\alpha_v}{\rho}\nabla\rho\right\}} = \frac{1}{2}\nabla\left\{\frac{\alpha_v}{\rho}\nabla\rho\right\} + \frac{1}{2}\left(\nabla\left\{\frac{\alpha_v}{\rho}\nabla\rho\right\}\right)^T - \frac{\mathbf{I}}{3}\nabla\cdot\left\{\frac{\alpha_v}{\rho}\nabla\rho\right\}. \tag{30}$$

The diffusive flux density of internal energy is given by,

$$\mathbf{j}_e = -k\nabla T - \alpha_v\frac{p}{\rho}\nabla\rho. \tag{31}$$

For a plane shock wave we have from (30) and (31) that,

$$\mathbf{P}_{xx} = p - \frac{4}{3}\mu\frac{du}{dx} - \frac{4}{3}\mu\frac{d}{dx}\left(\frac{\alpha_v}{\rho}\frac{d\rho}{dx}\right), \quad \mathbf{j}_{ex} = -k\frac{dT}{dx} - \alpha_v\frac{p}{\rho}\frac{d\rho}{dx}. \tag{32}$$

Substitution of (32) into (25) leads to,

$$\rho(x)u(x) = C_1,$$

$$\rho(x)u^2(x) + p - \frac{4}{3}\mu\frac{du}{dx} - \frac{4}{3}\mu\frac{d}{dx}\left(\frac{\alpha_v}{\rho}\frac{d\rho}{dx}\right) = C_2,$$

$$\left(\frac{3}{2}\frac{k_BT(x)}{m} + \frac{u^2(x)}{2}\right)\rho(x)u(x) + \left\{p - \frac{4}{3}\mu\frac{du}{dx} - \frac{4}{3}\mu\frac{d}{dx}\left(\frac{\alpha_v}{\rho}\frac{d\rho}{dx}\right)\right\}u(x)$$

$$-k\frac{dT}{dx} - \alpha_v\frac{p}{\rho}\frac{d\rho}{dx} = C_3. \tag{33}$$

This set of three equations is the one that we would like to solve; notice that the first equation can be used to express u in terms of ρ, for the shear viscosity and the thermal conductivity we will use the soft sphere model in which both transport coefficients are proportional to a power of the temperature, the pressure is given by the ideal gas equation of state, and we will use $\alpha = \alpha_v$. Then, we are led to two differential equations for T and ρ, there are first and second derivatives for ρ and a first derivative for T, then the system can be considered as a first order system in three dimensions as we will now see.

We now follow the same procedure as we did for the Navier–Stokes equations except that in order to simplify the calculations we assume an specific value of the distance by which we reduce the position, and also a specific temperature dependence of the transport coefficients. In the first instance we use

$$s = x \frac{3\rho_0 u_0^2}{4\mu_0},$$
(34)

where μ_0 is the shear viscosity at up-flow. This choice is not the same as the one used by Greenshields and Reese [14] for their reduced distance but, as was mentioned before, it gives the same reduced Navier–Stokes equations as those given by Holian et al. [41] for the rigid sphere case. Secondly, we use the soft sphere model – the interaction potential is proportional to an inverse power of the interatomic distance –, in which the shear viscosity is proportional to a power of the temperature and also use the first Sonine approximation to the transport coefficients – see (12) – so that the following equations are valid:

$$\mu_r \left(\tau \left(s \left(x \right) \right) \right) = \left(\tau \left(s \left(x \right) \right) \right)^{\Gamma}, \quad k_r \left(\tau \left(s \left(x \right) \right) \right) = \frac{15}{4} \frac{k_B \mu_0 \mu_r \left(\tau \left(s \left(x \right) \right) \right)}{k_0 m},$$

$$D \left(\mu_r \right) \left(\tau \left(s \left(x \right) \right) \right) = \frac{\left(\tau \left(s \left(x \right) \right) \right)^{\Gamma} \Gamma}{\tau \left(s \left(x \right) \right)},$$

$$D \left(k_r \right) \left(\tau \left(s \left(x \right) \right) \right) = \frac{15}{4} \frac{k_B \mu_0 D \left(\mu_r \right) \left(\tau \left(s \left(x \right) \right) \right)}{k_0 m},$$
(35)

where D means the derivative of the corresponding function.

Proceeding in much the same way as in the case of the Navier–Stokes equations we find that the integrated form of the momentum equation can be written in the form:[7]

$$\left(\rho_r \left(s \left(x \right) \right) \right)^{-1} + \rho_r \left(s \left(x \right) \right) \tau \left(s \left(x \right) \right) + \frac{\left(\tau \left(s \left(x \right) \right) \right)^{\Gamma} D \left(\rho_r \right) \left(s \left(x \right) \right)}{\left(\rho_r \left(s \left(x \right) \right) \right)^2}$$

$$- \frac{9}{8} \frac{\left(\left(\tau \left(s \left(x \right) \right) \right)^{\Gamma} \right)^2 \Gamma D \left(\tau \right) \left(s \left(x \right) \right) D \left(\rho_r \right) \left(s \left(x \right) \right)}{\tau \left(s \left(x \right) \right) \left(\rho_r \left(s \left(x \right) \right) \right)^2}$$

$$+ \frac{9}{4} \frac{\left(\left(\tau \left(s \left(x \right) \right) \right)^{\Gamma} \right)^2 \left(D \left(\rho_r \right) \left(s \left(x \right) \right) \right)^2}{\left(\rho_r \left(s \left(x \right) \right) \right)^3} - \frac{9}{8} \frac{\left(\left(\tau \left(s \left(x \right) \right) \right)^{\Gamma} \right)^2 \left(D^{(2)} \right) \left(\rho_r \right) \left(s \left(x \right) \right)}{\left(\rho_r \left(s \left(x \right) \right) \right)^2}$$

$$= 1 + \tau_0.$$
(36)

---

[7] While the procedure follows almost the same steps as the algebra for the Navier–Stokes equations, it is more demanding but still can be carried out by hand with patience. However, we found more convenient to do the algebra using the computer in order to avoid as much errors as possible. We can provide the interested readers, upon request, a Maple worksheet showing the calculations of Brenner's theory as well as some computations of the Navier–Stokes equations.

The Shock Wave Problem Revisited

and for the integrated form of the energy equation we find that

$$
\begin{aligned}
&\frac{5}{2}\tau\left(s\left(x\right)\right)+\frac{1}{2}\left(\rho_r\left(s\left(x\right)\right)\right)^{-2}+\frac{\left(\tau\left(s\left(x\right)\right)\right)^{\Gamma}D\left(\rho_r\right)\left(s\left(x\right)\right)}{\left(\rho_r\left(s\left(x\right)\right)\right)^3} \\
&-\frac{9}{8}\frac{\left(\left(\tau\left(s\left(x\right)\right)\right)^{\Gamma}\right)^2\Gamma D\left(\tau\right)\left(s\left(x\right)\right)D\left(\rho_r\right)\left(s\left(x\right)\right)}{\left(\rho_r\left(s\left(x\right)\right)\right)^3\tau\left(s\left(x\right)\right)} \\
&+\frac{9}{4}\frac{\left(\left(\tau\left(s\left(x\right)\right)\right)^{\Gamma}\right)^2\left(D\left(\rho_r\right)\left(s\left(x\right)\right)\right)^2}{\left(\rho_r\left(s\left(x\right)\right)\right)^4}-\frac{9}{8}\frac{\left(\left(\tau\left(s\left(x\right)\right)\right)^{\Gamma}\right)^2\left(D^{(2)}\right)\left(\rho_r\right)\left(s\left(x\right)\right)}{\left(\rho_r\left(s\left(x\right)\right)\right)^3} \\
&-\frac{45}{16}\left(\tau\left(s\left(x\right)\right)\right)^{\Gamma}D\left(\tau\right)\left(s\left(x\right)\right)-\frac{9}{8}\frac{\left(\tau\left(s\left(x\right)\right)\right)^{\Gamma}\tau\left(s\left(x\right)\right)D\left(\rho_r\right)\left(s\left(x\right)\right)}{\rho_r\left(s\left(x\right)\right)} \\
&=\frac{5}{2}\tau_0+1/2.
\end{aligned}
\tag{37}
$$

As mentioned before our strategy is to regard the previous differential system as a first order one in three dimensions, this can be achieved by using the following definitions:

$$
y_1\left(\xi\right)\equiv\rho_r\left(s\left(x\right)\right),\quad y_2\left(\xi\right)\equiv D\left(\rho_r\right)\left(s\left(x\right)\right)=D\left(y_1\right)\left(\xi\right),\quad y_3\left(\xi\right)\equiv\tau\left(s\left(x\right)\right),
\tag{38}
$$

so that

$$
D^{(2)}\left(\rho_r\right)\left(s\left(x\right)\right)=D\left(y_2\right)\left(\xi\right),\quad D\left(\tau\right)\left(s\left(x\right)\right)=D\left(y_3\right)\left(\xi\right).
\tag{39}
$$

Substitution of (38) and (39) into (36) and (37) leads to

$$
\begin{aligned}
&\left(y_1\left(\xi\right)\right)^{-1}+y_1\left(\xi\right)y_3\left(\xi\right)+\frac{\left(y_3\left(\xi\right)\right)^{\Gamma}y_2\left(\xi\right)}{\left(y_1\left(\xi\right)\right)^2}+\frac{9}{4}\frac{\left(\left(y_3\left(\xi\right)\right)^{\Gamma}\right)^2\left(y_2\left(\xi\right)\right)^2}{\left(y_1\left(\xi\right)\right)^3} \\
&-\frac{9}{8}\frac{\left(\left(y_3\left(\xi\right)\right)^{\Gamma}\right)^2\Gamma D\left(y_3\right)\left(\xi\right)y_2\left(\xi\right)}{y_3\left(\xi\right)\left(y_1\left(\xi\right)\right)^2}-\frac{9}{8}\frac{\left(\left(y_3\left(\xi\right)\right)^{\Gamma}\right)^2 D\left(y_2\right)\left(\xi\right)}{\left(y_1\left(\xi\right)\right)^2}=1+\tau_0,
\end{aligned}
$$

$$
\begin{aligned}
&\frac{5}{2}y_3\left(\xi\right)+\frac{1}{2}\left(y_1\left(\xi\right)\right)^{-2}+\frac{\left(y_3\left(\xi\right)\right)^{\Gamma}y_2\left(\xi\right)}{\left(y_1\left(\xi\right)\right)^3}+\frac{9}{4}\frac{\left(\left(y_3\left(\xi\right)\right)^{\Gamma}\right)^2\left(y_2\left(\xi\right)\right)^2}{\left(y_1\left(\xi\right)\right)^4} \\
&-\frac{9}{8}\frac{\left(\left(y_3\left(\xi\right)\right)^{\Gamma}\right)^2\left(\Gamma\right)D\left(y_3\right)\left(\xi\right)y_2\left(\xi\right)}{\left(y_1\left(\xi\right)\right)^3 y_3\left(\xi\right)}-\frac{9}{8}\frac{\left(\left(y_3\left(\xi\right)\right)^{\Gamma}\right)^2 D\left(y_2\right)\left(\xi\right)}{\left(y_1\left(\xi\right)\right)^3} \\
&-\frac{45}{16}\left(y_3\left(\xi\right)\right)^{\Gamma}D\left(y_3\right)\left(\xi\right)-\frac{9}{8}\frac{\left(y_3\left(\xi\right)\right)^{\Gamma}y_3\left(\xi\right)y_2\left(\xi\right)}{y_1\left(\xi\right)}=5/2\tau_0+1/2.
\end{aligned}
\tag{40}
$$

Finally, solving (40) for the derivatives we obtain the following first order system in three dimensions:

$$D(y_1)(\xi) = y_2(\xi) \equiv \mathbf{F}_1(y_1(\xi), y_2(\xi), y_3(\xi)),$$
$$D(y_2)(\xi) = \frac{2}{45}\frac{p_1(y_1(\xi), y_2(\xi), y_3(\xi))}{y_3(\xi)(y_1(\xi))^2} \equiv \mathbf{F}_2(y_1(\xi), y_2(\xi), y_3(\xi)),$$
$$D(y_3)(\xi) = -\frac{2}{45}\frac{p_2(y_1(\xi), y_2(\xi), y_3(\xi))}{(y_1(\xi))^2} \equiv \mathbf{F}_3(y_1(\xi), y_2(\xi), y_3(\xi)),$$
$$(41)$$

where

$$p_1(y_1, y_2, y_3) = 20(y_3)^{-2\Gamma+1}(y_1)^3 + 20(y_3)^{-2\Gamma+2}(y_1)^5$$
$$+20(y_3)^{-\Gamma+1}y_2(y_1)^2 - 12\Gamma(y_3)^{-\Gamma+1}y_2(y_1)^2 + 9\Gamma(y_2)^2 y_1 y_3$$
$$+20\Gamma(y_3)^{-\Gamma}y_2(y_1)^2\tau_0 + 4\Gamma(y_3)^{-\Gamma}y_2 + 4\Gamma(y_3)^{-\Gamma}y_2(y_1)^2$$
$$-8\Gamma(y_3)^{-\Gamma}y_2 y_1\tau_0 - 8\Gamma(y_3)^{-\Gamma}y_2 y_1 + 45(y_2)^2 y_3 y_1$$
$$-20(y_3(\xi))^{-2\Gamma+1}(y_1)^4 - 20(y_3)^{-2\Gamma+1}\tau_0(y_1)^4,$$
$$(42)$$

and

$$p_2(y_1, y_2, y_3) = -12y_3^{-\Gamma+1}y_1^2 + 9y_2 y_1 y_3 + 20y_3^{-\Gamma}y_1^2\tau_0 + 4y_3^{-\Gamma}$$
$$+4y_3^{-\Gamma}y_1^2 - 8y_3^{-\Gamma}y_1\tau_0 - 8y_3^{-\Gamma}y_1.$$
$$(43)$$

We now have to analyse the system:

$$\frac{d\mathbf{y}}{d\xi} = \mathbf{F}(\mathbf{y}(\xi), \tau_0),$$
$$(44)$$

where $\mathbf{y} = (y_1, y_2, y_3)$ and the components of the vector field $\mathbf{F}$ are given by (41). The shock wave problem consists in showing the existence of a solution orbit of (44) that joins the two equilibrium points given by the Rankine–Hugoniot jump conditions, with the present election of the variables, they correspond to the points:

$$(1, 0, \tau_0) \equiv \mathbf{y}^{up},$$
$$(\frac{4}{5\tau_0 + 1}, 0, -\frac{5}{16}\tau_0 + \frac{7}{8}\tau_0 + \frac{3}{16}) \equiv \mathbf{y}^{down},$$
$$(45)$$

that characterize up-flow and down-flow respectively. The question of structure for the shock problem is a difficult one and first we will analyze locally the dynamical system given by (44).

3.1 Local Analysis and Fortran Numerical Computations

The critical points (also called stationary or fixed points) of a vector field are the solutions of $\mathbf{F}(\mathbf{y}) = 0$; in our case the stationary points are the points $\mathbf{y}^{up}$ and $\mathbf{y}^{down}$ given above – see (45) –. The second goal is to see if they are hyperbolic,[8] as indeed they are. The algebra and computations are given in a Maple worksheet (available upon request), we thought it was not worth to describe the somewhat involved computations that the computer can handle easily. We did calculations for three interaction models: one is the rigid sphere case for which $\Gamma = 1/2$; another is the Maxwell model for which $\Gamma = 1$, and the other model corresponds to $\Gamma = 3/4$. We expect that the nature of the critical points to be the same for $\Gamma \in [1/2, 1]$. Up-flow is an unstable node (the dimension of its unstable manifold is three) and down-flow is a saddle (the dimension of the stable manifold is one and that of the unstable manifold is two). Then, we have the conditions to expect a "stable" heteroclinic orbit since, in order to have the existence, the stable manifold of one critical point must intersect the stable manifold of the other, and in order to survive under small perturbations (change for example τ_0 or Γ) the dimension of the stable manifold plus the dimension of the unstable manifold should be greater than the dimension of the space in which they are embedded (three in our case).

The ideas put forward by Gilbarg and Paolucci [18] for the Navier–Stokes equations to obtain an approximation to the heteroclinic trajectory[9] can then be used. The main point is to integrate the orbits in the negative sense (ξ is decreased), then up-flow becomes a stable node; starting at down-flow and integrating backwards there is a good chance that the orbit hits the stable manifold. Equivalently this amounts to consider the vector field $-\mathbf{F}$ instead of $\mathbf{F}$. We decided to try this using Adam's method to obtain the solution of the dynamical system numerically,[10] and the results are now described.

As was mentioned above, we start near down-flow and integrate in the negative sense, several computations were performed at different Mach numbers and different models, but we will describe the details only for $M = 2$. The actual initial values used for $M = 2$ ($\tau_0 = 0.15$) are,

$$s_i = 10.0, \quad y_1 = 2.2857142857142856, \quad y_2 = 0.0, \quad y_3 = 0.31171875,$$

for the values $\Gamma = 1$ (Maxwell model), $\Gamma = 1/2$ (rigid sphere), and $\Gamma = 3/4$. We also performed computations for the Navier–Stokes equations with practically the same initial conditions. The density profiles for Brenner's theory – (41) – and the

---

[8] A critical point is called hyperbolic if the eigenvalues of the jacobian matrix evaluated at the critical point have real parts different from zero, the interest in them comes from the fact that the Hartman–Grobman theorem [70] assures that the orbits of the nonlinear systems are homeomorphic to the orbits of the linearization of the vector field at the critical point.

[9] According to Lumpkin and Chapman [71] this idea can be traced back to von Mises.

[10] We used the Numerical Algorithms Group fortran library, generous information can be found at www.nag.co.uk. The code can be generated using Maple if needed.

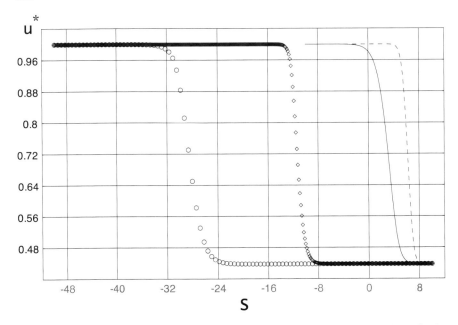

Fig. 1 Velocity profiles for $M = 2$. *Dashed line*; Navier–Stokes equations – see (21) – for the Maxwell model ($\Gamma = 1/2$), *solid line*; Navier–Stokes equations – see (21) – for the rigid sphere model ($\Gamma = 1$), *open circles*; Brenner's theory – see (41) – for the rigid sphere model, *open diamonds*; Brenner's theory – see (41) – for the Maxwell model

Navier–Stokes equations – (20) and (21) – are given in Fig. 1. All the integrations were done starting at $s_i = 10.0$ and ending at $s_f = -10.0$ for the Navier–Stokes equations, and $s_f = -50.0$ for Brenner's theory. The two theories and the models reported (rigid spheres and Maxwellian molecules) give what is expected for a shock.

In Fig. 2 the curves in the u^*–τ plane are given for $M = 2$. We provide results for the Navier–Stokes and Brenner's theory for the models mentioned above; also shown are the computations for Burnett equations and the DSMC method for rigid spheres are also shown. As can be seen, Brenner's theory results are above the Navier–Stokes but for the value of the Mach number considered the results are not as good as the Burnett equations when compared with the DSMC method.

The results for a mach number of value 134 are given in Fig. 3; even in this case we found evidence for the existence of an heteroclinic orbit. Another important feature is that near up-flow the projection of Brenner's solution shows a peculiar behavior; actually, looking in detail at the orbit it has regions in which for a given value of u^* there are at least three vales of τ; this is shown in Fig. 4 (but recall that Brenner's dynamical system is in three dimensions).[11] Our explanation of this

---

[11] A 3-D color figure of Brenner's orbits for $M = 2$ is available upon request.

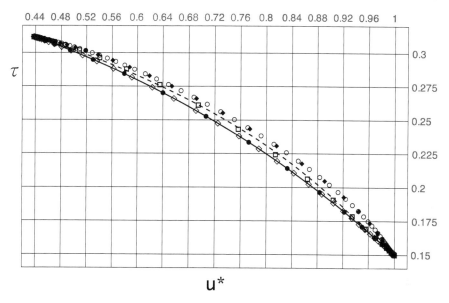

Fig. 2 Orbits or their projections in the τ–u^* plane for $M = 2$. *Solid line*; Navier–Stokes calculated with (20), *open circles*; Brenner's theory, see (41), for the rigid sphere model ($\Gamma = 1/2$), *solid diamonds*; Brenner's theory, see (41), for the Maxwell model ($\Gamma = 1$), *open diamonds*; Navier–Stokes equations, see (21), for the rigid sphere model, *solid circles*; Navier–Stokes equations, see (21), for the Maxwell model, *dashed line*; Burnett equations [34] for the rigid sphere case, *open squares*; DSMC method for the rigid sphere model [34]

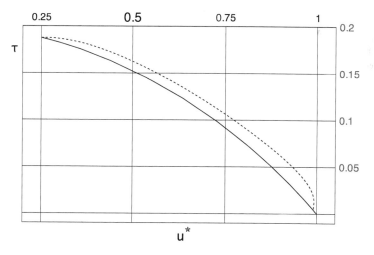

Fig. 3 Orbits or their projections in the τ–u^* plane for $M = 134$. *Solid line*; Navier–Stokes for the rigid sphere model calculated with (21), *dashed line*; Brenner's theory for the rigid sphere model calculated with (41)

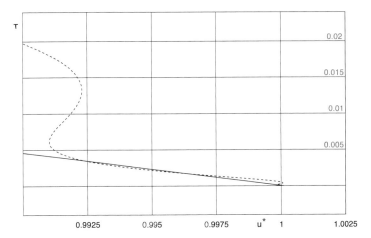

Fig. 4 Orbits or their projections in the τ–u^* plane for $M = 134$. *Solid line*; Navier–Stokes for the rigid sphere model calculated with (21), *dashed line*; Brenner's theory for the rigid sphere model calculated with (41)

behavior is that two of the eigenvalues of the jacobian matrix at up-flow are complex, therefore oscillations are to be expected. This behavior makes difficult to evaluate the asymmetry factor with confidence and this is why we do not report the values of the preliminary calculations done until now. Perhaps the behavior just described could explain the failure of the non-stationary code used by Greenshields and Reese [14] using OpenFoam for Mach numbers greater than or equal to eight.

4 Final Remarks

The important features of this work are the following;

- In terms of reduced variables the orbits of the Navier–Stokes equations are independent of the interaction potential. The "universal" differential equation given by (20) is not equivalent to the Navier–Stokes dynamical system given by (21) because the former one can be extended beyond the critical points in the case of the heteroclinic trajectory. It is interesting that the projected orbits of Brenner's dynamical system for the models considered also fall on one line.
- Using local analysis and numerical computations we have provided evidence that an heteroclinic orbit exists for Brenner's theory ($\alpha_v = \alpha$), this is in contradiction with what was stated by Greenshields and Reese [14]. Our approach leads to ordinary differential equations that are simpler than the partial differential equations used by them for their non-stationary analysis, but a mathematical proof of existence is still lacking.
- Brenner's theory points in the correct direction in the sense that the orbits are above the "universal" Navier–Stokes heteroclinic orbit, in accordance with

DSMC computations (see Fig. 1), but it is perhaps far above the DSMC data. In contrast the Burnett equations provide a better description for $M = 2$.

- The "universal" Navier–Stokes heteroclinic orbit is different when compared with the DSMC computations, this means that in order to improve on the Navier–Stokes equations the question of the interaction potential becomes irrelevant and a new theory is needed. However, our results rely on the use of the first Sonine expansion for the transport coefficients.

- For the shock wave problem Brenner's theory leads to a dynamical system in three dimensions which is simpler than the four dimensional one provided by the Burnett equations [34]. It seems worth to take a deeper look to the determination of α_v – as was done by Greenshields and Reese [14] that unfortunately found globally a too low asymmetry factor when compared with DSMC, and experiments [72] –, to see if Brenner's proposal ($\alpha_v = \alpha$) can be improved.

- Experiments and DSMC data shows that the trend in the asymmetry factor is relatively well described by the Burnett equations for Mach numbers lower than about 2.69, see Fig. 14 in reference [34] and Fig. 4 in reference [79]. The results by Greenshields and Reese [14] and our preliminary computations show that Benner's ideas so far do not provide a good asymmetry factor. Another theory that provides reasonable asymmetry factors is Struchtrup's regularized 13-moments method [17].[12] However, for the two theories that provide consistent asymmetry factors [17, 34], the problem of existence of structure for Mach numbers greater than about three remains open (not only mathematically but also numerically), which is in constrat with the numerical evidence given here in the case of Brenner's theory and, of course as is well known, with the Navier–Stokes equations for which also the mathematical aspects of existence and uniqueness of the heteroclinic orbit are well understood [80].

Brenner's two velocity hydrodynamics has certainly challenged our understanding of hydrodynamics; if it is correct or not is perhaps to soon to know, although arguments showing that is incorrect are available [67]. These arguments rely on what the authors think is the correct thermodynamics, this point is in our opinion also not resolved at present beyond LIT [66] (but see footnote 2). In one of his last works Brenner [12] has turn his attention to the kinetic theory of gases and in particular to the Burnett equations, actually to a particular case of them. Using linear constitutive principles based on them he provided arguments supporting the validity of his approach. However, if one has the full Burnett equations the question would be: why not to use them instead? The Burnett equations can be obtained with the Chapman–Enskog method for solving the Boltzmann equation [6] (which lacks rigor according to mathematicians) but also using the method of stretched fields that perhaps fulfills

---

[12] The situation with moments methods for the shock wave problem is in our opinion far from being well understood and there is an extensive literature on them. Apart from the local and global analysis by Grad [4] there are not many works dealing with these issues, the reason in part is that moment methods introduce spurious critical points that makes the analysis complicated. The interested reader may take a look at the several references; few, here, are provided [64, 73–77]. An interesting work by Santos dealing with the moment hierarchy has been published recently [78].

the requirement of mathematical rigor since the derivation was done by Truesdell [57], a discussion of some of the problems and good points associated with the Burnett equations is available [35]. As pointed out before, the phenomenological approach has the virtue of being of a more general character than studies based on the Boltzmann equation; this answers in part the question raised above. Apart from Brenner's phenomenological two velocity hydrodynamics there is also the two temperature phenomenological hydrodynamics proposed recently [68]. We end by expressing our believe that by bringing into discussion the fundamentals of fluid mechanics, the two velocity hydrodynamics has been important for the development of the difficult subject of fluid dynamics.

References

1. Bird, G. A.: Molecular Gas Dynamics and the Direct Simulation of Gas Flows. Clarendon, Oxford (1994)
2. Erwin, D. A., Pham-Van-Diep, G. C., Muntz, E. P.: Nonequilibrium gas flows I: A detailed validation of Monte Carlo direct simulation for monatomic gases. Phys. Fluids A **3** (1991) 697–705
3. Grad, H.: Principles of the kinetic theory of gases. In: Handbuch der Physik. Flügge, D. (ed.), Springer, Berlin (1958) 205–294
4. Grad, H.: The profile of a steady plane shock wave, Comm. Pure Appl. Math. **5** (1952) 257–300
5. Uribe, F. J., Velasco, R. M., García-Colín, L. S.: Hydrodynamics, Grad's thirteen moments method and the structure of shock waves. In: Developments in Mathematical and Experimental Physics, Volume C: Hydrodynamics and Dynamical Systems. Macias, A., Uribe, F. J., Díaz, E. (eds.), Kluwer, New York (2003) 53–77
6. Chapman, S., Cowling, T. G.: The Mathematical Theory of Non-Uniform Gases. Cambridge University Press, Cambridge (1970)
7. Caflisch, R. E., Nicolaenko, B.: Shock profile solutions of the Boltzmann-equation. Commun. Math. Phys. **86** (1982) 161–194
8. Caflisch, R. E., Nicolaenko, B.: Shock waves and the Boltzmann equation. In: Nonlinear Partial Differential Equations. Smoller, J. A. (ed.), American Mathematical Society, Providence R. I. (1983) 35–44
9. Brenner, H.: Navier–Stokes revisited. Physica A **349** (2005) 60–132
10. Brenner, H.: Fluid mechanics revisited. Physica A **370** (2006) 190–224
11. Brenner, H.: Kinematics of volume transport. Physica A **349** (2005) 11–59
12. Brenner, H.: Bi-velocity hydrodynamics. Physica A **388** (2009) 3391–3398
13. Brenner, H.: Bi-velocity transport processes. Single-component liquid and gaseous continua. Physica A **389** (2010) 1297–1316
14. Greenshields, C. J., Reese, J. M.: The structure of shock waves as a test of Brenner's modifications to the Navier–Stokes equations. J. Fluid Mech. **580** (2007) 407–429
15. Gatinol, R.: Kinetic theory for a discrete velocity gas and application for the shock structure, Phys. Fluids **18** (1975) 153–161
16. Struchtrup, H.: Macroscopic Transport Equations for Rarefied Gas Flows. Springer, Berlin (2005)
17. Torrilhon, M., Struchtrup, H.: Regularized 13-moment Equations: Shock structure calculations and comparison to Burnett models. J. Fluid Mech. **513** (2004) 171–198
18. Gilbarg, D., Paolucci, D.: The structure of shock waves in the continuum theory of fluids. J. of Rat. Mech. Anal. **2** (1953) 617–642
19. Yen, S. M.: Numerical solution of the nonlinear Boltzmann equation for nonequilibrium flows. Ann. Rev. Fluid. Mech. **16** (1984) 67–97

The Shock Wave Problem Revisited

20. Ohwada, T.: Structure of normal shock waves: Direct numerical analysis of the Boltzmann equation for hard-sphere molecules. Phys. Fluids A **5** (1993) 217–234
21. Gorban, A. N., Karlin, I. V.: Method of invariant manifolds and regularization of acoustic spectra. Transport. Theor. Stat. **23** (1994) 559–632
22. Gorban, A. N., Karlin, I. V.: Short-Wave Limit of Hydrodynamics: A soluble Example. Phys. Rev. Lett. **77** (1996) 282–285
23. Gorban, A. N., Karlin, I. V., Zinovyev, A. Y.: Constructive methods of invariant manifolds for kinetic problems. Phys. Rep. **396** (2004) 197–403
24. Ansumali, S., Karlin, I. V.: Stabilization of the lattice Boltzmann method by the H theorem: A numerical test. Phys. Rev. E **62** (2000) 7999–8003
25. Succi, S., Karlin, I. V., Chen, H.: Colloquium: Role of the H theorem in lattice Boltzmann hydrodynamic simulations. Rev. Mod. Phys. **74** (2002) 1203–1220
26. Benzi, R., Succi, S., Vergassola, M.: The Lattice Boltzmann-Equation – Theory and Applications. Phys. Rep. **222** (1992) 145–197
27. Weiss, W. : Continuous shock structure in extended thermodynamics. Phys. Rev. E **52** (1995) R5760–R5763
28. Burnett, D.: The distribution of velocities in a slightly non-uniform gas. Proc. Lond. Math. Soc. **39** (1935) 385–430
29. Burnett, D.: The distribution of molecular velocities and the mean motion in a non-uniform gas. Proc. Lond. Math. Soc. **40** (1936) 382–435
30. Foch Jr., J. D.: On higher order hydrodynamic theories of shock structure. Acta Phys. Austriaca, suppl. X (1973) 123–140
31. Foch Jr., J. D., Simon, C.E.: Numerical integration of the Burnett equations for shock structure in a Maxwell gas. In: Rarefied Gas Dynamics, Progress in Astronautics and Aeronautics. Potter, J. L. (ed.), vol. 51, AIAA, NewYork (1977) 493–500
32. Salomons, E.: Mareschal, M.: Usefulness of the Burnett description of strong shock-waves. Phys. Rev. Lett. **69** (1992) 269–272
33. Uribe, F. J., Velasco, R. M., García-Colín, L. S.: Burnett description of strong shock waves. Phys. Rev. Lett. **81** (1998) 2044–2047
34. Uribe, F. J., Velasco, R. M., García-Colín, L. S., Díaz-Herrera, E.: Shock wave profiles in the Burnett approximation. Phys. Rev. E **62** (2000) 6648–6666
35. García-Colín, L. S., Velasco, R. M., Uribe, F. J.: Beyond the Navier–Stokes equations: Burnett hydrodynamics. Phys. Rep. **465** (2008) 149–189
36. Bhatnagar, P. L., Gross, E.P., Krook, M.: A model for collision processes in gases. I: Small amplitude processes in charged and neutral one-component system. Phys. Rev. **94** (1954) 511–525
37. Liepmann, H. W., Narasimha, H., Chaine, M. T.: Structure of a Plane Shock Layer. Phys. Fluids. **5** (1962) 1313–1324
38. Xu, K., Josyula, E.: Continuum Formulation for Non-Equilibrium Shock Structure Calculation, Commun. Comput. Phys. **1** (2006) 425–450
39. K. Xu, X. He, and C. P. Cai, Multiple temperature kinetic model and gas-kinetic method for hypersonic non-equilibrium flow computations, J. Comput. Phys. **227** (2008) 6779–6794
40. Xu, K., Josyula, E.: Title: Multiple translational temperature model and its shock structure solution. Phys. Rev. E **71** (2005) 056308
41. Holian, B. L., Patterson, C. W., Mareschal, M., Salomons, E.: Modeling shock waves in an ideal gas: Going beyond the Navier–Stokes level. Phys. Rev. E **47** (1993) R24–R27
42. Hoover, W. G.: Smooth Particle Applied Mechanics: The State of the Art. World Scientific, Singapore (2006)
43. Müller, I., Ruggeri, T.: Extended Thermodynamics. Springer, New York (1993)
44. Cercignani, C., Frezzotti, A., Grosfils, P.: The structure of an infinitely strong shock wave. Phys. Fluids **11** (1999) 2757–2764
45. Mott-Smith, H. M.: The solution of the Boltzmann equation for a shock wave. Phys. Rev. **82** (1951) 885–892
46. Hoover, W. G.: Structure of a Wave Shock Front in a Liquid. Phys. Rev. Lett. **42** (1979) 1531–1534

47. Holian, B. L., Hoover, W. G., Moran, B., Straub, B.: Shock-wave structure via non-equilibrium molecular-dynamics and Navier–Stokes continuum mechanics. Phys. Rev. A **22** (1980) 2798–2808
48. Eu, B. C.: Nonequilibrium Statistical Methods. Kluwer, Dordrecht (1998)
49. Al-Ghoul, M., Eu, B. C.: Generalized hydrodynamics and shock saves. Phys. Rev. **56** (1997) 2981–2992
50. Fiscko, K. A., Chapman, D. R.: Comparison of Burnett, Super-Burnett and Monte Carlo Solutions for Hypersonic Shock Structures. In: Rareïņ̃Aed Gas Dynamics, Progress in Astronautics and Aeronautics Vol. 118. Muntz, E. P., Weaver, D. P., Campbell, D. H. (eds.), AAIA, Washington, D.C. (1989) 374
51. Zhong, X., MacCormack, R. W., Chapman, D.R.: AIAAJ **31** (1993) 1036–1043
52. Rosenau, P.: Extending hydrodynamics via the regularizations of the Chapman–Enskog expansion. Phys. Rev. A **40** (1989) 7193–7196
53. Jin, S., Slemrod, M.: Regularization of the Burnett equations via relaxation. J. Stat. Phys. **103** (2001) 1009–1033
54. Jin, S., Pareschi, L., Slemrod, M.: A relaxation scheme for solving the Boltzmann equation based on the Chapman–Enskog method. Manuscript provided to the authors by Professor Kun Xu. It has not been published yet according to the Institute for Scientific Information (ISI)
55. Soderholm, L. H.: Hybrid Burnett equations: A new method of stabilizing. Transport Theor. Stat. Phys. **36** (2007) 495–512
56. Bobylev, A. V.: Generalized Burnett Hydrodynamics. J. Stat. Phys. **132** (2008) 569–580
57. Truesdell, C., Muncaster, R. G.: Fundamentals of Maxwell's Kinetic Theory of a Simple MonatomicGas. Academic Press, New York (1980)
58. Villani, C.: A Review of Mathematical Topics in Collisional Kinetic Theory. In: Handbook of Mathematical Fluid Dynamics vol. 1. Friedlander, S., Serre, D. (eds.), Elsevier Science, Amsterdam (2002)
59. Cercignani, C., Illner, R., Pulvirenti, M.: The Mathematical Theory of Dilute Gases. Springer, NewYork (1994)
60. Agarwal, R. K., Yun, K. Y., Balakrishnan, R.: Beyond Navier–Stokes: Burnett equations for flows in the continuum-transition regime. Phys. Fluids **13** (2001) 3061–3085
61. Ohr, Y. G.: Iterative method to improve the Mott-Smith shock-wave structure theory. Phys. Rev E **57** (1998) 1723–1726
62. Ruggeri, T., Simic, S.: Average temperature and Maxwellian iteration in multitemperature mixture of fluids. Phys. Rev. E **80** (2009) 026317
63. Mason, E. A., McDaniel, E. W.: Transport Properties of Ions in Gases. Wiley, NY (1988)
64. García-Colín, L. S., Velasco, R. M., Uribe, F. J.: Inconsistency in the moment's method for solving the Boltzmann equation. J. Non-Equilib. Thermodyn. **29** (2004) 257–277
65. Lin, J., Shen, C., Fan, J.: IP Simulation of Micro Gas Flow under 3-D Head Sliders. In: Rarefied Gas Dynamics. Abe, T. (ed.), AIP Conference Proceedings 1084, AIP (2009)
66. de Groot, S., Mazur, P.: Nonequilibrium Thermodynamics. Dover, New York (1984)
67. Ottinger, H. C., Struchtrup, H., Liu, M.: Inconsistency of a dissipative contribution to the mass flux in hydrodynamics. Phys. Rev. E **80** (2010) 05303
68. Hoover, Wm. G., Hoover, C. G.: Well-posed two-temperature constitutive equations for stable dense fluid shock waves using molecular dynamics and generalizations of Navier–Stokes-Fourier continuum mechanics. Phys. Rev. E **81** (2010) 046302
69. Uribe, F. J.: Understanding dilute gases: Going beyond the Navier–Stokes equations. In: Computational Fluid Mechanics. Ramos, E., Cisneros, G., Fernández-Flores, R., Santillán-González, A. (eds.), World Scientific, Singapore (2001) 255–264
70. Glendining, P.: Stability, Instability, and Chaos. Cambridge University Press, New York (1994)
71. Lumpkin III, F. E., Chapman, D. R.: Accuracy of the Burnett equations for hypersonic real gas flow. J. Aeronautical Sci. **6** (1992) 419–425
72. Alsmeyer, H.: Density profiles in Argon and Nitrogen shock waves measured by the absorption of an electron beam. J. Fluid. Mech. **74** (1976) 497–513
73. Holway Jr., L. H.: Existence of kinetic theory solutions to the shock structure problem. Phys. Fluids **7** (1964) 911–913

74. Weiss, W.: Existence of kinetic theory solutions to the shock structure problem-Comment. Phys. Fluids **8** (1996) 1689–1690
75. Jou, D., Pavón, D.: Nonlocal and nonlinear effects in shock-waves. Phys. Rev. A **44** (1991) 6496–6502
76. Ruggeri, T.: Breakdown of shock-wave-structure solutions. Phys. Rev. E **47** (1993) 4135–4140
77. Ruggeri, T.: On the shock structure problem in non-equilibrium thermodynamics of gases. Transport Theor. Stat. Phys. **25** (1996) 567–574
78. Santos, A.: Solutions of the moment hierarchy in the kinetic theory of Maxwell models. Continuum Mech. Thermodyn. **21** (2009) 361–387
79. Pham-Van-Diep, G. C., Erwin, D. A., Muntz, E. P.: Testing continuum descriptions of low-Mach-number shock structures. J. Fluid Mech. **232** (1993) 403–413
80. Gilbarg, D.: The existence and limit behavior of the one-dimensional shock layer. Am. J. Math. **73** (1951) 256–274

Adaptive Simplification of Complex Systems: A Review of the Relaxation-Redistribution Approach

Eliodoro Chiavazzo and Ilya Karlin

Abstract We present a review of a recently introduced methodology for reducing complexity of large dissipative systems such as detailed reaction mechanisms of combustion and biochemical reactions whose dynamics is characterized by a hierarchy of slow invariant manifolds. Accurate reduced description is achieved by construction of slow invariant manifolds with an embarrassingly simple implementation in any dimension. The method is validated with the auto-ignition of the hydrogen-air mixture where a reduction to a cascade of slow invariant manifolds is observed.

1 Introduction

Detailed reaction mechanisms of biochemical processes in living cells and combustion of hydrocarbon fuels are prototypical examples of complex systems [1–3]. Modern research in the reactive flow phenomena has to cope with an increasing complexity, in two aspects. First, the number of species and reactions is usually quite large; second, their dynamics is characterized by a wide range of time-scales. As a result, the usage of detailed reaction mechanisms in the reactive flow simulation soon becomes intractable even for supercomputers, as for example in the turbulent combustion of even "simplest" fuels such as hydrogen [4]. Hence, there is a demand for reducing the complexity of detailed reaction systems without a loss of accuracy and predictive power (commonly known as model reduction; see, e.g. a recent collection of modern approaches [5]).

E. Chiavazzo
Department of Energetics, Politecnico di Torino, 10129 Torino, Italy
e-mail: eliodoro.chiavazzo@gmail.com

I. Karlin (✉)
Aerothermochemistry and Combustion Systems Lab, ETH Zurich, 8092 Zurich, Switzerland
e-mail: karlin.ilya@gmail.com
and
School of Engineering Sciences, University of Southampton, SO17 1BJ Southampton, UK

A.N. Gorban and D. Roose (eds.), *Coping with Complexity: Model Reduction and Data Analysis*, Lecture Notes in Computational Science and Engineering 75, DOI 10.1007/978-3-642-14941-2_11, © Springer-Verlag Berlin Heidelberg 2011

Let the detailed description be given by an autonomous system in terms of the state ψ on a phase space U,

$$\frac{d\psi}{dt} = f(\psi). \tag{1}$$

Important example of (1) to be addressed below is the reaction kinetics where $\psi = (\psi^1, \dots, \psi^n)$ is a n-dimensional vector of concentrations of various species while the vector field f is constructed according to the detailed reaction mechanism and (usually) the mass action law.

A good, rigorous approach to model reduction is provided by the Method of Invariant Manifold (MIM) [6] which we first briefly review. In MIM, the problem of model reduction is identified it with the construction of a slow invariant manifold (SIM) Ω_{SIM}. A sub-manifold Ω (not necessarily a SIM) is embedded in the phase space U and is represented by a function $F(\xi)$ which maps a macroscopic variables space Ξ into U. Introducing a projector P onto the tangent space T of a manifold Ω (not necessarily invariant), the reduced dynamics on it is defined by the projection $Pf(\Omega) \in T$. A manifold Ω is termed invariant (but not necessarily *slow*) if the vector field f is tangent to the manifold at every point: $f(F(\xi)) - Pf(F(\xi)) = 0$. While the notion of a manifold's invariance is relatively straightforward, a definition of slowness is more delicate as it necessarily compares a (faster) approach towards the SIM with a (slower) motion along SIM. In MIM, slowness is understood as stability, and SIM is a stable fixed point $F_{\text{SIM}}(\xi)$ of the *film equation* defined on the space of maps F [6],

$$\frac{dF(\xi)}{dt} = f(F(\xi)) - Pf(F(\xi)), \tag{2}$$

Projector P in (2) reflects the separation of motions in the vicinity of SIM; slow motions along SIM are locked in the image, $\text{im} P = T$, while the null-space spans the fibres of fast motion transversal to SIM. For thermodynamically consistent systems (1) equipped with a thermodynamic potential G, MIM offers a projector whose construction is based on the tangent space T and the gradient of the thermodynamic potential, $\partial G/\partial \psi$, at every point of SIM (explicit formulae for this thermodynamic projector are not necessary for our presentation and can be found in [6]). For finite-dimensional systems, a more "conventional" projector based on the spectral analysis of the Jacobian of the vector field f or its symmetrized forms can also be applied. Rigorous proofs of existence and uniqueness of SIM, by the film equation (2), were recently given for linear systems [7].

Finally, a computationally advantageous realization is provided by a *grid* representation of MIM [6], where grid nodes in the phase space are defined by a discrete set of macroscopic variables ξ_i while finite difference operators are used to compute the tangent space at every node $F(\xi_i)$. Thanks to locality of MIM constructions, we further make no distinction between manifolds and grids.

A natural approach to the construction of SIM's is a direct numerical solution of the film equation (2) starting with an initial manifold (grid). For that, both the initial condition as well as implicit or semi-implicit schemes were developed.

However, the latter approaches are hindered by severe numerical (Courant) instabilities. Furthermore, usually the construction of invariant manifolds is attempted in the whole phase space, and their dimensionality is pre-assigned somewhat arbitrarily. Construction of high-dimensional homogeneous (fixed dimensionality) grids is problematic, in addition to the significant difficulties of storing and accessing a large amount of data. Finally, the dimension of SIM generally changes in the phase-space, and adaptive methods capable to take this into account are requested.

A new approach to model reduction introduced recently (the relaxation-redistribution method, RRM hereafter [8,9]) and reviewed below in this contribution to the present volume allows for a highly efficient, instability-free construction of heterogeneous slow invariant manifolds (grids) with the dimension adaptively varying from one region of the phase space to another. The key idea of our approach is to abandon an attempt of *solving* the film equation (2) in favor of a *simulation* of the physics behind this equation, in a spirit similar to Monte Carlo methods. Application to hydrogen-air combustion mechanism convincingly shows that the reduced dynamics thus obtained can replace the complex detailed kinetics without loosing any accuracy.

2 RRM: Global Formulation

In order to introduce our method, we consider reaction kinetics and assume that a slow dynamics of (1) evolves on a q-dimensional SIM in the n-dimensional concentration space (this assumption will be relaxed in a sequel). Inspection of the right-hand side of (2) reveals a composition of two motions: The first term, $f(F(\xi))$, is the relaxation of the initial approximation to SIM due to the detailed kinetics, while the second term, $-Pf(F(\xi))$ is the motion *antiparallel* to the slow dynamics. Let a time stepping δt and a numerical scheme (e.g. Euler, Runge–Kutta, etc.) be chosen for solving the system of kinetic equations: All grid nodes relax towards the SIM under the full dynamics f during δt. Fast component of f leads any grid node closer to the SIM while at the same time, the slow component causes a shift towards the steady state. As a result, while keeping on relaxing, the grid shrinks towards the steady state (we term this a "shagreen effect" per *de Balzac*'s famous novel [10]). Subtraction of the slow component therefore prevents the shagreen effect to occur, and it is precisely the difficulty in the numerical realization: explicit evaluation of the projector P on the approximate SIM does not always balance the effect of shrinking. This leads to instabilities, and results in a drastic decreasing of the time step.

The key idea here is to neutralize the slow component of motion by a simple *redistribution* of the points on the manifold after the relaxation step. Specifically, using linear interpolation, the set of relaxed points ψ_i' is replaced with a new set of points ψ_i corresponding to the nodes $\xi_x = (\xi_x^1, \ldots, \xi_x^q)$ of a fixed q-dimensional grid in the parameter space $\varXi$:

$$\psi_x = \left(1 - \sum_{i=1}^{q} w_i\right) \psi_0' + \sum_{i=1}^{q} w_i \psi_i', \tag{3}$$

where $\psi_0', \psi_1', \ldots, \psi_q'$ are relaxed grid points whose macroscopic states are the vertexes of a q–simplex $\mathcal{T}$, such that $\psi_x = F(\xi_x)$ and $\xi_x \in \mathcal{T} \subset \Xi$. First order Lagrange interpolation requires:

$$\begin{bmatrix} \xi_1^1 - \xi_0^1 & \cdots & \xi_q^1 - \xi_0^1 \\ \vdots & \ddots & \vdots \\ \xi_1^q - \xi_0^q & \cdots & \xi_q^q - \xi_0^q \end{bmatrix} \begin{bmatrix} w_1 \\ \vdots \\ w_q \end{bmatrix} = \begin{bmatrix} \xi_x^1 - \xi_0^1 \\ \vdots \\ \xi_x^q - \xi_0^q \end{bmatrix}, \tag{4}$$

where ξ_i^j is the jth macroscopic variable corresponding to ψ_i'. Finally, the above procedure should be supplemented by the boundary conditions applied at the edges of the grid. Here we use a simple linear extrapolation: Grid nodes at the boundary ψ_{bound} are reconstructed by extrapolation after the relaxation step. The formula (3) is used where $\psi_x = \psi_{\text{bound}} \notin \mathcal{T}$ is located in the vicinity of a simplex $\mathcal{T}$ with vertexes $\psi_0, \psi_1, \ldots, \psi_q$: In general, $\mathcal{T}$ can be chosen such that ψ_0 is the relaxed node of ψ_{bound}.

In other words (see a cartoon, Fig. 1), in the redistribution step, the slow motion is subtracted by stretching the macroscopic variables to the nodes of the initial grid ξ_x. Note that all intermediate grids are, by construction, regular in the parameter space Ξ (if the initial grid was regular) and, for grids in the vicinity of the slow invariant manifold, the effects due to relaxation and redistribution tend to compensate each other. Hence, invariant grids are stable fixed points of the described procedure, here termed *relaxation redistribution method* (RRM), which can be terminated when the overall displacement of any node (relaxation+redistribution) is small compared to that due to the relaxation alone. In effect, the choice of the time stepping δt for the RRM with imposed boundary conditions is not longer restricted by Courant instability because no explicit evaluation of projector P is required in the "anti-shagreen" redistribution step.

In order to test the RRM, we first consider a simple benchmark suggested by Davis and Skodje [11] (a two-dimensional system with a one-dimensional SIM known in a closed analytical form). The RRM was performed for a variety of initial grids with different spacing, instability was never observed. Results are presented in Fig. 2 for two different initial grids, a regular linear and a randomly generated grid. Convergence to SIM is even striking in Fig. 2 given the fact that both initial grids are far from SIM. Note that, limiting the sources of a spurious (numerical) instability of iterative methods is highly desirable. In RRM, grid convergence indeed confirms the existence of a reduced description with a fixed number of degrees of freedom q (existence of q-dimensional slow invariant manifold). On the contrary, no convergence in RRM indicates that more degrees of freedom are needed to recover the detailed system dynamics. This concept shall be used below for adaptively choosing the invariant grid dimension.

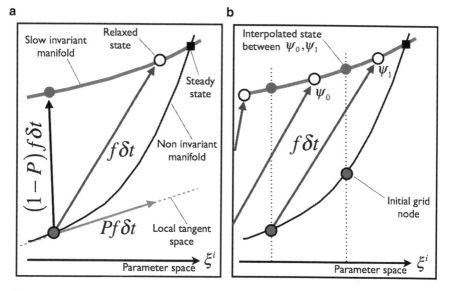

Fig. 1 Explanation of the relaxation-redistribution method. (**a**) The relaxation due to (1) of a non-invariant manifold. Fast dynamics drives it towards the slow invariant manifold, however the concurrent action of the slow dynamics causes a shift towards the steady state (shagreen effect, *open circles*). On the contrary, relaxation due to the film equation (2) allows a non-invariant manifold to move only in the fast subspace (*full circles*); (**b**) RRM: The displacement in the slow subspace, generated by the relaxation step, is compensated by the subsequent redistribution step

3 RRM: Local Formulation

Both construction and usage of a global reduced description soon become impracticable as the dimension q increases. In fact, computing and storage of high dimensional SIM's may be problematic already at $q \geq 3$. Above all that, data retrieval by interpolation on such large arrays is computationally intensive, and sometimes full construction of manifolds can be useless: For example, in combustion applications, regions with a high concentration of radicals are unlikely to be visited. In general, what is required in model reduction is the mapping of a point $\xi = (\xi^1, \ldots, \xi^q)$ describing the system macro-state into a micro-state $F(\xi) \in U$: In the following, this is referred to as the closure problem of the reduced model. Therefore, *local* approaches with an adaptive choice of the dimension q are much more appealing, because they provide a closure without generating the whole of a high-dimensional manifold (or any significant portions of it).

Importantly, the RRM allows for a straightforward local formulation, where only a small patch of an invariant grid is constructed (see a cartoon, Fig. 3). Let $\bar{\xi} = (\bar{\xi}^1, \ldots, \bar{\xi}^q)$ and $F(\bar{\xi})$ be a fixed macroscopic state and the corresponding microscopic state, respectively. Starting from $F(\bar{\xi})$ (termed *pivot* in the following), it is possible to construct a q-dimensional simplex $\mathcal{T}$ by linking the pivot to q

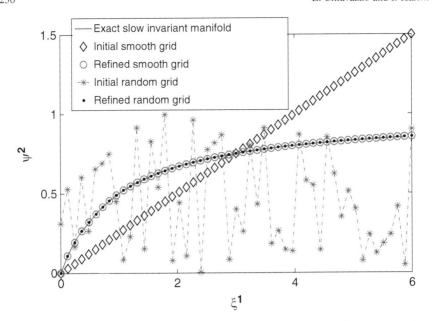

Fig. 2 The Davis–Skodje system [11], $d\psi^1/dt = -\psi^1$, $d\psi^2/dt = -\gamma\psi^2 + [(\gamma - 1)\psi^1 + \gamma\psi^1\psi^1]/(1 + \psi^1)^2$, has unique stable steady state $\psi^1 = \psi^2 = 0$, and a one-dimensional SIM, $\psi^2 = \psi^1/(1+\psi^1)$. Macroscopic variable is $\xi^1 = \psi^1$. Two different initial grids are refined using explicit Euler scheme for the relaxation ($\delta t = 10^{-2}$). Results after 50 RRM iterations are reported, $\gamma = 50$

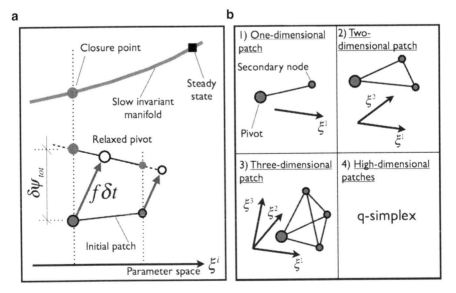

Fig. 3 (**a**) Relaxation redistribution method: Local construction. Only a small portion (patch) of the slow invariant manifold is constructed. (**b**) Simplexes can be conveniently used for the patch-wise description of slow invariant manifolds in any dimension

secondary nodes $F\left(\bar{\xi}_1\right), \ldots, F\left(\bar{\xi}_q\right)$ chosen in a neighborhood of $F\left(\bar{\xi}\right)$ such that $\bar{\xi}_i = \left(\bar{\xi}^1, \ldots, \bar{\xi}^i + \delta\bar{\xi}^i, \ldots, \bar{\xi}^q\right)$, with $\delta\bar{\xi}^i$ being a small deviation along the ith macroscopic direction. A sequence of relaxation and redistribution (by linear extrapolation) sub-steps can be easily applied to the nodes of $\mathcal{T}$ in any dimension q: This realizes indeed the simplest instance of the RRM method, where refinements end as soon as a norm of the total displacement of the pivot $|\delta\psi_{\text{tot}}|$ is sufficiently small compared to the displacement caused by the relaxation alone $|\delta\psi_{\text{rel}}|$.

A convenient (but not the only possible) initialization of the RRM procedure for dissipative systems of chemical kinetics is achieved by existing techniques. For a chemical reaction with n species, and r conservation laws (of elements) and a thermodynamic potential G, we use spectral variables [6], $\xi = B\psi$, where the $q \times n$ matrix $B = [b_1, \ldots, b_q]^T$ and $b_1, \ldots, b_{n-r}$ are the eigenvectors of the Jacobian $J = \partial f/\partial\psi$ at the equilibrium, corresponding to non-zero eigenvalues λ_i and numbered in the order of increase of $|\lambda_i|$. The pivot of the initial simplex $\mathcal{T}$ is defined as the quasi-equilibrium point ψ^*, corresponding to $\bar{\xi} = \left(\bar{\xi}^1, \ldots, \bar{\xi}^q\right)$, as the minimizer of thermodynamic potential G: $G\left(\psi\right) \to \min$, $B\psi = \bar{\xi}$, $C\psi = c$, where C is the $r \times n$ matrix formed by conservation laws. Secondary nodes of the simplex are calculated by expansion of the minimization problem about the quasi-equilibrium as suggested in [12]: A secondary node ψ^k is found as $\psi^k = \psi^* + \sum_{i=1}^{n-r} \delta_k^i \rho_i$, with $(\rho_1, \ldots \rho_{n-r})$ and $\delta_k = \left(\delta_k^1, \ldots, \delta_k^{n-r}\right)$ being a vector basis spanning the null space of C and the solution of a linear algebraic system

$$
\begin{cases}
\sum_{i=1}^{n-r} \left(t_j, H^*\rho_i\right) \delta_i = -\nabla G^* t_j, \quad j = 1, \ldots, n-r-q, \\
\sum_{i=1}^{n-r} \left(b^1, \rho_i\right) \delta_i = 0, \\
\ldots \\
\sum_{i=1}^{n-r} \left(b^k, \rho_i\right) \delta_i = \varepsilon_k, \\
\ldots \\
\sum_{i=1}^{n-r} \left(b^q, \rho_i\right) \delta_i = 0.
\end{cases}
$$

The vector basis $\left(t_1, \ldots t_{n-q-r}\right)$ spans the kernel of the linear space defined by the rows b^i of B and the rows of C, H^* and ∇G^* are the second derivative matrix and the gradient of the function G at the pivot respectively, while ε_k defines the length of the edge of $\mathcal{T}$ along the kth direction.

Setting an upper limit to both the number of refinements $N_{\max}$ and the tolerance $\epsilon = |\delta\psi_{\text{tot}}|/|\delta\psi_{\text{rel}}|$, the local RRM can be adaptively performed starting with $q = 1$. If the latter requirements are not fulfilled, the dimension is updated to $q = 2$ and the procedure repeated. Upon convergence with $q = \bar{q}$, the closure $F_{\text{RRM}}\left(\xi^1, \ldots, \xi^{\bar{q}}\right)$ is provided by the coordinates of the pivot. The above local RRM fully alleviates any assumption about the dimensionality of SIM, the local dimension is found automatically and if no reduced description is possible at all, no convergence at any $q < n - r$ will clearly indicate this.

4 Auto-Ignition of Hydrogen-Air Mixture

We considered the hydrogen-air mixture at stoichiometric proportion reacting according to the realistic detailed mechanism of Li et al. [3] (nine chemical species ($n = 9$), three elements conservation ($r = 3$), twenty-one reversible reactions; this mechanism is universally used in turbulent combustion simulations) in an batch reactor under fixed enthalpy ($1,000[kJ/kg]$) and pressure ($1[atm]$). Figure 4 shows a part of the heterogeneous SIM constructed by the local RRM where one-, two- and three-dimensional patches are clearly visible.

Finally, the RRM method provides the reduced system,

$$\frac{d\xi}{dt} = BPf(F_{\text{RRM}}(\xi)).$$

Results in terms of basic variables (concentrations of species) are obtained upon a post-processing of the spectral variables which amounts to a linear transformation. A typical problem, where dynamics evolves along a *cascade* of slow invariant manifolds with progressively lower dimensions, is the auto-ignition of a fuel-air mixture.

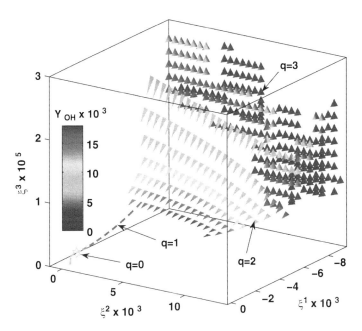

Fig. 4 Heterogeneous slow invariant manifold of hydrogen-air combustion mechanism by local RRM. Three-dimensional projection of the six-dimensional phase space onto spectral variables (see text). The two-dimensional patch ("kite", *triangles*) is tight by a one dimensional "thread" (*line*) to the zero-dimensional equilibrium and merges with the three-dimensional "cloud" (*tetrahedra*). Legend: mass fraction of OH. Explicit Euler scheme with $\delta t = 5 \times 10^{-8}[s]$ was used for the relaxation of simplex. RRM convergence criteria: $N_{\max} = 2,000$, $\epsilon = 10^{-4}$

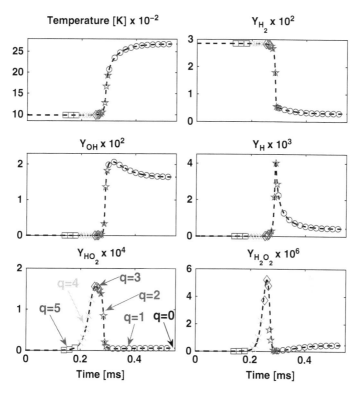

Fig. 5 Auto-ignition of homogeneous stoichiometric mixtures of hydrogen and air: Histories of the temperature and of the mass fraction of chemical species. *Line*: detailed reaction; *Symbol*: local RRM method by adaptively following a cascade of reduced models of various dimension q: $q = 5$ (*square*), $q = 4$ (*cross*), $q = 3$ (*diamond*), $q = 2$ (*star*), $q = 1$ (*circle*) and $q = 0$ (steady state)

In Fig. 5, results of integration of the reduced system as obtained by RRM of Fig. 4 is compared with the integration of the detailed reaction mechanism. Results are in excellent agreement for all the chemical species and the temperature. Note that, although one- and two-dimensional SIM's are able to recover the most of the dynamics of major species and of the temperature, the minority species (such as radicals HO_2 and H_2O_2) do require high dimensional manifolds ($q \geq 3$) to be correctly predicted.

5 Conclusion

To conclude, we have reviewed a novel approach to simplification of complex systems. Our approach is based on a simulation (instead of a solution) of the fundamental film equation (2). In that respect, the RRM is similar in its spirit (but certainly not in the implementation) to other successful simulation strategies such as the Direct Simulation Monte Carlo method [13] which replaces the solution of the Boltzmann equation by a stochastic simulation of "collisions" or the lattice Boltzmann method [14] which replaces discretization of the Navier-Stokes equations by a simulation of fictitious "particles". In the RRM, we replaced the difficult part of solving the film equation (subtraction of slow motions) by an embarrassingly simple and computationally efficient redistribution of the slow invariant manifold points which simulates this subtraction. Examples presented above convincingly show that RRM achieves all the objectives of model reduction in terms of accuracy and efficiency, the resulting adaptively reduced systems can be used *instead* of the detailed reaction. The computationally efficient, fully adaptive construction of heterogeneous slow invariant manifolds as in the case of hydrogen-air mixture is difficult if at all possible with any other model reduction technique [5], most notable is the full adaptivity of RRM which makes it unnecessary to guess the number of essential variables. Finally, while we focused on the important class of dissipative systems arising in combustion, we look forward to generalization of the above technique of simplification to other dissipative systems such as master and Fokker–Planck equations and other complex dynamics.

References

1. Jamshidi, N., Palsson, B. O.: Formulating genome-scale kinetic models in the post-genome era. Mol. Syst. Biol. **4** (2008) 171
2. Endy, D., Brent, R.: Modelling cellular behaviour. Nature **409** (2001) 391–395
3. Li, J., Zhao, Z., Kazakov, A., Dryer, F.L.: An updated comprehensive kinetic model of hydrogen combustion. Int. J. Chem. Kinet. **36** (2004) 566–575
4. Nur eine kleine Flamme und schon so kompliziert. Interview with Prof. K. Boulouchos, NZZ Neue Zürcher Zeitung (Jul. 05, 2009)
5. Model Reduction and Coarse-graining Approaches for Multiscale Phenomena, A.N. Gorban et al. (eds.) Springer, Berlin (2006)
6. Gorban, A.N., Karlin, I.V.: Invariant Manifolds for Physical and Chemical Kinetics. Springer, Berlin (2005)
7. Ramm, A.G.: Invariant manifolds for dissipative systems. J. Math. Phys. **50** (2009) 042701
8. Chiavazzo, E.: Ph. D. thesis, ETH Zurich (2009)
9. Chiavazzo, E., Karlin, I. preprint (2010)
10. Honore de Balzac, La Peau de chagrin (1831); English translation: The Wild Ass's Skin. Trans. Herbert J. Hunt. Penguin Books: New York (1977)
11. Davis, M.J., Skodje, R.T.: Geometric investigation of low-dimensional manifolds in systems approaching equilibrium. J. Chem. Phys. **111** (1999) 859–874
12. Chiavazzo, E., Karlin, I.V.: Quasi-equilibrium grid algorithm: Geometric construction for model reduction. J. Comput. Phys. **227** (2008) 5535–5560
13. Bird, G.A.: Molecular Gas Dynamics and the Direct Simulation of Gas Flows. Clarendon Press, Oxford (1994)
14. Benzi, R., Succi, S., Vergassola, M.: The lattice Boltzmann-equation – theory and applications. Phys. Rep. **222** (1992) 145–197

Geometric Criteria for Model Reduction in Chemical Kinetics via Optimization of Trajectories

Dirk Lebiedz, Volkmar Reinhardt, Jochen Siehr, and Jonas Unger

Abstract The need for reduced models of chemical kinetics is motivated by the fact that the simulation of reactive flows with detailed chemistry is generally computationally expensive. In dissipative dynamical systems different time scales cause an anisotropically contracting phase flow. Most kinetic model reduction approaches explicitly exploit this and separate the dynamics into fast and slow modes. We propose an implicit approach for the approximation of slow attracting manifolds by computing trajectories as solutions of an optimization problem suggesting a variational principle characterizing trajectories near slow attracting manifolds. The objective functional for the identification of suitable trajectories is supposed to represent the extent of relaxation of chemical forces along the trajectories which is proposed to be minimal on the slow manifold. Corresponding geometric criteria are motivated via fundamental concepts from differential geometry and physics. They are compared to each other through three kinetic reaction mechanisms.

D. Lebiedz (✉)
Interdisciplinary Center for Scientific Computing (IWR), University of Heidelberg,
Im Neuenheimer Feld 368, 69120 Heidelberg, Germany
e-mail: dirk.lebiedz@biologie.uni-freiburg.de
and
Center for Systems Biology (ZBSA), University of Freiburg, Habsburgerstraße 49, 79104 Freiburg,
Germany

V. Reinhardt and J. Siehr
Interdisciplinary Center for Scientific Computing (IWR), University of Heidelberg,
Im Neuenheimer Feld 368, 69120 Heidelberg, Germany
e-mail: Volkmar.Reinhardt@iwr.uni-heidelberg.de, Jochen.Siehr@iwr.uni-heidelberg.de

J. Unger
Center for Systems Biology (ZBSA), University of Freiburg, Habsburgerstraße 49, 79104 Freiburg,
Germany
e-mail: jonas.unger@zbsa.de

A.N. Gorban and D. Roose (eds.), *Coping with Complexity: Model Reduction and Data Analysis*, Lecture Notes in Computational Science and Engineering 75, DOI 10.1007/978-3-642-14941-2_12, © Springer-Verlag Berlin Heidelberg 2011

1 Introduction

In dissipative ordinary differential equation (ODE) systems modeling chemical reaction kinetics different time scales cause anisotropic phase volume contraction along trajectories. This leads to a bundling of trajectories on or near "manifolds of slow motion" of successively lower dimension. Many model reduction methods exploit this for simplifying chemical kinetics via a time scale separation into fast and slow modes.

The background of numerical model reduction of such systems lies mainly in the fact that the computational effort for a full simulation of reactive flows, e.g., of fluid transport involving multiple time scale chemical reaction processes, is extremely high [1]. However, model reduction is often also of general interest for theoretical purposes in modeling and simulation [2]. In this context reduced models should describe some essential characteristics as the full dynamical system while omitting less relevant issues.

Reaction trajectories in phase space as solutions of an ODE model are uniquely determined by their initial values and bear information on the structure of the phase space. Following Lebiedz' idea searching for an extremum principle that distinguishes special trajectories on or near slow attracting manifolds, an optimization approach for computing such trajectories is applied and discussed in [3–5]. In [6] we discuss three geometrically motivated objective criteria for the identification of those trajectories lying on slow attracting manifolds.

The present overview shortly discusses the method and shows results of its application to different kinetic mechanisms: a six component model mechanism, a six component simplified combustion mechanism, and a realistic nine species hydrogen combustion mechanism.

2 Theoretical Background

In this section the theoretical basis of the method is discussed. Different objective functionals can be used in the general optimization problem for the identification of trajectories on or near the slow invariant manifold (SIM).

2.1 Optimization Problem

The basic idea of the method is the formulation of an optimization problem for identifying special trajectories. Trajectories in phase space converge with progress of time towards the SIM to be identified and are invariant by definition. The formulation of the method as finding the numerical solution of an optimization problem assures the existence of a reduced model and sophisticated optimization software can be used for the numerical solution.

Model Reduction via Optimization of Trajectories

The general optimization problem can be formulated as

$$\min_{c(t)} \int_{t_0}^{t_f} \Phi\left(c(t)\right) \, dt \tag{1a}$$

subject to

$$\frac{dc(t)}{dt} = f\left(c(t)\right) \tag{1b}$$

$$0 = g\left(c(t_*)\right) \tag{1c}$$

$$c_j(t_*) = c_j^{t_*}, \quad j \in I_{\text{fixed}}, \tag{1d}$$

with $t_0 \leqslant t_* \leqslant t_f$. The vector $c = (c_i)_{i=1}^n$ denotes e.g. concentrations of chemical species, and I_{fixed} is an index set that contains the indices of variables with fixed values at time t_* (so-called reaction progress variables) chosen to parameterize the reduced model, i.e., points lying on a slow attracting manifold to be computed. Thus, $c_j(t_*), j \notin I_{\text{fixed}}$ represent the degrees of freedom in the optimization problem further reduced by the function g in (1c). The process of determining $c_j^{t_*}, j \notin I_{\text{fixed}}$ from $c_j^{t_*}, j \in I_{\text{fixed}}$ is known as *species reconstruction*. The system dynamics (e.g., chemical kinetics determined by the reaction mechanism) are described by (1b) and are introduced as equality constraints. Hence an optimal solution of (1) always satisfies the system dynamics of the full ODE system. Conservation relations due to the physical background of the model are collected in the function g in (1c). The concentrations at time t_* of the reaction progress variables are fixed via the equality constraint (1d). The objective functional $\Phi(c(t))$ in (1a) will be discussed in the next section.

In previous publications [3, 4, 6] the general optimization problem (1) is solved with $t_0 = t_* = 0$ and for the computations t_f is chosen large ensuring sufficient approximation of equilibrium. In contrast, in the present article the formulation $t_f = t_* = 0$ is also used. This means that for the determination of a point in phase space as a solution of the optimization problem the course of the trajectory through that point "backwards" in time is taken into account. The idea behind this is the fact that for a point on a slow invariant manifold the trajectory through that point should be slow also backwards in time.

We will refer to the first case with $t_0 = t_* = 0$ as *forward* mode of the method and to the latter $(t_f = t_* = 0)$ as *reverse* mode. The general formulation (1) includes both modes as special cases. While solving the reverse mode problem one has to deal with the instabilities of the trajectories backwards in time. We employ a numerically robust collocation approach (see Sect. 3).

2.2 Optimization Criteria

Of course, the choice of the criterion $\Phi(c(t))$ is important for both success and degree of accuracy of the computed approximations of the slow attracting manifold. In [6]

$$\Phi_A(c) = \frac{\|J_f(c)\,f\|_2}{\|f\|_2}$$

is proposed with $J_f(c)$ being the Jacobian of the right hand side f evaluated at $c(t)$ using the Euclidean norm. The resulting objective functional is

$$\int_{t_0}^{t_f} \|J_f(c)\,f\|_2 \, dt. \tag{2}$$

This criterion is rooted in the idea that trajectories on the SIM undergo a "smaller" change of velocity along their way as those still relaxing to the SIM. We propose the integrated velocity change

$$D_v \dot{c} := \left.\frac{d}{d\alpha} f(c + \alpha v)\right|_{\alpha=0} = J_f(c) \cdot \frac{f}{\|f\|_2} \tag{3}$$

in direction of the phase flow $v := \frac{\dot{c}}{\|\dot{c}\|_2} = \frac{f}{\|f\|_2}$ to be minimal. The whole term (3) is evaluated in the Euclidean norm and the norm of f cancels out during reparameterization from arclength to real time.

A slightly modified objective functional is

$$\int_{t_0}^{t_f} \|J_f(c)\,f\|_W \, dt = \int_{t_0}^{t_f} f^T\,J_f^T(c)\,\mathrm{diag}(1/c_i)\,J_f(c)\,f\,dt, \tag{4}$$

with $W = \mathrm{diag}(1/c_i)$ being the diagonal matrix with diagonal elements $1/c_i$ as a weighting function. This criterion represents the Riemannian geometry induced by the second differential of the Lyapunov function G

$$G = \sum_{i=1}^{n} c_i[\log(c_i/c_i^{\mathrm{eq}}) - 1], \quad W = \mathrm{Hess}(G)$$

which is derived from the Helmholtz free energy for a perfect system. The corresponding metric has been discussed in the context of an entropic scalar product [7] which has already been used by Weinhold in [8]. This norm is also known as Shahshahani norm [9] and is employed for model reduction purposes in [7, 10].

3 Numerical Methods

The optimization problem (1) can be solved after discretrization as a standard nonlinear optimization problem (NLP) via the sequential quadratic programming (SQP) method [11] or interior point (IP) methods, e.g., [12]. In case of unstable dynamics, such as the backwards dynamics considered here, it is useful to have an "all at once" approach coupling simulation and optimization via discretization of the ODE constraint. This simultaneous approach introduces more freedom into the optimiza-

Model Reduction via Optimization of Trajectories 245

tion problem and is generally more stable. In the collocation method used for the computation of the results in the next section polynomials on a predefined grid are constructed fulfilling the differential equation at a certain number of nodes depending on the polynomial's degree. For the solutions presented in the following we use a Radau-method with linear, quadratic, and cubic polynomials, respectively, see e.g., [13]. The resulting NLP is solved using the interior point software package IPOPT [14] including linear algebra solvers of the HSL routines [15]. Derivatives necessary for the Newton steps of the optimizer are computed using the open source package CppAD [16]. Plots are generated using MATLAB®.

4 Numerical Results

In this section results for three different mechanisms are presented: a six component model mechanism, a six component simplified hydrogen combustion mechanism, and a realistic nine species hydrogen combustion mechanism. By means of these mechanisms the different criteria and modes of the method are utilized and results are visualized.

4.1 Model Mechanism

The first example is a six species model mechanism representing the characteristics of a simple hydrogen combustion (see Table 1). Element mass conservation for H and O ($2c_{H_2} + 2c_{H_2O} + c_H + c_{OH} = 2.0$, $2c_{O_2} + c_{H_2O} + c_O + c_{OH} = 1.0$) restricts the degrees of freedom in the optimization problem to four.

The different optimization criteria and formulations of our method are analyzed using this mechanism. Figure 1 shows results for the problem applying the forward mode of the method with the Euclidean norm objective functional (2). All free species concentrations are plotted vs. the reaction progress variable c_{H_2O}. The open circles mark different initial values of solutions to the optimization problems for $c_i(t_0 = t_*)$ and the solid lines are trajectories starting from these and converging to equilibrium (filled dot). Solving the optimization problem for different values of c_{H_2O} obviously yields a manifold as a collection of points (open circles) which approximates the SIM but violates invariance significantly.

Table 1 Simple test mechanism from [7]. Forward and backward rate coefficients are assumed to be temperature-independent

Reaction		k_+	k_-
H_2	$\rightleftharpoons 2H$	2.0	216.0
O_2	$\rightleftharpoons 2O$	1.0	337.5
H_2O	$\rightleftharpoons H + OH$	1.0	1400.0
$H_2 + O$	$\rightleftharpoons H + OH$	1000.0	10800.0
$O_2 + H$	$\rightleftharpoons O + OH$	1000.0	33750.0
$H_2 + O$	$\rightleftharpoons H_2O$	100.0	0.7714

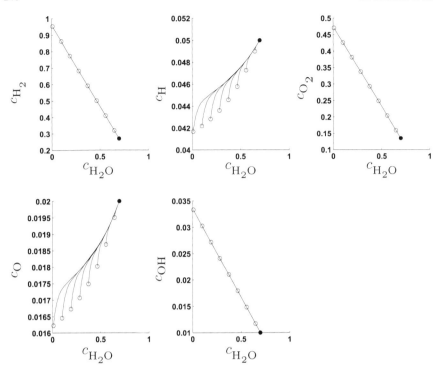

Fig. 1 Result of a one-dimensional manifold of the model mechanism with c_{H_2O} as progress variable and (2) as objective functional in the forward mode formulation ($t_0 = t_* = 0$) of problem (1)

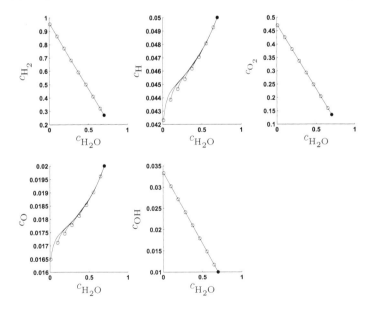

Fig. 2 Result of a one-dimensional manifold of the model mechanism with c_{H_2O} as progress variable and (4) as objective functional in the forward mode formulation

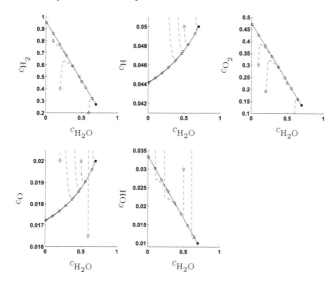

Fig. 3 Result of a one-dimensional manifold of the model mechanism with c_{H_2O} as progress variable and (2) as objective functional in the reverse mode formulation

In contrast, Fig. 2 shows results with better invariance properties. Here, the only difference in the problem formulation is the modification of the objective functional to (4) including the thermodynamic metric. The corresponding norm can also be understood as a weighting of different species concentrations. Small-valued radical species (as c_O and c_H, where the results in Fig. 1 particularly show deviations from invariance of the slow manifold) get a larger weight in the objective functional value. A drawback of (4) is the numerical difficulty caused by the weighting matrix. As concentrations of radical species can become very small one has to deal with very large weighting factors. This can lead to severe numerical instabilities during the computations.

Further improvement (avoiding the weighting terms) can be achieved using the reverse mode. Figure 3 depicts corresponding results. Here (2) serves as objective functional. The results of the reverse formulation are very close to invariance. Here the open circles represent final values $c_i(t_*) = c_i(t_f)$ of the optimal trajectory. Trajectories (continuous lines) starting from these values are added for visualization. Arbitrary trajectories being attracted by the computed slow manifold are also added in Fig. 3 (dashed lines).

4.2 Simplified Reaction Mechanism

As a second example a simplified hydrogen combustion mechanism is used. Ren et al. derive this mechanism as part of the detailed realistic one of Li et al. [17] and use it for testing the ICE-PIC method in [18]. It consists of six chemical species (including the inert gas N_2) and one third body M which are reacting in twelve elementary

Table 2 Adapted version of the simplified mechanism [18]. Rate coefficients k are computed as $k(T) = AT^b \exp(-E_a/RT)$, where R is the universal gas constant. In the mechanism M represents a third body being any species with collision efficiency $f_H = 1$, $f_{H_2} = 2.5$, $f_{OH} = 1$, $f_O = 1$, $f_{H_2O} = 12$, and $f_{N_2} = 1$

Reaction		A/cm, mol, s	b	$E_a / \frac{kJ}{mol}$
$O + H_2$	$\to H + OH$	5.08×10^{04}	2.7	26.3
$H + OH$	$\to O + H_2$	2.24×10^{04}	2.7	18.5
$H_2 + OH$	$\to H_2O + H$	2.16×10^{08}	1.5	14.4
$H_2O + H$	$\to H_2 + OH$	9.62×10^{08}	1.5	77.7
$O + H_2O$	$\to 2\,OH$	2.97×10^{06}	2.0	56.1
$2\,OH$	$\to O + H_2O$	2.94×10^{05}	2.0	-15.1
$H_2 + M$	$\to 2\,H + M$	4.58×10^{19}	-1.4	436.7
$2\,H + M$	$\to H_2 + M$	1.18×10^{19}	-1.4	0.7
$O + H + M$	$\to OH + M$	4.71×10^{18}	-1.0	0.0
$OH + M$	$\to O + H + M$	8.07×10^{18}	-1.0	428.2
$H + OH + M$	$\to H_2O + M$	3.80×10^{22}	-2.0	0.0
$H_2O + M$	$\to H + OH + M$	6.57×10^{23}	-2.0	499.4

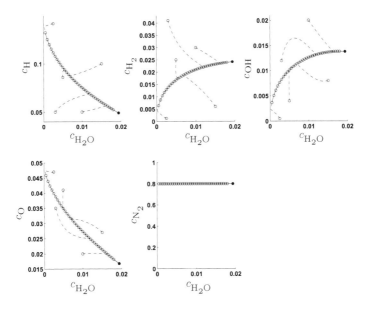

Fig. 4 Result of a one-dimensional manifold of the simplified hydrogen combustion mechanism with c_{H_2O} as progress variable and (2) as objective functional in the forward mode formulation

reactions. We use an adapted version given in Table 2 with three element mass conservation relations: $c_H + 2c_{H_2} + c_{OH} + 2c_{H_2O} = 0.15$, $c_{OH} + c_O + c_{H_2O} = 0.05$, and $2c_{N_2} = 1.6$. We compare forward and reverse mode of our method at a fixed temperature $T = 3000$ K.

Figure 4 shows results for the forward mode with (2) as objective functional. The value of c_{H_2O} serves as reaction progress variable and the plots are arranged as

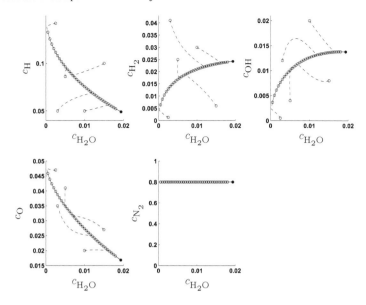

Fig. 5 Result of a one-dimensional manifold of the simplified hydrogen combustion mechanism with c_{H_2O} as progress variable and (2) as objective functional in the reverse mode formulation

in the example presented in the previous section. Here, the forward mode already yields a manifold with good invariance properties. Arbitrary (dashed) trajectories demonstrate the attraction property of the slow manifold. The reverse mode yields comparable results, these are depicted in Fig. 5.

4.3 Realistic Reaction Mechanism

As a final example a realistic detailed mechanism for hydrogen combustion is used. Originally published (with the coefficients used here) as part of a larger one in [19] it consists of nine species (including the inert gas N_2), two third bodies, and 20 (forward and reverse) reactions. These reactions include two pressure-dependent *Troe*-reactions [20, 21], where the rate coefficients are fitted to $p = 1$ bar in our case. The mass conservation relations reduce the degrees of freedom to six. A fixed temperature at $T = 1{,}500$ K is used. We show results for one- and two-dimensional manifolds applying the forward mode and objective functional (2).

Figure 6 depicts results of a one-dimensional manifold. The progress variable c_{H_2O} is varied again for checking the invariance of the solution. Arbitrary trajectories converging to the computed manifold illustrate the good quality of SIM approximation. The course of these trajectories suggests the geometry of a two-dimensional manifold.

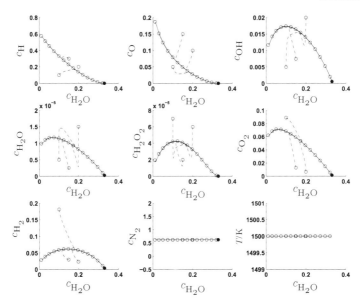

Fig. 6 Result of a one-dimensional manifold of the detailed hydrogen combustion mechanism with c_{H_2O} as progress variable and (2) as objective functional in the forward mode formulation. Additional arbitrary (*dashed*) trajectories are plotted showing relaxation on the one-dimensional manifold

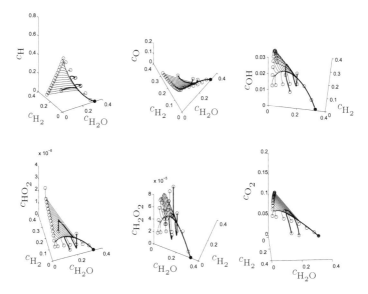

Fig. 7 Result of a two-dimensional manifold of the detailed hydrogen combustion mechanism with c_{H_2O} and c_{H_2} as progress variables and the same problem formulation as in Fig. 6. The same three arbitrary trajectories are plotted again (*thicker ones*)

Model Reduction via Optimization of Trajectories

Results for the computation of a two-dimensional slow manifold are shown in Fig. 7. The concentrations c_{H_2O} and c_{H_2} are chosen to parameterize the manifold and are varied along the edge of the physically feasible domain. The constant species N_2 and the constant temperature $T = 1,500\,K$ are omitted in the plot. The manifold can be regarded as being spanned by "optimal" trajectories in our approach. The same arbitrary trajectories as in Fig. 6 are plotted (thicker ones) relaxing onto the manifold.

Acknowledgements This work was supported by the German Research Foundation (DFG) through the Collaborative Research Center (SFB) 568.

References

1. Warnatz, J., Maas, U., Dibble, R.W.: Combustion: Physical and Chemical Fundamentals, Modeling and Simulation, Experiments, Pollutant Formation. Springer, Berlin (2006)
2. Lebiedz, D., Kammerer, J., Brandt-Pollmann, U.: Automatic network coupling analysis for dynamical systems based on detailed kinetic models. Phys. Rev. E, **72** (2005) 041911
3. Lebiedz, D.: Computing minimal entropy production trajectories: An approach to model reduction in chemical kinetics. J. Chem. Phys. **120** (2004) 6890–6897
4. Lebiedz, D., Reinhardt, V., Kammerer, J.: Novel trajectory based concepts for model and complexity reduction in (bio)chemical kinetics. In Gorban, A.N., Kazantzis, N., Kevrekidis I. G., Theodoropoulos C. (eds.): Model reduction and coarse-graining approaches for multi-scale phenomena. Springer, Berlin (2006) 343–364
5. Reinhardt, V., Winckler, M., Lebiedz, D.: Approximation of Slow Attracting Manifolds in Chemical Kinetics by Trajectory-Based Optimization Approaches. J. Phys. Chem. A **112** (2008) 1712–1718
6. Lebiedz, D., Reinhardt, V., Siehr, J.: Minimal curvature trajectories: Riemannian geometry concepts for model reduction in chemical kinetics. J. Comput. Phys. **229** (2010) 6512–6533
7. Gorban, A.N., Karlin, I.V., Zinovyev, A.Y.: Constructive methods of invariant manifolds for kinetic problems. Phys. Rep. **396** (2004) 197–403
8. Weinhold, F.: Metric geometry of equilibrium thermodynamics. J. Chem. Phys. **63** (1975) 2479–2483
9. Shahshahani, S.: A new mathematical framework for the study of linkage and selection. Mem. Am. Math. Soc. **17** (1979)
10. Chiavazzo, E., Karlin, I.V., Gorban, A.N., Boulouchos, K.: Combustion simulation via lattice Boltzmann and reduced chemical kinetics. J. Stat. Mech. (2009) P06013
11. Powell, M.J.D.: A fast algorithm for nonlinearly constrained optimization calculations. Vol. 630 of Lecture Notes in Mathematics, Springer, Berlin (1978) 144–157
12. Forsgren, A., Gill, P.E., Wright, M.H.: Interior Methods for Nonlinear Optimization. SIAM Rev. **44** (2002) 525–597
13. Ascher, U., Petzold, L.: Computer methods for ordinary differential equations and differential-algebraic equations. SIAM, Philadelphia (1998)
14. Wächter, A., Biegler, L.T.: On the implementation of a primal-dual interior point filter line search algorithm for large-scale nonlinear programming. Math. Prog. **106** (2006) 25–57
15. HSL: A collection of fortran codes for large-scale scientific computation. See URL http://www.hsl.rl.ac.uk (2007)
16. Bell, B.M., Burke, J.V.: Algorithmic differentiation of implicit functions and optimal values. In Bischof, C.H., Bücker, H.M., Hovland, P.D., Naumann U., Utke, J. (eds.): Advances in Automatic Differentiation. Springer, Berlin (2008) 67–77

17. Li, J., Zhao, Z., Kazakov, A., Dryer, F.L.: An updated comprehensive kinetic model of hydrogen combustion. Int. J. Chem. Kinet. **36** (2004) 566–575
18. Ren, Z., Pope, S.B., Vladimirsky, A., Guckenheimer, J.M.: The invariant constrained equilibrium edge preimage curve method for the dimension reduction of chemical kinetics. J. Chem. Phys. **124** (2006) 114111
19. Hegheş, C.I.: C_1-C_4 Hydrocarbon Oxidation Mechanism. PhD thesis, University of Heidelberg (2006)
20. Troe, J.: Theory of Thermal Unimolecular Reactions in the Fall-off Range. I. Strong Collision Rate Constants. Ber. Bunsenges. Phys. Chem. **87** (1983) 161–169
21. Gilbert, R., Luther, K., Troe, J.: Theory of Thermal Unimolecular Reactions in the Fall-off Range. II. Weak Collision Rate Constants. Ber. Bunsenges. Phys. Chem. **87** (1983) 169–177

Computing Realizations of Reaction Kinetic Networks with Given Properties

Gábor Szederkényi, Katalin M. Hangos, and Dávid Csercsik

Abstract The solution to the problem of finding the reaction kinetic realization of a given system obeying the mass action law containing the minimal/maximal number of reactions and complexes is shown in this paper. The proposed methods are based on Mixed Integer Linear Programming where the mass action kinetics is encoded into the linear constraints. Although the problems are NP-hard in the current setting, the developed algorithms give a usable answer to some of the questions first raised in [1].

1 Introduction

Positive systems are characterized by the property that all state variables remain nonnegative if the trajectories start in the nonnegative orthant. Thus, positive systems play an important role in fields such as chemistry, economy, population dynamics or even in transportation modeling where the state variables of the models are often physically constrained to be nonnegative [2]. It is remarked that most non-positive systems can be transformed into the positive class, where the distortion of the phase-space can be kept minimal in the region of interest [3].

An important class of positive nonlinear dynamical systems is the class of chemical reaction networks (shortly CRNs) obeying the mass-action law [4]. Such

---

K.M. Hangos (✉)
Process Control Research Group, Systems and Control Laboratory, Computer and Automation Research Institute, Hungarian Academy of Sciences 1518, P.O. Box 63, Budapest, Hungary
e-mail: hangos@sztaki.hu
and
Department of Electrical Engineering and Information Systems, University of Pannonia, 8200, Egyetem u. 10, Veszprém, Hungary

G. Szederkényi and D. Csercsik
Process Control Research Group, Systems and Control Laboratory, Computer and Automation Research Institute, Hungarian Academy of Sciences 1518, P.O. Box 63, Budapest, Hungary
e-mail: szeder@scl.sztaki.hu, csercsik@scl.sztaki.hu

A.N. Gorban and D. Roose (eds.), *Coping with Complexity: Model Reduction and Data Analysis*, Lecture Notes in Computational Science and Engineering 75, DOI 10.1007/978-3-642-14941-2_13, © Springer-Verlag Berlin Heidelberg 2011

networks can be used to describe pure chemical reactions, but they are also widely used to model the dynamics of intracellular processes, metabolic or cell signalling pathways [5]. Thus, reaction kinetic systems are able to describe key mechanisms both in industrial processes and living systems.

Moreover, CRNs represent a wide class of polynomial systems and therefore they are actively studied in the framework of general nonlinear systems' theory [6]. This practical and theoretical significance has triggered a renewed interest in the investigation of CRNs recently in the control community [7]. Even theorists working in quite distant fields from chemistry such as aerospace engineering consider CRNs as a very promising candidate for modeling and control of nonlinear systems [8].

The so-called inverse problem of reaction kinetics (i.e., the characterization of those polynomial differential equations which are kinetic) was solved in [1]. It is known from the "fundamental dogma of chemical kinetics" that different reaction networks can produce the same kinetic differential equations [9]. Naturally, this property has a fundamental impact on the identifiability of reaction rate constants [10]. However, the task of finding realization(s) with given prescribed properties, e.g., realizations with minimal number of reactions or with a given deficiency, has not attracted a lot of attention. This task will be called the *realization problem* of reaction kinetic networks, and it is the subject of our paper.

Recent studies indicate [11] that it is possible and quite useful to formulate and solve realization problems of different kinds in the framework of mixed integer linear programming (MILP) [12], where the continuous optimization variables are the nonnegative reaction rate coefficients, and the corresponding integer variables ensure the finding of the realization with the minimal or maximal number of reactions. Although the general MILP problem is NP-hard, the above approach is applicable to realizations problems of realistic size, thus it gives an alternative to model reduction of CNRs.

In addition to the analysis of CRNs, several authors studied the possibilities of *dimension reduction* for large chemical networks (see, e.g., [13,14]) which is of vital importance in the case of networks with several hundred (or even more) complexes and reactions that can often be found in natural or technological systems. Some of the approaches in this area treat the dimension reduction task as an optimal dynamic approximation problem.

The aim of the present paper is to further generalize the above MILP-based optimal realization construction results [11] by developing methods for finding realizations with the minimal or maximal number of complexes and/or reactions.

2 Basic Notions and Tools

The mathematical description of reaction kinetic networks together with the most important notions of mixed integer linear programming problems are given in this section following [15].

2.1 Mass Action Reaction Networks and their Representation

The original physical picture underlying the reaction kinetic system class is a closed system under isothermal and isobaric conditions, where chemical species X_i, $i = 1, \ldots, n$ take part in r chemical reactions. The system is perfectly stirred, i.e., concentrated parameter in the simplest case. The concentrations x_i, $i = 1, \ldots, n$ form the state vector the elements of which are non-negative by nature. For the sake of simplicity, physico-chemical properties of the system are assumed to be constant.

The origin of mass action law lies in the *molecular collision picture* of chemical reactions. Here the reaction occurs when either two reactant molecules collide, or a reactant molecule collides with an inactive (e.g., solvent) molecule. Clearly, the probability of having a reaction is proportional to the probability of collisions, that is proportional to the concentration of the reactant(s).

A straightforward generalization of the above molecular collision picture is when we allow to have multi-molecule collisions to obtain *elementary reaction steps* in the following form [16]:

$$\sum_{i=1}^{n} \alpha_{ij} X_i \to \sum_{i=1}^{n} \beta_{ij} X_i, \quad j = 1, \ldots, r \tag{1}$$

where α_{ij} is the so-called *stoichiometric coefficient* of component X_i in the jth reaction, i.e., the number of colliding X_i molecules, and $\beta_{i\ell}$ is the stoichiometric coefficient of the product X_ℓ. The linear combinations of the species in eq. (1), namely $\sum_{i=1}^{n} \alpha_{ij} X_i$ and $\sum_{i=1}^{n} \beta_{ij} X_i$ for $j = 1, \ldots, r$ are called the complexes and are denoted by $C_1, C_2, \ldots, C_m$. Note that *the stoichiometric coefficients are always non-negative integers in classical reaction kinetic systems*.

According to the extended molecular picture, the reaction rate of the above reactions can be described as

$$\rho_j = k_j \prod_{i=1}^{n} [X_i]^{\alpha_{ij}} = k_j \prod_{i=1}^{n} x_i^{\alpha_{ij}}, \quad j = 1, \ldots, r \tag{2}$$

where $[X_i] = x_i$ is the concentration of the component X_i, and $k_j > 0$ is the *reaction rate constant* of the jth reaction, that is always positive.

If the reactions $C_i \to C_j$ and $C_j \to C_i$ take place at the same time in a reaction network for some i, j then this pair of reactions is called a reversible reaction (but it will be treated as two separate elementary reactions).

Similarly to [17], we can assign the following directed graph (see, e.g., [18]) to the reaction network (1) in a straightforward way. The directed graph $D = (V_d, E_d)$ of a reaction network consists of a finite nonempty set V_d of vertices and a finite set E_d of ordered pairs of distinct vertices called directed edges. The vertices correspond to the complexes, i.e., $V_d = \{C_1, C_2, \ldots C_m\}$, while the directed edges represent the reactions, i.e., $(C_i, C_j) \in E_d$ if complex C_i is transformed to C_j in the reaction network. The reaction rates k_j for $j = 1, \ldots, r$ in (2) are assigned

Fig. 1 Kinetic realization of the classical 2 dimensional Lotka–Volterra predator-prey model

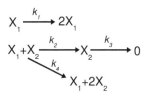

as positive weights to the corresponding directed edges in the graph. A *walk* in the reaction graph is an alternating sequence $W = C_1 E_1 C_2 E_2 \ldots C_{k-1} E_{k-1} C_k E_k$ where $C_i \in V_d$, $E_i \in E_d$ for $i = 1, \ldots, k$. W is a *directed path* if all the vertices in it are distinct. P is called a *directed cycle* if the vertices $C_1, C_2, \ldots, C_{k-1}$ are distinct, $k \geq 3$ and $C_1 = C_k$. A set of complexes $\{C_1, C_2, \ldots, C_k\}$ is a *linkage class* of a reaction network if the complexes of the set are linked to each other in the reaction graph but not to any other complex [19]. As an illustration, the CRN realization of the classical Lotka–Volterra predator-prey model is shown in Fig. 1, where the concentrations of species X_1 and X_1 correspond to the number of preys and predators, respectively.

There are several possibilities to represent the dynamic equations of mass action systems (see, e.g., [17, 20], or [10]). The most advantageous form for our purposes is the one that is used e.g., in Lecture 4 of [17], i.e.,

$$\dot{x} = Y \cdot A_k \cdot \psi(x) \tag{3}$$

where $x \in \mathbb{R}^n$ is the concentration vector of the species, $Y \in \mathbb{R}^{n \times m}$ stores the stoichiometric composition of the complexes, $A_k \in \mathbb{R}^{m \times m}$ contains the information corresponding to the weighted directed graph of the reaction network, and $\psi : \mathbb{R}^n \mapsto \mathbb{R}^m$ is a monomial-type vector mapping defined by

$$\psi_j(x) = \prod_{i=1}^{n} x_i^{y_{ij}}, \quad j = 1, \ldots, m \tag{4}$$

where $y_{ij} = [Y]_{ij}$. The exact structure of Y and A_k is the following. The ith column of Y contains the composition of complex C_i, i.e., Y_{ji} is the stoichiometric coefficient of C_i corresponding to the specie $\mathbf{X}_j$. A_k is a column conservation matrix (i.e., the sum of the elements in each column is zero) defined as

$$[A_k]_{ij} = \begin{cases} -\sum_{l=1}^{m} k_{il}, & \text{if } i = j \\ k_{ji}, & \text{if } i \neq j \end{cases} \tag{5}$$

In other words, the diagonal elements $[A_k]_{ii}$ contain the negative sum of the weights of the edges starting from the node C_i, while the off-diagonal elements $[A_k]_{ij}$, $i \neq j$ contain the weights of the directed edges (C_j, C_i) coming into C_i. Based on the above properties, it is appropriate to call A_k the *Kirchhoff matrix* of a reaction network.

To handle the exchange of materials between the environment and the reaction network, the so-called "zero-complex" can be introduced and used which is a special

Computing Realizations of Reaction Kinetic Networks with Given Properties 257

complex with the all stoichiometric coefficients zero i.e., it is represented by a zero vector in the Y matrix (for the details, see, e.g., [17] or [21]).

It is not difficult to show that all positive linear systems are kinetic (i.e., a mass-action reaction network exactly realizes their dynamics). The situation is a bit more complex in the nonlinear case, but vast majority of polynomial positive systems are also kinetic [22].

We can associate an n-dimensional vector with each reaction in the following way. For the reaction $C_i \rightarrow C_j$, the corresponding reaction vector denoted by e_k is given by

$$e_k = [Y]_{.,j} - [Y]_{.,i} \tag{6}$$

where $[Y]_{.,i}$ denotes the ith column of Y. Any convention can be used for the numbering of the reaction vectors (e.g., the indices i and j in (6) can be treated as digits in a decimal system). The *rank* of a reaction network denoted by s is defined as the rank of the vector set $\{e_1, e_2 \ldots, e_r\}$ where r is the number of reactions.

The deficiency δ of a reaction network is defined as [17], [19]

$$\delta = m - l - s \tag{7}$$

where m is the number of complexes in the network, l is the number of linkage classes and s is the rank of the reaction network.

A reaction network is called *reversible*, if each of its reactions is a reversible reaction. A reaction network is called weakly reversible, if each complex in the reaction graph lies on at least one directed cycle (i.e., if complex C_j is reachable from complex C_i on a directed path in the reaction graph, then C_i is reachable from C_j on a directed path).

The deficiency combined with reversibility is a very useful tool for studying the dynamical properties of reaction networks and for establishing parameter-independent global stability conditions [19, 23]. The *deficiency zero* property for a sufficiently reversible (more precisely: "weakly reversible") reaction network guarantees the global stability of the solutions of the system with a known Lyapunov function [19].

Using the notation

$$M = Y \cdot A_k, \tag{8}$$

Equation (3) can be written in the compact form

$$\dot{x} = M \cdot \psi(x) \tag{9}$$

2.2 Mixed Integer Linear Programming (MILP) and Its Solution

A special subset of optimization problems is the class of Mixed Integer Linear Programs (MILPs) where the objective function and the constraints are linear functions of the decision variables. A mixed integer linear program with k variables (denoted by $y \in \mathbb{R}^k$) and p constraints can be written as [24]:

$$
\begin{aligned}
&\text{minimize } c^T y \\
&\text{subject to:} \\
&A_1 y = b_1 \\
&A_2 y \leq b_2 \\
&l_i \leq y_i \leq u_i \text{ for } i = 1, \ldots, k \\
&y_j \text{ is integer for } j \in I, \ I \subseteq \{1, \ldots, k\}
\end{aligned}
\tag{10}
$$

where $c \in \mathbb{R}^k$, $A_1 \in \mathbb{R}^{p_1 \times k}$, $A_2 \in \mathbb{R}^{p_2 \times k}$, and $p_1 + p_2 = p$.

If all the variables can be real, then (10) is a simple linear programming problem that can be solved in polynomial time. However, if any of the variables is integer, then the problem becomes NP-hard. In spite of this, there exist a number of free (e.g., YALMIP [25] or the GNU Linear Programming Kit [26]) and commercial (such as CPLEX or TOMLAB [27]) solvers that can efficiently handle many practical problems.

3 Using MILP to Compute Realizations with Required Properties

This section is devoted to show how the problems of finding realizations of CNRs with minimal or maximal number of reactions and complexes can be cast as a MILP and how to solve them effectively.

Consider the polynomial system (9). We will call the matrix M in (9) *admissible*, if the polynomial differential equations describe a mass-action reaction network. Conditions for this were first given in [1] but not through the properties of M. The matrix pair (Y, A_k) is called a *realization* of an admissible matrix M if $Y \cdot A_k = M$, the elements of Y are non-negative integers, and A_k is a column conservation matrix with non-positive diagonal and non-negative off-diagonal elements. This way, we can define the alternative realizations of a reaction network, since M is computable from the structure and parameters of a given reaction system.

3.1 Representation of Mass Action Kinetics as Linear Equality Constraints

The starting point for the forthcoming calculations is that a reaction network is given with its reaction graph or equivalently with its realization (Y, A_k) and we want to compute its sparsest or densest realization denoted by (Y^s, A_k^s) and (Y^d, A_k^d), respectively. Furthermore, we make the restriction that the complexes in the newly found realizations form a subset of the original complexes, i.e., $\mathrm{col}(Y^s) \subseteq \mathrm{col}(Y)$ and $\mathrm{col}(Y^d) \subseteq \mathrm{col}(Y)$. In principle, the alternative realizations may contain such

Computing Realizations of Reaction Kinetic Networks with Given Properties

complexes that do not appear in the original reaction network, but we won't elaborate on this case. It is assumed that a maximal possible set of complexes for the reaction network is given in advance. We remark, that obviously, the sparsest or densest realization may not be unique (parametrically and/or structurally), but here our goal is to find one possible solution.

For the computations, let us represent the Kirchhoff matrix of a reaction network containing m complexes as

$$A_k = \begin{bmatrix} -a_{11} & a_{12} & \cdots & a_{1m} \\ a_{21} & -a_{22} & \cdots & a_{2m} \\ \vdots & & & \vdots \\ a_{m1} & a_{m2} & \cdots & -a_{mm} \end{bmatrix} \tag{11}$$

Keeping in mind the properties of A_k, the negative sign in (11) for the diagonal elements a_{ii} for $i = 1, \ldots, m$ will allow us to set a uniform nonnegativity (or identically tractable lower and upper bound) constraint for all a_{ij} in the later computations.

Let us denote the ith row and ith column of a matrix W by $[W]_{i,\cdot}$ and $[W]_{\cdot,i}$, respectively. Using (11), the individual linear equations of the matrix equation (8) can be written as

$$-y_{11}a_{11} + y_{12}a_{21} + \cdots + y_{1m}a_{m1} = [M]_{11} \tag{12}$$

$$\vdots$$

$$-y_{n1}a_{11} + y_{n2}a_{21} + \cdots + y_{nm}a_{m1} = [M]_{n1} \tag{13}$$

$$y_{11}a_{12} - y_{12}a_{22} + \cdots + y_{1m}a_{m2} = [M]_{12} \tag{14}$$

$$\vdots$$

$$y_{n1}a_{12} - y_{n2}a_{22} + \cdots + y_{nm}a_{m2} = [M]_{n2} \tag{15}$$

$$\vdots$$

$$y_{11}a_{1m} + y_{12}a_{2m} + \cdots - y_{1m}a_{mm} = [M]_{1m} \tag{16}$$

$$\vdots$$

$$y_{n1}a_{1m} + y_{n2}a_{2m} + \cdots - y_{nm}a_{mm} = [M]_{nm} \tag{17}$$

The property that A_k is a column conservation matrix can also be expressed in the form of linear equations:

$$-a_{11} + a_{21} + a_{31} + a_{41} = 0 \tag{18}$$

$$a_{12} - a_{22} + a_{32} + a_{42} = 0 \tag{19}$$

$$\vdots$$

$$a_{1m} + a_{2m} + \cdots - a_{mm} = 0 \tag{20}$$

Equations (12)–(20) can be written in the following more compact form:

$$\begin{bmatrix} \bar{Y}^1 & 0 & 0 & \cdots & 0 \\ 0 & \bar{Y}^2 & 0 & \cdots & 0 \\ & & \vdots & & \\ 0 & 0 & 0 & \cdots & \bar{Y}^m \end{bmatrix} \begin{bmatrix} [A_k]_{\cdot,1} \\ [A_k]_{\cdot,2} \\ \vdots \\ [A_k]_{\cdot,m} \end{bmatrix} = \begin{bmatrix} [\bar{M}]_{\cdot,1} \\ [\bar{M}]_{\cdot,2} \\ \vdots \\ [\bar{M}]_{\cdot,m} \end{bmatrix} \tag{21}$$

where the zeros denote zero matrix blocks of size $(n+1) \times m$ and

$$\bar{Y}^i = \begin{bmatrix} [Y]_{\cdot,1} & [Y]_{\cdot,2} & \cdots & [Y]_{\cdot,i-1} & -[Y]_{\cdot,i} & [Y]_{\cdot,i+1} & \cdots & [Y]_{\cdot,m} \\ 1 & 1 & \cdots & 1 & -1 & 1 & \cdots & 1 \end{bmatrix} \in \mathbb{R}^{(n+1) \times m}, \tag{22}$$

$$\bar{M} = \begin{bmatrix} M \\ 0 \cdots 0 \end{bmatrix} \in \mathbb{R}^{(n+1) \times m} \tag{23}$$

3.2 Computing Realizations with the Minimal/Maximal Number of Reactions

It is visible from (21) that the optimization variable will contain the reaction rate coefficients, i.e., the elements of A_k as the matrix Y is known and fixed by the problem statement. For the sake of simplicity, let us use the notation

$$z = \begin{bmatrix} z^{(1)} \\ z^{(2)} \\ \vdots \\ z^{(m)} \end{bmatrix} = \begin{bmatrix} [A_k]_{\cdot,1} \\ [A_k]_{\cdot,2} \\ \vdots \\ [A_k]_{\cdot,m} \end{bmatrix} \tag{24}$$

where obviously, $z^{(i)} \in \mathbb{R}^m$, $i = 1, \ldots, m$.

When we seek the sparsest realization of the original reaction network (A_k, Y) then we are searching for the sparsest solution of (21), i.e., the one containing the maximal number of zeros (or the minimal number of zeros, if the densest realization is to be computed). For this, let us associate logical variables $\delta_j^{(i)}$ with the continuous variables $z_j^{(i)}$ for $i, j = 1, \ldots, m$. Then the optimization variable previously denoted by y is

$$y = \begin{bmatrix} z \\ \delta \end{bmatrix}. \tag{25}$$

Following from the problem statement and construction, the lower bound for the continuous variables is zero. For the solvability of the MILP problem, also an upper

Computing Realizations of Reaction Kinetic Networks with Given Properties 261

bound is introduced for z, i.e.,

$$0 \leq z_i \leq u_i, \ u_i > 0, \ i = 1, \ldots, m^2 \tag{26}$$

To minimize (or maximize) the number of nonzeros in the continuous solution part z, the following compound statement have to be translated to linear inequalities

$$\delta_i = 1 \leftrightarrow z_i > 0, \ i = 1, \ldots, m^2 \tag{27}$$

To be able to numerically distinguish between practically zero and nonzero solutions, (27) is modified to

$$\delta_i = 1 \leftrightarrow z_i > \epsilon, \ i = 1, \ldots, m^2 \tag{28}$$

where $0 < \epsilon \ll 1$ (i.e., solutions below ϵ are treated as zero). Taking into consideration (26), the linear inequalities corresponding to (28) are

$$0 \leq z_i - \epsilon \delta_i, \ i = 1, \ldots, m^2 \tag{29}$$

$$0 \leq -z_i + u_i \delta_i, \ i = 1, \ldots, m^2 \tag{30}$$

Now, the MILP problem for finding the sparsest realization can be constructed as

$$\text{minimize} \ \sum_{m^2+1}^{2m^2} y_i \tag{31}$$

subject to:

$$\begin{bmatrix} \bar{Y}^1 & 0 & 0 & \ldots & 0 \\ 0 & \bar{Y}^2 & 0 & \ldots & 0 \\ \vdots & & & & \vdots \\ 0 & 0 & 0 & \ldots & \bar{Y}^m \end{bmatrix} \begin{bmatrix} y_1 \\ y_2 \\ \vdots \\ y_{m^2} \end{bmatrix} = \begin{bmatrix} [\bar{M}]_{\cdot,1} \\ [\bar{M}]_{\cdot,2} \\ \vdots \\ [\bar{M}]_{\cdot,m} \end{bmatrix} \tag{32}$$

$$0 \leq y_i \leq u_i \ \text{for} \ i = 1, \ldots, m^2 \tag{33}$$

$$0 \leq y_i - \epsilon y_{i+m^2}, \ i = 1, \ldots, m^2 \tag{34}$$

$$0 \leq -y_i + u_i y_{i+m^2}, \ i = 1, \ldots, m^2 \tag{35}$$

$$y_j \ \text{is integer for} \ j = m^2 + 1, \ldots, 2m^2 \tag{36}$$

In the case when the densest realization is searched for, the optimization task (31) is simply changed to

$$\text{minimize} \ \left(- \sum_{m^2+1}^{2m^2} y_i \right) \tag{37}$$

Remark 1. By setting the lower and upper bounds for y_i differently from what is given in (33), the presence or omission of certain reactions can be forced during the optimization.

Remark 2. The block-diagonal structure of the coefficient matrix in (32) and the independence of the inequalities (33)–(35) allow us to partition the optimization variable y to m partitions and thus to solve the m resulting MILP subproblems paralelly which is a significant advantage from a computational point of view [28].

Remark 3. The block-diagonal structure mentioned in the previous remark makes it possible to combine different objective functions for different source complexes (since column i of A_k contains the rate coefficients corresponding to the reactions starting from complex C_i). E.g., the number of reactions starting from certain complexes can be minimized while it can be maximized for other complexes.

Remark 4. We note that the sparsest solution of certain sets of underdetermined linear equations can be obtained in polynomial time using linear programming (LP) [29, 30]. However, the applicability conditions of this LP solution are not fulfilled for many reaction networks.

3.3 Computing Realizations with the Minimal/Maximal Number of Complexes

To solve this case, we need the following simple observation. A complex disappears from the reaction network's graph, if both the corresponding column and row in A_k contain only zeros. This means that no directed edges start from or point to this complex in the graph and therefore it becomes an isolated vertex that can be omitted. Therefore, the m integer (or more precisely, boolean) variables are now assigned to row and column sums of A_k in the following way:

$$\delta_i = 1 \leftrightarrow \sum_{j_1=1}^{m} A_k(i, j_1) + \sum_{j_2=1}^{m} A_k(j_2, i) > \epsilon, \quad i = 1, \dots, m \qquad (38)$$

The linear constraints corresponding to mass action characteristics are the same as in (32). Equation (38) can also be translated into appropriate linear equalities. The objective function to be minimized is now

$$\pm \sum_{i=1}^{m} \delta_i \qquad (39)$$

We remark that this problem is not paralellizable in a straightforward way like the previous one, but the number of discrete variables is only m (compared to m^2 in the previous case).

4 Case Studies

Two simple yet interesting case studies illustrate the use of the proposed methods for finding the realizations of a CNR with minimal/maximal reactions and/or complexes.

4.1 A Simple 2-Dimensional Lotka–Volterra Model

Consider the simple oscillating Lotka–Volterra model shown in Fig. 1. The differential equations of the reaction mechanism are well-known:

$$\begin{aligned}\dot{x}_1 &= k_1 x_1 - k_2 x_1 x_2 \\ \dot{x}_2 &= -k_3 x_2 + k_4 x_1 x_2\end{aligned} \quad (40)$$

where x_i denote the concentrations of species X_i, and all constants are positive. The Y and A_k matrices characterizing the system are the following:

$$Y = \begin{bmatrix} 0 & 1 & 2 & 1 & 0 & 1 \\ 0 & 0 & 0 & 1 & 1 & 2 \end{bmatrix}, \quad A_k = \begin{bmatrix} 0 & 0 & 0 & 0 & k_3 & 0 \\ 0 & -k_1 & 0 & 0 & 0 & 0 \\ 0 & k_1 & 0 & 0 & 0 & 0 \\ 0 & 0 & 0 & -k_2 - k_4 & 0 & 0 \\ 0 & 0 & 0 & k_2 & -k_3 & 0 \\ 0 & 0 & 0 & k_4 & 0 & 0 \end{bmatrix} \quad (41)$$

For the numerical computations, let us choose the reaction rates $k_1, \ldots, k_4$ to be uniformly 1. The algorithm described in Sect. 3.2 gives that if the set of complexes are given by matrix Y in (41), then the sparsest realization is the original network shown in Fig. 1 with a deficiency of 2. However, the densest realization contains eight reactions instead of the original 4 as it is shown in Fig. 2. It is visible that this realization contains only one linkage class and therefore its deficiency is 3.

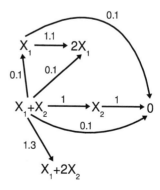

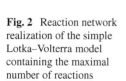

Fig. 2 Reaction network realization of the simple Lotka–Volterra model containing the maximal number of reactions

Fig. 3 Realization of the simple Lotka–Volterra model with a new complex and only three reactions

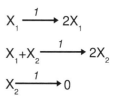

Fig. 4 Initial reaction network of example 4.2. All reaction rates are chosen to be 1

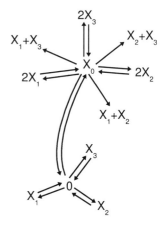

An interesting case occurs, if a new complex, namely $2X_2$ is added to the system that was previously not present in Y. The updated matrices of the system are then the following

$$Y' = \begin{bmatrix} 0 & 1 & 2 & 1 & 0 & 1 & 0 \\ 0 & 0 & 0 & 1 & 1 & 2 & 2 \end{bmatrix}, \quad A'_k = \begin{bmatrix} A_k & & 0 \\ & \vdots & \\ 0 & \ldots & 0 \end{bmatrix}. \tag{42}$$

If we now search for the realization with the minimal number of reactions, we find one with only three reactions, three linkage classes and a deficiency of 1 as it can be seen in Fig. 3. After mass-balancing this network, we find that it is identical to the one that is proved to be the smallest reaction network that produces oscillations [31, 32]. It has to be added that a parametric condition for the existence of this minimal realization is $k_2 = k_4$ in (40).

4.2 An Interesting Literature Example

Let us revisit the reaction network that was shown in Fig. 6 in Sect. 6 of [10]. The graph of the system is reproduced in Fig. 4. For convenience, let us select each

Fig. 5 Reaction network of example 4.2 containing the minimal number of complexes

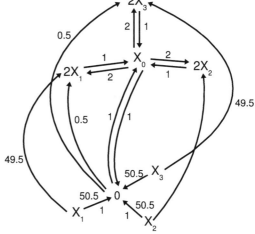

Fig. 6 Reaction network of example 4.2 obtained by first minimizing the number of complexes and then minimizing the number of reactions

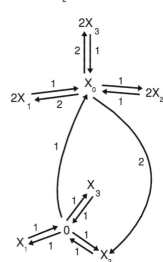

reaction rate to be 1 again. The sparsest realization of this network with nine complexes and 12 reactions was shown in [15]. The algorithm described in Sect. 3.3 shows that complexes $X_1 + X_3$, $X_2 + X_3$ and $X_1 + X_2$ can be omitted from the network. The graph of the resulting network containing the minimal number of complexes can be seen in Fig. 5. It is visible that this realization has eight complexes and 15 reactions.

Therefore, it makes sense to search for a sparse realization (containing the minimal number of reactions) of this system using the reduced complex set. The result with eight complexes and 14 reactions is visible in Fig. 6. It is important to remark that firstly reducing the number of complexes made the minimization of the number of reactions less effective.

5 Summary

The solution to the problem of finding the reaction kinetic realization of a given system obeying the mass action law containing the minimal/maximal number of reactions and complexes were presented in this paper. The proposed methods are based on Mixed Integer Linear Programming where the mass action kinetics is encoded into the linear constraints.

The problem of finding realization with minimal/maximal number of reactions was shown to be effectively parallelizable. It was also illustrated that firstly reducing the number of complexes made the minimization of the number of reactions less effective.

The proposed methods were illustrated in two case studies: in finding optimal realizations of a simpe oscillating Lotka-Volterra model, and on the example taken from [10].

Acknowledgements This work was supported by the Hungarian National Research Fund (OTKA K-67625). The first author is a grantee of the Bolyai scholarship of the Hungarian Academy of Sciences.

References

1. Hárs, V., Tóth, J.: On the inverse problem of reaction kinetics. In: Qualitative Theory of Differential Equations, Farkas, M., Hatvani L. (eds.). North-Holland, Amsterdam (1981) 363–379
2. Farina, L., Rinaldi, S.: Positive Linear Systems: Theory and Applications. Wiley, NY (2000)
3. Samardzija, N., Greller, G.D., Wassermann, E.: Nonlinear chemical kinetic schemes derived from mechanical and electrical dynamical systems. Journal of Chemical Physics **90(4)** (1989) 2296–2304
4. Horn, F., Jackson, R.: General mass action kinetics. Archive for Rational Mechanics and Analysis **47** (1972) 82–116
5. Haag, J., Wouver, A., Bogaerts, P.: Dynamic modeling of complex biological systems: a link between metabolic and macroscopic description. Mathematical Biosciencs **193** (2005) 25–49
6. Sonta, E.: Structure and stability of certain chemical networks and applications to the kinetic proofreading model of T-cell receptor signal transduction. IEEE Trans. Autom. Control **46** (2001) 1028–1047
7. Angeli. D.: A tutorial on chemical network dynamics. European Journal of Control **15** (2009) 398–406
8. Chellaboina, V., Bhat, S.P., Haddad, W.M., Bernstein., D.S.: Modeling and Analysis of Mass-Action Kinetics – Nonnegativity, Realizability, Reducibility, and Semistability. IEEE Control Systems Magazine **29** (2009) 60–78
9. Schnell, S., Chappell, M.J., Evans, N.D., Roussel, M.R.: The mechanism distinguishability problem in biochemical kinetics: The single-enzyme, single-substrate reaction as a case study. Comptes Rendus Biologies **329** (2006) 51–61
10. Craciun, G., Pantea, C.: Identifiability of chemical reaction networks. Journal of Mathematical Chemistry **44** (2008) 244–259
11. Szederkényi, G.: Comment on Identifiability of chemical reaction networks by G. Craciun and C. Pantea. Journal of Mathematical Chemistry **45** (2009) 1172–1174

12. Floudas, C.A.: Nonlinear and mixed-integer optimization. Oxford University Press, London (1995)
13. Gorban, A.N., Karlin I.V.: Method of invariant manifold for chemical kinetics. Chemical Engineering Science **58** (2003) 4751–4768
14. Hardin, H.M., can Schuppen, J.H.: System reduction of nonlinear positive systems by linearization and truncation. Lecture Notes in Control and Information Sciences **341** (2006) 431–438
15. Szederkenyi, G.: Computing sparse and dense realizations of reaction kinetic systems. J. Math. Chem. **47**(2) (2010) 551–568
16. Hangos, K.M., Szederkényi, G.: Special Positive Systems: the QP and the Reaction Kinetic System Class. In: Preprints of the Workshop on Systems and Control Theory in honor of József Bokor on his 60th Birthday, Hungarian Academy of Sciences (2008)
17. Feinberg, M.: Lectures on chemical reaction networks, Notes of lectures given at the Mathematics Research Centre, University of Wisconsin (1979)
18. Bang-Jensen, J., Gutin, G.: Digraphs: Theory, Algorithms and Applications. Springer, Berlin (2001)
19. Feinberg, M.: Chemical reaction network structure and the stability of complex isothermal reactors – I. The deficiency zero and deficiency one theorems. Chemical Engineering Science **42**(10) (1987) 2229–2268
20. Gorban A.N., Karlin, I.V., Zinovyev, A.Y.: Invariant grids for reaction kinetics. Physica A **33** (2004) 106–154
21. Craciun, G., Feinberg, M.: Multiple equilibria in complex chemical reaction networks: I. The injectivity property. SIAM Journal on Applied Mathematics **65**(5) (2005) 1526–1546
22. Erdi, P., Tóth, J.: Mathematical Models of Chemical Reactions. Theory and Applications of Deterministic and Stochastic Models. Manchester University Press, Manchester; Princeton University Press, Princeton (1989)
23. Feinberg, M.: Chemical reaction network structure and the stability of complex isothermal reactors – II. Multiple steady states for networks of deficiency one. Chemical Engineering Science **43** (1988) 1–25
24. Nemhauser, G.L., Wolsey, L. A.: Integer and Combinatorial Optimization. Wiley, NY (1988)
25. Löfberg, J.: YALMIP : A Toolbox for Modeling and Optimization in MATLAB In: Proceedings of the CACSD Conference, Taipei, Taiwan (2004)
26. Makhorin, A.: GNU Linear Programming Kit. Reference Manual. Version 4.10 (2006)
27. Homlström, K., Edvall, M.M., Göran, A.O.: TOMLAB for large-scale robust optimization. In: Nordic MATLAB Conference (2003)
28. Aykanat, C., Pinar, A.: Permuting sparse rectangular matrices into block-diagonal form. SIAM J. Scientific Computing **25** (2004) 1860–1879
29. Donoho, D.L., Tanner, J.: Sparse nonnegative solution of underdetermined linear equations by linear programming. Proc. of the National Academy of Sciences of the USA (PNAS) **102** (2005) 9446–9451
30. Donoho, D.L.: For most large undetermined systems of linear equations the minimal l1-norm solution is also the sparsest solution. Communications on Pure and Applied Mathematics **59** (2006) 903–934
31. Lotka, A.J.: Undamped oscillations derived from the law of mass action. J. Am. Chem. Soc. **42** (1920) 1595–1599
32. Wilhelm, T.: The smallest chemical reaction system with bistability. BMC Systems Biology **3** (2009) 90

A Drift-Filtered Approach to Diffusion Estimation for Multiscale Processes

Yves Frederix and Dirk Roose

Abstract This paper deals with parameter estimation in the context of so-called multiscale diffusions. The aim is to describe the coarse dynamics of a multiscale system by an approximation in the form of a stochastic differential equation with the coarse (homogenized) drift and diffusion coefficients estimated from numerical simulations of the multiscale system. Recently, it was found that for general stochastic homogenization problems a minimal time interval between consecutive observations must be respected for the estimators to be able to "see" the homogenized behavior of the system. If this *subsampling* interval is large compared to the coarse dynamics of the system, the standard estimators for these coefficients cannot be used to obtain accurate estimates as their convergence behavior strongly depends on the used subsampling. In this work, we focus on diffusion estimation and propose a procedure based on maximum-likelihood estimation that addresses this problem. First, the contributions due to the drift are filtered from the input data, after which the transformed data is used in the estimation. Due to this preprocessing step, the behavior of the standard estimator changes and it is possible to find good estimates for the homogenized diffusion coefficient as long as the subsampling interval is larger than the minimal required value. Various numerical examples are presented for both known and unknown coarse drift.

1 Introduction

Approximating complex multiscale systems by simpler models is a useful way to study its dynamics without having to deal with the various fine-scale effects that are present in the full description. Much research effort has gone into finding good approximation strategies to describe the relevant dynamics by adequate

---

Y. Frederix (✉) and D. Roose
Department of Computer Science, K.U.Leuven, Celestijnenlaan 200A, 3001 Leuven, Belgium
e-mail: yves.frederix@cs.kuleuven.be, Dirk.Roose@cs.kuleuven.be

A.N. Gorban and D. Roose (eds.), *Coping with Complexity: Model Reduction and Data Analysis*, Lecture Notes in Computational Science and Engineering 75, DOI 10.1007/978-3-642-14941-2_14, © Springer-Verlag Berlin Heidelberg 2011

low-dimensional effective models. Application areas are ubiquitous with examples in molecular dynamics, finance, ocean sciences and computational biology.

In this paper, we focus on systems for which the coarse dynamics is an effective diffusion model described by a stochastic differential equation (SDE) of the form

$$dX(t) = \gamma(X)dt + \sigma dW_t, \tag{1}$$

in which $X \in \mathbb{R}^N$ and W_t is an N-dimensional Wiener process. The predominant dynamics of the multiscale system is described by the drift $\gamma(X)$, while unresolved small scale effects are represented by the diffusion σ via a stochastic noise term. If it can be assumed that the coarse dynamics of the multiscale system can indeed be described by (1), the main challenge is to find appropriate expressions for the unknown drift and diffusion. In some cases, it is possible to derive (1) from the full system description by making certain assumptions or approximations, so that closed expressions can be obtained. Often, however, this cannot be done and they have to be estimated by matching observations from the multiscale system to the effective model. Typical strategies are based on the ideas of maximum-likelihood, see, e.g., [1, 8]. An alternative spectral approach was proposed by Crommelin and Vanden-Eijnden and is based on matching data from time series to a discrete-time Markov chain model [3].

Recently, an increasing amount of attention has been given to complex systems with a strong multiscale character. For that case, Pavliotis and Stuart indicated that standard estimation techniques might fail because the effective behavior is not visible on small time scales, see [6, 7]. The authors prove that, to avoid biasing and to enable the estimators to capture the effective dynamics, the data should be *subsampled*, i.e., a minimal interval between observations should be respected in the estimation procedure.

The present work builds on the observations made in [7]. We consider systems that fit in the framework of homogenization for SDEs,

$$\begin{aligned} dx^\epsilon(t) &= \left(\frac{1}{\epsilon} f_0(x^\epsilon, y^\epsilon) + f_1(x^\epsilon, y^\epsilon) \right) dt + f_2(x^\epsilon, y^\epsilon) dU_t, \\ dy^\epsilon(t) &= \frac{1}{\epsilon^2} g_0(x^\epsilon, y^\epsilon) dt + \frac{1}{\epsilon} g_1(x^\epsilon, y^\epsilon) dV_t, \end{aligned} \tag{2}$$

with $x^\epsilon \in \mathbb{R}^N$, $y^\epsilon \in \mathbb{R}^M$ and mutually independent N- and M-dimensional Wiener processes U_t and V_t. If $y^\epsilon(t)$ is ergodic for fixed x^ϵ and $f_0(x^\epsilon, y^\epsilon)$ averages to zero against the invariant measure generated by y^ϵ for fixed x^ϵ, it is well-known that in the limit $\epsilon \to 0$, the solution $x^\epsilon(t)$ of (2) converges weakly to the solution of (1), with known expressions that define the drift and diffusion coefficients [2]. Computing drift and diffusion from these expressions might, however, be difficult, so that numerical estimation of the unknown parameters is appropriate. Pavliotis et al. [7] commented that, to estimate the effective drift and diffusion for systems of the form (2), a minimal subsampling interval between used observations must be respected to be able to capture the correct homogenized, effective behavior. This interval is both necessary and unavoidable. The time scale on which the homogenized behavior becomes apparent is directly related to the time scale separation in the system (2). Weaker time scale separation ("large" ϵ) corresponds to larger subsampling intervals. As in practice ϵ must be seen as a model parameter and can typically not be controlled, large difficulties arise in the estimation of drift and diffusion if ϵ is only moderately small. In this case, the estimation procedure should operate on a time scale that is not small compared to the effective dynamics. Combined with the fact that the accuracy of standard parameter estimation procedures improves for decreasing time interval between observations, this indicates that there is a conflict between the accuracy of the estimators and the accuracy of the effective model (with respect to the observed simulation data).

Here, we will primarily focus on diffusion estimation in the context of problems for which subsampling is required, assuming that the drift coefficient is known. Moreover, we consider examples that do not exhibit a very large time scale separation, and, hence, for which the required subsampling interval is not small with respect to the time scale of the effective dynamics. It will be shown with numerical experiments that the classic estimators fail in this case because on a larger time scale, the drift can no longer be ignored and starts to influence their accuracy. We propose to use a better suited maximum-likelihood based diffusion estimator, in which we compensate for this effect by filtering out the contributions of the drift at each discrete time step. Numerical examples are provided both for the case that the coarse drift is known and for the case it has to be estimated from numerical simulations.

This paper is organized as follows. In Sect. 2, we introduce the general setting and the standard diffusion estimator. We also give an example to illustrate both the need for subsampling and the shortcomings of the standard estimator. We then derive a better suited estimator in Sect. 3 and present numerical results in Sect. 4. Finally, Sect. 5 summarizes and gives conclusions.

2 Setup

In this section, we sketch the general setting of this work and formulate some assumptions about the considered multiscale systems. We then briefly discuss the estimation of the diffusion coefficient and present the standard estimator in the context of multiscale homogenization problems. Next, we provide a numerical example in which this estimator is used to illustrate the fact that for a general homogenization problem, its naive application does not result in correct estimation of the homogenized diffusion and appropriate subsampling [7] should be used. However, also *with* subsampling, it is found from the example that the standard estimator has a number of shortcomings and is not always usable in practice. This observation then serves as a motivation to develop a better suited diffusion estimator.

As already pointed out in the introduction, the final goal is to find a coarse low-dimensional stochastic model that captures the coarse dynamic behavior of an (often) high-dimensional multiscale system. We restrict ourselves to coarse models that can be described by (1). For certain classes of multiscale models, such as, e.g., systems of the form (2), this coarse model can be derived rigorously and equations exist from which the unknown drift and diffusion coefficients can be computed [2]. If these equations are difficult to solve, an often better alternative lies in the numerical estimation of the unknown coefficients from simulation data. Moreover, if the multiscale system has a general form for which it is expected that the coarse behavior can indeed be modeled by this equation, but for which no closed derivation can be found, numerical estimation is the only alternative.

Although we provide numerical examples for systems of the form (2), we keep the multiscale model as general as possible, with only a few basic assumptions.

Assumption 1 *We dispose of a numerical simulator for the multiscale system in which we have full control over the initial values of all relevant variables v^ϵ of the system.*

The above assumption does not imply that the equations describing the multiscale system are known. It merely requires that we have a time stepper $\phi_{\Delta t}(v^\epsilon(t))$ that computes the state of the multiscale system at time $t + \Delta t$ starting from $v^\epsilon(t)$,

$$v^\epsilon(t + n\Delta t) = \phi^n_{\Delta t}(v^\epsilon(t)). \tag{3}$$

Situations in which the model is described by a legacy code or a black box simulator for a complex system are thus also covered, as long as the approximation of the coarse dynamics by (1) is valid.

Assumption 2 *If the multiscale model exhibits slow-fast dynamic behavior, the separation in slow variables x^ϵ and fast variables y^ϵ is explicitly known. Furthermore, it is assumed that the invariant measure $\mu^X(\mathrm{d}y^\epsilon)$ generated by $y^\epsilon(t)$ for fixed $x^\epsilon \equiv X$, can be found (numerically or analytically).*

Assumption 3 *The dynamics of the slow variables of the considered multiscale system can indeed be approximated by (1).*

Here, we deal with multiscale homogenization problems (2). For this type of systems, it is well-known that, under certain assumptions on its parameters (see introduction) and in the limit $\epsilon \to 0$, the time series $x^\epsilon(t)$ converges weakly to the solution of an effective SDE (1). Hence, for this type of systems, all above assumptions apply.

As the main focus of this paper lies on the estimation of the diffusion coefficient, we make an additional assumption on the drift coefficient to avoid unrelated issues.

Assumption 4 *Let the homogenized drift be of the form $\gamma(X; A)$. It is assumed that, given simulation data $\mathbf{X}_k = [X(t_0), \ldots, X(t_k)]$ from the multiscale system, a good estimator $\Gamma_{\Delta t}(\mathbf{X}_k)$ for A is available.*

Note that the latter assumption implies that a good drift estimator is available also for the case that it involves subsampling of the input data.

2.1 Diffusion Estimation

In the context of multiscale homogenization problems, the standard way to estimate the diffusion σ is via the mean-squared displacement of a particle [7]. Typically, estimators based on maximum-likelihood from observations of a single time series are used,

$$\overline{\sigma}_{\text{ts}}^2 := \frac{1}{n\delta\Delta t} \sum_{k=1}^{n} \left(x^\epsilon(t_{k\delta}) - x^\epsilon(t_{(k-1)\delta}) \right)^2, \tag{4}$$

in which $t_k = t_0 + k\Delta t$ and $x^\epsilon(t_0) = X_0$.

In this paper, however, we focus on estimation techniques that enable us to estimate the unknown coefficients from short *localized* simulations with the multiscale model, which could then be used to compute, e.g., the evolution of the probability density of X directly on a coarser level [5]. Time series based estimators do not fulfill this requirement, so that we will concentrate on estimators based on ensemble averaging. The standard estimator related to (4) has the form

$$\overline{\sigma}^2 := \frac{1}{R\delta\Delta t} \sum_{r=1}^{R} \left(x_r^\epsilon(t_\delta) - x^\epsilon(t_0) \right)^2, \tag{5}$$

with $x^\epsilon(t_0) = X_0$ for all realizations. To compute $x_r^\epsilon(t_\delta)$, an ensemble of R initial conditions of $y^\epsilon(t_0)$ must be generated. These have to be chosen so that $y_r^\epsilon(t_0)$ samples the invariant measure of y^ϵ for fixed $x^\epsilon(t)$ [2], which was assumed to be possible by Assumption 3.

The estimators (4) and (5) are in some sense equivalent for data without multiscale character. In the context of multiscale systems, however, their convergence behavior might differ significantly. We will illustrate and discuss these differences in the last example of Sect. 4.

Note that (5) is in fact derived assuming that the discrete data originates from observations of a discretization of (1) (without multiscale character), for which it converges in the limit $\delta\Delta t \to 0$. To estimate the effective coarse diffusion of the multiscale system, we still use this estimator, but replace the input data by observations of (2) instead. This has, however, consequences for the performance of the estimators, which is discussed below in Sect. 2.2.

2.2 Subsampling

Pavliotis and Stuart [7] noted that in approximating the coarse dynamics of multiscale systems (2) by an effective model (1), the effective (homogenized) diffusion is not visible on very small time scales. If one wants to estimate these coefficients numerically, this means that the parameter δ in (5) should be sufficiently large. Otherwise, the effective dynamic behavior cannot be captured by the estimator and, as a result, it is expected to converge to the *unhomogenized* value of the diffusion. It was proven for a subclass of (2) that for too small δ the estimator is indeed biased; see [7]. In the context of drift estimation, Papavasiliou et al. [6] also gave a rigorous proof of this biasing for the general case of (2).

Example 1. We provide an example to illustrate the need for subsampling in a more general multiscale setting. The results, however, will also serve as an illustration of the limitations of the estimator (5). Consider a homogenization problem of the form (2) described by a simple bistable slow mode coupled to a fast Ornstein–Uhlenbeck (OU) process:

$$\dot{x} = x - x^3 + \frac{B}{\epsilon} y, \tag{6a}$$

$$dy = -\frac{y}{\epsilon^2} dt + \frac{1}{\epsilon} dW_t, \tag{6b}$$

with W_t a Wiener process. The invariant measure $\mu^X(dy)$ for the fast variable at fixed $x = X$ can easily be computed and is the standard Gaussian distribution. In the limit $\epsilon \to 0$, the time series $x(t)$ converges weakly to the solution $X(t)$ of (1) with the exact drift and diffusion given by

$$\gamma(X) = X - X^3 \qquad \sigma = B. \tag{7}$$

The system (6) is solved numerically using the Euler–Maruyama scheme with $\Delta t = 0.001$ and system parameters $B = 0.3$ and $\epsilon = 0.1$. We then apply (5) and compute the dependence of the diffusion estimate on the subsampling time $\delta \Delta t$ for different initial x; see Fig. 1. For small $\delta \Delta t$, the estimator recovers (within the limits of the discretization) the unhomogenized value of the diffusion coefficient, which in this case is zero. In other words, the estimator is biased and we have illustrated that for this example, subsampling is indeed needed. Furthermore, it can be observed that, although for increasing subsampling interval there is always evolution of the estimates towards zero, depending the initial condition, the exact dependence on $\delta \Delta t$ may differ significantly.

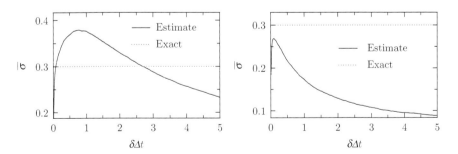

Fig. 1 Diffusion estimate using (5) as a function of the subsampling interval $\delta \Delta t$ for a slow-fast SDE homogenization problem computed with $R = 10,000$ realizations. *Left*: $x_0 = 0.5$. *Right*: $x_0 = 0.9$

2.2.1 Remarks

The fact that subsampling is unavoidable does not mean that (5) can be used to obtain an *accurate* estimate of the effective diffusion coefficient. To underpin this statement, we now formulate some observations and remarks with respect to the usage of this "standard" estimator in a multiscale homogenization setting. Note that if we speak about *convergence* of the estimator, we mean that in certain limits the estimator converges to the homogenized value of the coefficient we are trying to estimate.

- For some subclass of (2), it was proven that if $\delta \Delta t$ is too small, the standard estimator is biased and that there is *no convergence* to the homogenized value [7]. A similar result was obtained for drift estimation in the general case of (2); see [6]. It is believed, however, that also in the context of diffusion estimation for the general (non-degenerate) case of (2) the subsampling interval cannot be too small. As a result, to be able to compute the homogenized coefficient, $\delta \Delta t$ will always be bounded from below by a required minimal subsampling interval $\delta_c \Delta t$.
- If subsampling cannot be avoided, the question remains how large the subsampling interval should be to obtain convergence to the correct value of the diffusion coefficient. For (5) to be usable in practice, it should be possible to a priori find a value of δ for which the estimator converges to the correct value. Preferably, there should even be a range of acceptable values for which the estimate does not depend too strongly on δ. Unfortunately, this is not necessarily the case, as could clearly be observed in Fig. 1. For instance, for $x_0 = 0.9$ (right), the estimator does not converge for any choice of δ. As our objective is to obtain accurate estimates of the diffusion coefficient, the estimator (5) is thus of little practical use in the context of a multiscale setting.
- Although it is not apparent from the figure, the speed of the initial relaxation of the estimate towards the homogenized value depends on the time scale separation ϵ. If $\epsilon \to 0$, then also the required minimal subsampling interval $\delta_c \Delta t \to 0$ and the standard estimators should perform better. In practice, however, the scale separation is a property of the studied system and cannot be changed. In this work, we pay special attention to situations for which ϵ is relatively large. This was the case in the numerical example above, where we used $\epsilon = 0.1$. As a result, it should not be surprising that the minimal subsampling interval was relatively large compared to the coarse time scale, which in turn caused (5) to perform badly.

3 Drift-Filtered Diffusion Estimation

It was illustrated in the previous section that subsampling is indeed necessary if the input data comes from a general multiscale system. Furthermore, for finite and relatively large ϵ, the minimal subsampling interval can be relatively large compared to the coarse time scale of the multiscale system, which can conflict with the requirement that $\delta \Delta t$ should be small for the estimators to be accurate. If the time between observations used in (5) is small, we are in fact computing a good approximation of the wrong (unhomogenized) diffusion using an estimator that performs increasingly bad. This effect, however, is unrelated to subsampling and the multiscale nature of the input data, and should thus be addressed separately.

In this section, we pinpoint the exact reason for the bad performance of the standard estimator and offer an alternative diffusion estimator that is better suited for a multiscale setting. To this end, we consider a simple SDE example without multiscale effects and illustrate that the performance of the estimator (5) for $\delta \Delta t$ away from zero is similar to its behavior in the context of the multiscale example of the previous section.

Example 2. Consider a one-dimensional SDE of the form

$$dX(t) = \gamma(X)dt + \sigma dW_t, \qquad (8)$$

in which $\gamma(X) = X - X^3$ and $\sigma = 0.3$. The equation is solved with the Euler–Maruyama scheme with $\Delta t = 0.001$. We then estimate the diffusion coefficient using (5) and compute its evolution as a function of $\delta \Delta t$; see Fig. 2 (solid). This figure should be compared with Fig. 1. On the one hand, we see for this example that, for small $\delta \Delta t$, the estimate converges to the exact diffusion, whereas in the multiscale example we found convergence to the unhomogenized value. This is not surprising as, for this example, the diffusion is visible also on the smallest scales, so that $\delta \Delta t$ can be chosen small. In fact, one could argue that in both cases the estimator (5) does allow to compute a good approximation of the diffusion coefficient, but only of the diffusion that is visible *on that time scale* $\delta \Delta t$. On the other hand, we observe identical qualitative behavior when comparing Figs. 1 and 2 for larger $\delta \Delta t$, i.e., the standard estimator performs equally bad, converging towards zero in both cases. As a result, we can focus on (8) to improve the convergence behavior of the estimators on longer time scales so that multiscale related issues are avoided.

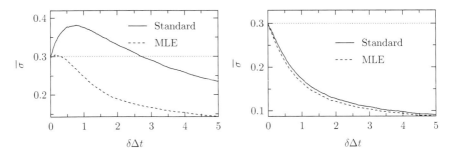

Fig. 2 Diffusion estimate using (5) as a function of the subsampling interval $\delta \Delta t$ for an SDE computed with $R = 10{,}000$ realizations. *Left*: $X_0 = 0.5$. *Right*: $X_0 = 0.9$

A Drift-Filtered Approach to Diffusion Estimation

3.1 Maximum-Likelihood Estimation

If the SDE (8) is approximated by a first order (Euler–Maruyama) discretization [4], the maximum-likelihood estimator (MLE) for the diffusion coefficient is given by

$$\overline{\sigma}^2 := \frac{1}{R\delta\Delta t} \sum_{r=1}^{R} (X_r(t\delta) - X(t_0) - \gamma\,(X(t_0))\,\delta\Delta t)^2\,, \tag{9}$$

which takes into account contributions of the drift. The estimator (5) can thus be seen as an approximation of the above maximum-likelihood estimator in the limit $\delta\Delta t \to 0$.

As in our setting $\delta\Delta t$ cannot be too small, one would expect (9) to perform better than its approximation (5). A numerical experiment indicates, however, that this is not the case. In Fig. 2 (dashed), it can be observed that, although the behavior of (9) on longer time scales can be significantly different compared to this of the standard estimator, it suffers from similar problems.

The solution to this problem readily presents itself from the following observation. While (9) is indeed a better estimator than (5), it is derived for the situation in which *each* observation can be used in the estimation. In a multiscale setting, this is no longer the case as the data should then be subsampled, which implies that only every δth observation is used. As a result, much information is lost or left unused by the estimation formula. As will be shown below, it is in fact this last effect that causes the observed bad convergence behavior of the estimator. Instead of merely subsampling the time series $X_r(t_k)$, we propose to use a more suited diffusion estimator that includes *all* information from the computed time series. In fact, this estimator will be derived as the MLE for the diffusion coefficient (given a time series $\{X_0, \ldots, X_\delta\}$), while taking into account a subsampling interval $\delta\Delta t$.

Theorem 5. *Let* (8) *be approximated by the Euler–Maruyama discretization scheme,*

$$X_{k+1} = X_k + \gamma(X_k)\Delta t + \sigma\sqrt{\Delta t}\,\Delta W_k, \tag{10}$$

with initial value $X_0 = X(t_0)$, $k = 0, \ldots, n$ *and* ΔW_k *independent standard Gaussian distributed variables. The MLE for the diffusion* σ *using the time series* $\{X_0, \ldots, X_\delta\}$ *and subsampling time* $\delta\Delta t$ *is given by*

$$\overline{\sigma}_\delta^2 = \frac{\left(X_\delta - X_0 - \sum_{k=0}^{\delta-1} \gamma\,(X_k)\right)\Delta t\right)^2}{\delta\Delta t}. \tag{11}$$

Proof. Equation (11) can be derived in a straightforward matter. First, consider a discrete-time approximation of (8) using the Euler–Maruyama scheme as given by (10). We can then express X_δ in terms of X_0,

$$X_\delta = X_0 + \left(\sum_{k=0}^{\delta-1} \gamma(X_k)\right) \Delta t + \sigma \left(\sum_{k=0}^{\delta-1} \sqrt{\Delta t} \, \Delta W_k\right),$$

$$= X_0 + \left(\sum_{k=0}^{\delta-1} \gamma(X_k)\right) \Delta t + \sigma \sqrt{\delta \Delta t} \, \Delta V_\delta, \tag{12}$$

in which ΔV_δ is a standard Gaussian distributed variable. From (12), we have that the variable $X_\delta - X_0 - \left(\sum_{k=0}^{\delta-1} \gamma(X_k)\right) \Delta t$ is normally distributed with mean zero and variance $\delta \Delta t \sigma^2$. For this distribution the log-likelihood function is known and, after applying the principle of maximum-likelihood, we find that the MLE for the diffusion coefficient is given by (11). $\qquad\square$

Using the result of Theorem 5, we can now define an improved ensemble averaging based estimator for the diffusion,

$$\overline{\sigma}_\delta^2 := \frac{1}{R\delta \Delta t} \sum_{r=1}^{R} \left(X_r(t_\delta) - X(t_0) - \sum_{k=0}^{\delta-1} \gamma\left(X_r(t_k)\right) \Delta t\right)^2. \tag{13}$$

For completeness, we also provide the improved time series based estimator corresponding to (4),

$$\overline{\sigma}_{\mathrm{ts},\delta}^2 := \frac{1}{n\delta \Delta t} \sum_{k=1}^{n} \left(X(t_{k\delta}) - X(t_{(k-1)\delta}) - \sum_{j=(k-1)\delta}^{k\delta-1} \gamma\left(X(t_j)\right) \Delta t\right)^2. \tag{14}$$

Note that the sum over j in (13) can be seen as a compensation for the contribution of the drift in each discrete time step. This observation enables us to give another interpretation to the estimator. First, define the net time series $X_r^f(t_k)$, that eliminates all effects due to the drift.

Definition 1. Let $X(t_k)$ be the discrete solution of (8) starting at time t_0. The drift-filtered solution $X^f(t_k)$ is then given by

$$X^f(t_k) := X(t_k) - \sum_{j=0}^{k-1} \gamma\left(X(t_j)\right) \Delta t. \tag{15}$$

This filtering is a preprocessing step that should be carried out for all R time series X_r in the ensemble. If we then apply the standard estimator (5) to $X_r^f(s_k)$, we obtain exactly the same result as in (13).

It should be noted that, as this filtered time series only contains contributions of the diffusive part of the equation, it is expected that the estimate does not deteriorate (nor decrease towards zero) for increasing $\delta \Delta t$.

A Drift-Filtered Approach to Diffusion Estimation 279

3.2 Estimated Drift

The diffusion estimate (13) can directly be computed as long as the coarse drift coefficient $\gamma(X)$ is explicitly known. If this is not the case, an approximation $\overline{\gamma}(X)$ should be constructed, which can then be used instead of $\gamma(X)$.

Here, we limit ourselves to the case for which no subsampling is required for the drift estimation. The coarse drift is then visible also on the smallest time scales, which greatly facilitates its numerical drift estimation. In Sect. 2, we assumed that the drift coefficient is of the form $\gamma(X; A)$ and that a good estimator $\overline{A} = \Gamma_{\Delta t}(\mathbf{X}_k)$ is available with $\mathbf{X}_k = [X(t_0), \ldots, X(t_k)]$. The drift can then be estimated as

$$\overline{A}_\delta = \frac{1}{R} \sum_{r=1}^{R} \Gamma_{\Delta t}(\mathbf{X}_\delta^r). \tag{16}$$

Note that the drift estimation does not necessarily involve an increased computational cost as there is no need to generate separate simulation data for the drift estimation. The data required for the estimation of the diffusion coefficient can simply be reused for the drift.

For homogenization problems of the form (2), it should be expected that the homogenized drift is only visible over sufficiently large time scales and that an appropriate subsampling interval should be used. However, as also standard drift estimators only offer accurate estimates in the limit $\delta \Delta t \rightarrow 0$, it may come as no surprise that to estimate the drift coefficient in practice, issues similar to the ones reported in Sect. 2.2 can arise. Unfortunately, no general estimators that perform well in a multiscale settings are available for the drift coefficient, hence the above restriction.

3.3 Remarks

The drift function $\gamma(X)$, which is required in (13), should in principle be the homogenized drift. If this is the case, then the estimator converges to the correct homogenized diffusion as a function of $\delta \Delta t$ for $\Delta t \rightarrow 0$ and $R \rightarrow \infty$. However, even if $\gamma(X)$ is analytically known, the use of an ensemble average based estimator can have consequences on the convergence speed, as it might be affected by initial transient behavior (which could be seen as the ensemble evolving towards the homogenized dynamics). This will be discussed below in Sect. 4.3.

It is pointed out that (13) takes into account all information from the computed time series, which in practice results in better performance of the estimator. This does, however, not mean that subsampling is avoided altogether. As was already pointed out in the introduction, this cannot be done, as the effective diffusion *is* only visible on larger time scales. If the objective is thus to compute the diffusion coefficient, it also intuitively makes sense to filter the accumulated drift up to t_δ

from each of the time series X_r and to take into account only diffusive contributions to the computed time series.

4 Numerical Results

In this section, we present numerical examples to illustrate the performance of the improved estimator (13). As this formula was derived from the discretization of a one-dimensional SDE (8), we first provide results for this case. Afterwards it is applied to a variety of multiscale homogenization examples. For each example, we first compare its convergence behavior for increasing $\delta\Delta t$ with the behavior of the standard estimator (5) using the known analytical expression for the homogenized drift. Afterwards, we compare the latter with the case in which the coarse drift is unknown and is estimated from numerical simulations.

4.1 One-Dimensional SDE

As a first illustration, we return to Example 2 and assess the performance of (13) for data generated from numerical simulation of (8). It is expected that the estimator will in some sense perform optimally as the discrete data fully corresponds to the proposed coarse model. We consider the evolution of the diffusion estimate as a function of $\delta\Delta t$, both for known and unknown drift coefficients.

4.1.1 Known Drift

To compute the diffusion estimate (13), we use the knowledge of the exact effective drift $\gamma(X) = X - X^3$. Figure 3 (left) shows the resulting diffusion estimate as a function of $\delta\Delta t$ for some initial values of X_0, each time computed for an ensemble of $R = 10,000$ initial conditions. It can be observed that, whereas the standard estimator deteriorates fast for increasing $\delta\Delta t$, the drift-filtered estimate provides good results for all $\delta\Delta t$. Moreover, although not shown, we found that the behavior of the estimator is now independent of the choice of X_0. This could be expected as for this example σ is a constant and thus independent of X while all space dependent effects of the drift are fully compensated for.

4.1.2 Estimated Drift

Instead of using the analytical expression for the drift, it is now computed from numerical simulations by applying maximum-likelihood estimation. First, we assume that the drift has the form

A Drift-Filtered Approach to Diffusion Estimation

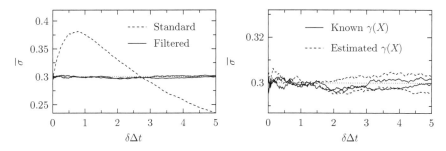

Fig. 3 Dependence of the diffusion estimate (13) on the subsampling interval $\delta \Delta t$ for a one-dimensional SDE; $R = 10{,}000$. *Left*: Comparison between the standard estimate (*dashed*) and the drift-filtered estimate using the analytical expression for the drift (*solid*). Two realizations of the experiment are shown. *Right*: Comparison of the drift-filtered estimate using the analytical (*solid*) and numerically estimated drift (*dashed*). For both cases, two realizations of the experiment are shown

$$\gamma(X; A) = A\left(X - X^3\right). \tag{17}$$

The MLE for A (only using simulation data $X_r(t_1)$) becomes

$$\overline{A} := \frac{1}{R \Delta t} \sum_{r=1}^{R} \frac{(X_r(t_1) - X(t_0))}{X(t_0) - X(t_0)^3}. \tag{18}$$

The diffusion estimate is then computed via (13) with $\gamma(X) = \gamma(X; \overline{A})$. Figure 3 shows a comparison between the behavior of the estimator with known (solid) and estimated (dashed) drift for $R = 10{,}000$. It can be observed that both estimates behave similarly, although the variance on the estimate for numerically computed drift is larger.

4.2 Slow-Fast SDE System

As a second illustration, we consider a multiscale homogenization problem of the form (2) and retake the system of Example 1, which is described by (6). For the numerical solution, identical simulation parameters are chosen as in Sect. 2.2.

4.2.1 Known Drift

The exact coarse drift can easily be computed and is given by $\gamma(X) = X - X^3$. This expression is then used in (13) to compute the diffusion estimate. Figure 4 (left) shows a comparison with the standard estimator. In both cases, an ensemble of $R = 10{,}000$ realizations initialized with $X_0 = 0.5$ was used. As in the

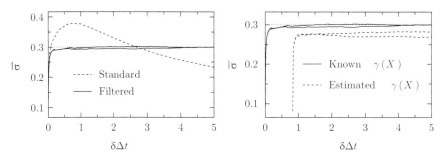

Fig. 4 Dependence of the diffusion estimate (13) on the subsampling interval $\delta \Delta t$ for a slow-fast stochastic homogenization problem; $R = 10{,}000$. *Left*: Comparison between the standard estimate (*dashed*) and the drift-filtered estimate using the analytical expression for the drift (*solid*). Two realizations of the experiment are shown. *Right*: Comparison of the drift-filtered estimate using the analytical (*solid*) and numerically estimated drift (*dashed*). For both cases, two realizations of the experiment are shown

previous example, we find that the improved estimator easily outperforms the standard estimator. For small $\delta \Delta t$, it can be observed that both estimators recover the unhomogenized diffusion $\sigma = 0$. For increasing $\delta \Delta t$, the standard estimate deteriorates, while (13) converges to the correct, homogenized diffusion.

4.2.2 Estimated Drift

We repeat the above experiment but now numerically estimate the drift $\gamma(X)$. As for this example no subsampling is required for the drift estimation and, as in the previous example, the drift is chosen to be of the form (17), we can use (18) to compute $\overline{A}$. The diffusion estimate is then computed by applying (13) with $\gamma(X) = \gamma(X; \overline{A})$. The results for known (solid) and estimated (dashed) drift are shown in Fig. 4 (right). Again, as before, we see that the convergence behavior of both estimators is similar, with slightly larger variance on the result when using the estimated drift function.

4.3 *Movement in Multiscale Potential with Thermal Noise*

This example (taken from [7]) serves as an illustration for the case that subsampling is required for the estimation of both drift and diffusion. As already commented in Sect. 3.2, a good drift estimator is then not necessarily available and our discussion will inevitably be restricted to the performance of (13) using known drift. Furthermore, for this type of more general homogenization problems, it will become clear that our focus on ensemble averaging based estimators has consequences on the convergence speed of the estimator.

A Drift-Filtered Approach to Diffusion Estimation 283

Consider the first order Langevin equation in one dimension of the form

$$dx^\epsilon(t) = -\nabla V\left(x^\epsilon(t), \frac{x^\epsilon}{\epsilon}; \alpha\right) dt + \sigma dW_t,$$
(19)

with W_t a Wiener process and σ the diffusion. This equation describes the movement of a particle in a multiscale potential subject to thermal noise. For this example, the (unhomogenized) drift is given by the gradient of a two-scale potential $V(x, y; \alpha) = \alpha V(x) + p(y)$, with p chosen to be an oscillatory periodic function. Under certain conditions on p it is known that the solution $x^\epsilon(t)$ of (19) converges to the solution of (1), with known expressions for the homogenized drift and diffusion; see [7] for more details.

Here, we consider a potential of the form

$$V\left(x, \frac{x}{\epsilon}\right) = \frac{\alpha}{2}x^2 + \cos\left(\frac{x}{\epsilon}\right),$$
(20)

and solve (19) using the Euler–Maruyama scheme, with sufficiently small time step to resolve the fast oscillations in the equation, i.e., Δt small with respect to ϵ^2. For this experiment, we used $\alpha = 1, \sigma = 0.5, \epsilon = 0.1, \Delta t = 0.0001$ and $R = 10,000$.

4.3.1 Known Drift

It was demonstrated in [7] that for system (19), subsampling is unavoidable both to estimate the homogenized drift and diffusion. A good numerical estimator for the drift (for fixed finite ϵ) is not available, so we only consider the case in which an analytical expression for the drift is known. With V as in (20), the homogenized drift can be computed and is given by

$$\gamma(X) = -\alpha X \frac{L^2}{Z^2}, \qquad Z = \int_0^L e^{\frac{\cos(y/\epsilon)}{\sigma}} dy.$$
(21)

Using the above $\gamma(X)$ in (13) to estimate the diffusion, we then compare with the standard estimator (5). The results are shown in Fig. 5. It can be observed that the standard estimator (dashed) does not converge, and although there indeed *exists* an optimal subsampling interval, it cannot be determined in advance. On the other hand, the diffusion estimate using the improved estimator (solid) evolves from the unhomogenized diffusion towards its correct, homogenized value.

As already pointed out above, the use of an ensemble average based estimator might have an effect on the convergence behavior of the estimator. The reason for this is that after initialization of the system, all realizations see the *same* unhomogenized drift. Hence, during an initial transitory phase, the effective drift (i.e., the drift seen by the ensemble) *is* the unhomogenized drift. In fact, during this transient phase the drift evolves from its unhomogenized towards its homogenized value.

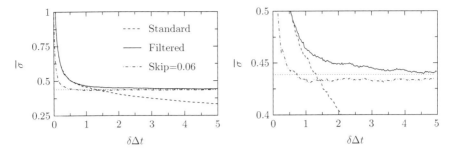

Fig. 5 *Left*: Dependence of the diffusion estimate (13) on the subsampling interval $\delta\Delta t$ for a stochastic homogenization problem with fast oscillations on the drift, using the analytical expression of the homogenized drift. *Right*: zoom around the exact diffusion

Fig. 6 Comparison of the evolution of the empirical mean of R trajectories initialized at $X_0 = 0.5$; $R = 10{,}000$ (*solid*) and the analytical expected value of the solution $X(t)$ of the effective SDE (*dashed*)

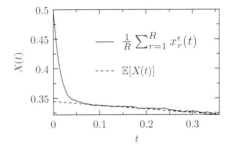

To apply (13), the drift function therein should therefore take into account this evolution if the estimator is to converge quickly. To illustrate, we first show the transient behavior by plotting the evolution of $1/R \sum_{r=1}^{R} x_r^\epsilon(t)$ with $x_r^\epsilon(t_0) = 0.5$; see the solid line of Fig. 6. In the same figure, we also plotted in dashed line the mean of the solution of the homogenized equation, i.e., $\mathbb{E}[X(t)] = X_0^h \exp\left(-\alpha L^2/Z^2 t\right)$, for appropriately chosen X_0^h. It can be seen that for $t > \delta\Delta t \approx 0.06$, the transient behavior has disappeared and the homogenized dynamics is recovered. Hence, to improve the convergence speed of (13), we could skip the first observations and only use data for which $t > 0.06$. It is found that this indeed speeds up the convergence of the estimator, as can be seen from the dash-dotted line in Fig. 5. Note that the presence of this transient behavior depends on the initial condition (via the unhomogenized drift in X_0) and cannot always be observed. If this is not the case, it should also not be expected that the convergence speed can be improved by simply skipping the initial observations. The time series based diffusion estimator equivalent to (13) does not suffer from this behavior.

5 Conclusions

In this paper, we dealt with the approximation of the coarse dynamics of a complex multiscale system by an effective diffusion model in the form of an SDE. Although many good procedures are proposed in the literature to match stochastic coarse models to simulation data, the estimation of drift and diffusion still poses a challenge if the input data has a multiscale character. It was indicated by Pavliotis and Stuart [7] that the effective dynamics of the system might not always be visible on the smallest time scales. Hence, the simulation data should be *subsampled*, i.e., a minimal interval between observations should be respected when using (standard) estimation formulas. However, we found that the standard estimators from SDE literature are not appropriate for use in a multiscale setting and that their performance rapidly deteriorates for increasing subsampling interval.

Our objective was to find better suited estimators that would also offer accurate estimates if this minimal subsampling interval is relatively large, as is the case when the time scale separation in the multiscale system is not very pronounced. With available estimators, the error in approximating a multiscale system by an SDE model, was dominated by errors unrelated to the multiscale nature of the problem, whereas one would hope that only the error due to the "fit" of the data to the model would play a role. Although the above-mentioned problems are apparent for both drift and diffusion coefficients, this work mainly focused on diffusion estimation.

It was first demonstrated that the reason for which the standard diffusion estimator does not perform well, was that it failed to take into account the effect of changing drift over longer time intervals. As a solution, we proposed to first filter out the contributions of the drift in every discrete time step, after which the standard estimator could be applied. This procedure was derived as a MLE for the diffusion, but taking into account the larger time interval between used data points. Also, although it is a relatively simple procedure, it proved to be highly effective and accurate diffusion estimates could be obtained in practice. This was illustrated with various examples, where we treated both the case of analytically known and numerically estimated effective drift coefficient. However, as also the estimation of the drift coefficient suffers from accuracy problems when multiscale input data is used, we assumed that, if this is the case, an analytical expression of the effective drift is available, which was done to avoid practical issues unrelated to the diffusion estimation.

Although we now dispose of a procedure to compute accurate estimates of the diffusion coefficient, various problems remain to be addressed. First, we only dealt with the estimation of a constant diffusion coefficient. An important question is how the ideas presented in this paper can be extended to derive a suitable estimator for the (more difficult) case in which the diffusion is space dependent. Second, it is crucial that also for the drift better estimators are developed when dealing with multiscale data. This will be the topic of future work.

Acknowledgements This paper presents research results of the Belgian Network DYSCO (Dynamical Systems, Control, and Optimization), funded by the Interuniversity Attraction Poles Programme, initiated by the Belgian State, Science Policy Office. The scientific responsibility rests with its authors.

References

1. Ait-Sahalia, Y.: Maximum likelihood estimation of discretely sampled diffusions: A closed-form approximation approach. Econometrica. **70(1)** (2002) 223–262
2. Bensoussan, A., Lions, J.-L., Papanicolaou, G.: Asymptotic analysis for periodic structures (1978)
3. Crommelin, D. T., Vanden-Eijnden, E.: Reconstruction of diffusions using spectral data from timeseries. Communications in Mathematical Sciences **4(3)** (2006) 651–668
4. Florens-Zmirou, D.: Approximate discrete-time schemes for statistics of diffusion processes. Statistics: A Journal of Theoretical and Applied Statistics **20(4)** (1989) 547–557
5. Frederix, Y., Samaey, G., Roose, D.: An analysis of noise propagation in the multiscale simulation of coarse Fokker–Planck equations. ESAIM, Math. Model. Numer. Anal. (accepted); Technical Report TW549, Department of Computer Science, K.U.Leuven (2009)
6. Papavasiliou, A., Pavliotis, G. A., Stuart, A. M.: Maximum likelihood drift estimation for multiscale diffusions. Stochastic Processes and their Applications **119(10)** (2009) 3173–3210
7. Pavliotis, G. A., Stuart, A. M.: Parameter estimation for multiscale diffusions. Journal of Statistical Physics **127(4)** (2007) 741–781
8. Pokern, Y., Stuart, A. M., Vanden-Eijnden, E.: Remarks on drift estimation for diffusion processes. SIAM Multiscale Modeling and Simulation **8(1)** (2009) 69–95

Model Reduction of a Higher-Order KdV Equation for Shallow Water Waves

Tassos Bountis, Ko van der Weele, Giorgos Kanellopoulos, and Kostis Andriopoulos

Abstract We present novel results on a non-integrable generalized KdV equation proposed by Fokas [A.S. Fokas, Physica D87, 145 (1995)], aiming to describe unidirectional solitary water waves with greater accuracy than the standard KdV equation. The profile of the solitary wave solutions is determined via a reduction of the partial differential equation (PDE) to a set of ordinary differential equations (ODEs). Subsequently, we study the stability of the wave using this profile as initial condition for the PDE. In the case of the standard KdV equation it is well-known that the solitary wave solutions are always stable, irrespective of their height. However, in the case of our higher-order KdV equation we find that the stability of the solutions breaks down beyond a certain critical height, just like solitary waves in real water experiments.

1 Introduction

The Korteweg–de Vries (KdV) equation, written in a stationary frame as

$$\eta_t + c_0 \eta_x + \tfrac{3}{2} \frac{c_0}{h_0} \eta\eta_x + \tfrac{1}{6} c_0 h_0^2 \eta_{xxx} = 0, \tag{1}$$

represents a first-order approximation in the study of waves of small amplitude and long wavelength propagating in one direction over the surface of a shallow

---

T. Bountis (✉) and K. Andriopoulos
Department of Mathematics and Centre for Research and Applications of Nonlinear Systems,
University of Patras, 26500 Patras, Greece
e-mail: bountis@math.upatras.gr, kand@aegean.gr
and
Centre for Education in Science, 26504 Platani – Rio, Greece

K. van der Weele and G. Kanellopoulos
Department of Mathematics and Centre for Research and Applications of Nonlinear Systems, University of Patras, 26500 Patras, Greece
e-mail: weele@math.upatras.gr, giorgoskan@lycos.gr

A.N. Gorban and D. Roose (eds.), *Coping with Complexity: Model Reduction and Data Analysis*, Lecture Notes in Computational Science and Engineering 75, DOI 10.1007/978-3-642-14941-2_15, © Springer-Verlag Berlin Heidelberg 2011

layer of an inviscid and incompressible fluid [1–8]. In this equation h_0 denotes the height of the undisturbed fluid layer, $\eta(x,t)$ the elevation of the wave above h_0, and $c_0 = \sqrt{gh_0}$ the nondispersive phase velocity.

A good example of a situation described by the KdV equation is the famous solitary wave first sighted by John Scott Russell in 1834 [7, 9, 10], while observing the motion of a boat, rapidly drawn along a canal near Edinburgh by a pair of horses. Through the work of Scott Russell and many others after him, it became clear that similar solitary waves can always be observed on the surface of a thin sheet of water (with velocity exceeding c_0) over a smooth floor. Certainly the most celebrated property of the solitary-wave solutions of (1) is that they collide elastically: when two of them cross each other, they emerge unscathed from the interaction. It is because of this particle-like property that they are called solitons [11].

We should mention that the study of solitary water waves can also be pursued at the level of the governing fluid-dynamical equations [12–14]. However, the KdV equation is often preferred – notwithstanding the fact that it is only an approximation – because it is completely integrable and admits a great variety of analytical solutions.

There are, however, two important drawbacks from which the KdV equation suffers: First, since it is integrable, its solutions are generally linearly stable with respect to small changes in their initial conditions. Furthermore, the parameters in front of the nonlinear and diffusion terms in (1) can be scaled out and hence do not significantly affect the spatial and temporal shape of the solutions. In the particular case of the solitary waves, it has been experimentally verified that, as the parameters vary, these shapes noticeably differ from what the KdV equation predicts and are not even observed beyond certain threshold values [15].

By contrast, the generalized KdV equation studied here exhibits solitary waves whose shapes are much closer to reality and *do* show a stability threshold. As the height and speed of the wave increase, we find that it becomes *unstable* under small perturbations of its profile and breaks down at threshold values close to the experimental ones, thus providing a mathematical explanation for the fact that they are not observed in these regimes.

Equation (1) can be brought into dimensionless form by a change of variables: $t \to \tau = t/t_c$ (where $t_c = \ell/c_0$ represents a typical time scale), $x \to \xi = x/\ell$ and $\eta \to u = \eta/a$. Here ℓ is a characteristic length scale in the horizontal direction, say the wavelength of a typical (non-solitary) wave solution of (1), and a a characteristic length in the vertical direction, say the height of the typical wave. Applying this change of variables we arrive at the following form of the KdV:

$$u_\tau + u_\xi + \alpha u u_\xi + \beta u_{\xi\xi\xi} = 0, \tag{2}$$

where $\alpha = \frac{3}{2}a/h_0$ and $\beta = \frac{1}{6}(h_0/\ell)^2$. In the derivation of the KdV equation from physical principles both α and β are required to be small compared to 1. Moreover, it is usually assumed that $a/h_0 \approx (h_0/\ell)^2$; this is a natural assumption, expressing the balance between nonlinearity and dispersion that plays a crucial role in the soliton solution for which the KdV equation is famous [16]. It is not obligatory,

Model Reduction of a Higher-Order KdV Equation

however, since a and ℓ are two independent quantities and formal considerations support the view that the KdV equation is a valid approximation to the underlying hydrodynamical equations also outside the regime where a/h_0 and $(h_0/\ell)^2$ have roughly the same value [17].

When α and β are not too small (or when a more accurate description is required), higher-order terms in these parameters should be included. The resulting equation up to second order terms has been studied extensively by Kodama and others [18–24]; see [25] for more details.

The next equation, containing all terms up to third order in α and β, was studied recently by Marchant [26, 27]. An interesting reduced version of this equation – first introduced by Fokas [28] – arises when $O(a/h_0) > O((h_0/\ell)^2)$, i.e., when the waves are higher and/or wider than is usually assumed. In this case the terms of order $O(\beta^2)$, $O(\alpha\beta^2)$ and $O(\beta^3)$ may be neglected, and the third-order equation reduces to [28]:

$$
\begin{aligned}
u_\tau + u_\xi &+ \alpha u u_\xi + \beta u_{\xi\xi\xi} \\
&+ \alpha^2 \rho_1 u^2 u_\xi + \alpha\beta \left(\rho_2 u u_{\xi\xi\xi} + \rho_3 u_\xi u_{\xi\xi}\right) + \alpha^3 \sigma_1 u^3 u_\xi \\
&+ \alpha^2\beta \left(\sigma_2 u^2 u_{\xi\xi\xi} + \sigma_3 u u_\xi u_{\xi\xi} + \sigma_4 u_\xi^3\right) = 0.
\end{aligned}
\tag{3}
$$

This is the generalized KdV equation we now proceed to study.

The constant coefficients in the above equation (determined from the hydrodynamic equations for an inviscid, incompressible fluid) are given by $\rho_1 = -\frac{1}{6}$, $\rho_2 = \frac{5}{3}$, $\rho_3 = \frac{23}{6}$, $\sigma_1 = \frac{1}{8}$, $\sigma_2 = \frac{7}{18}$, $\sigma_3 = \frac{79}{36}$ and $\sigma_4 = \frac{45}{36}$ [28]. Unlike the standard KdV equation, (3) with the given values for the coefficients is *not integrable* and its solitary wave solutions are not true "solitons" since their collisions are not perfectly elastic: During the interaction, a dispersive wave train of small amplitude is generated. Only for special choices of the coefficients ρ_1 to σ_4 – different from the above values, and *not* representing the hydrodynamic situation – does (3) admit exact soliton solutions with a sech^2 profile [29, 30].

For our analysis, we shall take $\alpha = 0.60$ and $\beta = 0.01$. The choice of α implies that $a/h_0 = \frac{2}{3}\alpha = 0.40$, which is not much smaller than 1, hence there is indeed good reason to include higher order terms in α. The choice of β means that $(h_0/\ell)^2 = 6\beta = 0.06$, in accordance with the assumption $O(a/h_0) > O((h_0/\ell)^2)$, which lies at the basis of (3).

It should be noted that our choice of $\alpha = 0.60$ (or $a/h_0 = 0.40$) is large enough to justify the higher-order approach, while also being *small* enough to lie within the realistic range for solitary water waves over a horizontal bottom. The maximum amplitude (height above the quiescent water surface) that is actually measured in practice is somewhere around $0.55h_0$ [15], while the theoretical upper limit is $0.78h_0$ [31–33]. For higher amplitudes the velocity of the water particles on the wave crest becomes larger than the propagation speed, hence the wave breaks. In our equation this means that $u = \eta/a$ should not exceed $0.55h_0/a = 1.38$ if we wish to compare our results with experimental data, or $0.78h_0/a = 1.95$ if we compare them with theory. Nevertheless, we also consider larger values of u to gain insight in the mathematical structure of the problem.

The rest of the paper is organized as follows: In Sect. 2 we determine the shape of the solitary wave solution of (3) using the fact that a unidirectional wave travelling at constant speed c does not depend on ξ and τ independently, but only on the combined variable $\xi - c\tau(=\zeta)$. This makes it possible to reduce the PDE (3) to an ODE in terms of ζ alone, which is solved numerically. The shape thus found is taken, in Sect. 3, as initial condition for (3). Applying then a small perturbation to the initial profile and following its time evolution, we determine the stability of the solitary wave. It is found that the waves become unstable beyond a critical height of $u \approx 4$. This is larger than what is observed for real water waves, but already much better than for the standard KdV equation, for which such stability threshold does not exist. Finally, Sect. 4 contains concluding remarks.

2 Reduction to an ODE: Shape of the Solitary Waves

2.1 The Standard KdV Equation

To demonstrate the method we are going to use, and for reasons of comparison with the higher-order equation, we briefly review the analysis of the unperturbed KdV equation (2), i.e., the first line of (3). Since we are interested in waves travelling to the right, we consider solutions of the form $u(\xi, \tau) = u(\xi - c\tau)$, which depend only on the combined variable $\zeta = \xi - c\tau$. Now, with $u_\xi = u_\zeta \cdot d\zeta/d\xi = u_\zeta$ and $u_\tau = u_\zeta \cdot d\zeta/d\tau = -cu_\zeta$, the KdV equation reduces to the following ODE (the prime denotes the derivative with respect to ζ):

$$(1 - c)u' + \alpha uu' + \beta u''' = 0. \tag{4}$$

Integrating this equation once yields:

$$(1 - c)u + \tfrac{1}{2}\alpha u^2 + \beta u'' = 0, \tag{5}$$

where the integration constant (the right hand side of this equation) is zero because of the vanishing boundary conditions $u(\zeta) = u'(\zeta) = u''(\zeta) = 0$ at $\zeta = \pm\infty$. In order to analyze (5) by means of a phase-plane analysis, we rewrite this second-order ODE in the form of two first-order ODEs:

$$
\begin{aligned}
u' &= y \\
u'' = y' &= -\frac{1}{\beta}\left[(1 - c)u + \tfrac{1}{2}\alpha u^2\right].
\end{aligned}
\tag{6}
$$

The corresponding (u, u') phase-plane is depicted in Fig. 1. We observe two fixed points, a saddle in $(0, 0)$ and a centre in $(2\alpha^{-1}(c - 1), 0) = (3.33, 0)$. The homoclinic orbit (separatrix) emanating from the saddle point $(0, 0)$ represents the

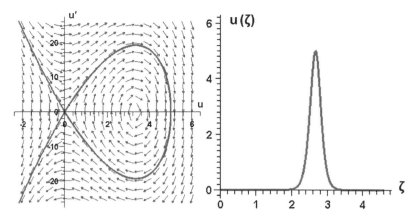

Fig. 1 *Left*: The (u, u')-plane for (6) for the values $\alpha = 0.6$, $\beta = 0.01$ and $c = 2$. The homoclinic orbit emanating from the saddle point $(0, 0)$ represents a solitary wave. *Right*: The corresponding solitary-wave profile $u(\zeta)$

well-known soliton solution

$$u_{\text{sol}}(\xi, \tau) = \frac{3(c-1)}{\alpha} \text{sech}^2 \left[\sqrt{\frac{c-1}{4\beta}} (\xi - c\tau) \right]. \tag{7}$$

Note that this solution only exists for $c > 1$, in agreement with our earlier observation that solitary waves occur when the speed exceeds the nondispersive phase velocity $c_0 = \sqrt{gh_0}$, which has been implicitly rescaled to 1 by the nondimensionalization.

The solitary wave solution (7) divides the unbounded orbits in the outer region of the (u, u')-plane from the periodic orbits within the inner region. These latter solutions are known as the cnoidal waves of the KdV equation [6, 7] and correspond to the Stokes wave solutions of the governing hydrodynamical equations [2, 7, 34, 35].

2.2 The Generalized KdV Equation

We now consider the full (3). When we again restrict ourselves to travelling-wave solutions of the form $u = u(\xi - c\tau) = u(\zeta)$, (3) takes the form of the following third order ODE:

$$(1-c)u' + \alpha u u' + \beta u''' \\ + \alpha^2 \rho_1 u^2 u' + \alpha \beta (\rho_2 u u''' + \rho_3 u' u'') + \alpha^3 \sigma_1 u^3 u' \\ + \alpha^2 \beta (\sigma_2 u^2 u''' + \sigma_3 u u' u'' + \sigma_4 u'^3) = 0. \tag{8}$$

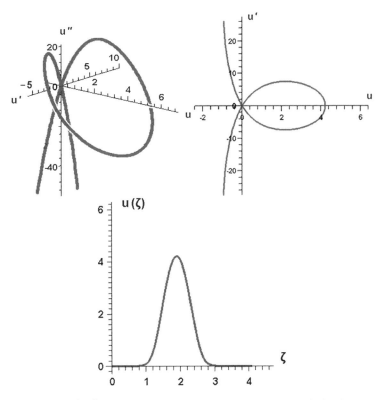

Fig. 2 *Top*: The (u, u', u'') phase-space for the third-order KdV system (9) for the same values $\alpha = 0.60$, $\beta = 0.01$, and $c = 2$ as in Fig. 1. The homoclinic orbit corresponding to the solitary wave solution is seen to lie on a curved surface. *Down, left*: Projection of the homoclinic orbit on the (u, u')-plane to facilitate the comparison with Fig. 1. *Down, right*: The solitary wave profile $u(\zeta)$. Its height has decreased, and its width increased, with respect to the soliton solution (the $\text{sech}^2(\zeta)$-profile) of the standard KdV equation in Fig. 1

Equation (8) is, as far as we can tell, not integrable – meaning that it cannot be reduced to an ODE of order less than three. We therefore transform it directly to a system of three first-order ODEs, as follows:

$$\begin{aligned}
u' &= y \\
u'' &= y' = z \\
u''' &= z' = \frac{-1}{\beta(1 + \alpha\rho_2 u + \alpha^2\sigma_2 u^2)} \{(1-c)y + \alpha u y + \alpha^2 \rho_1 u^2 y \\
&\quad + \alpha\beta\rho_3 yz + \alpha^3 \sigma_1 u^3 y + \alpha^2 \beta \sigma_3 u y z + \alpha^2 \beta \sigma_4 y^3\}.
\end{aligned} \quad (9)$$

This system can be solved numerically in the (u, u', u'') phase space, and its solution is depicted in Fig. 2. For all speeds $c > 1$ we observe a homoclinic orbit – or solitary wave – emanating from the unstable equilibrium point in the origin $(0, 0, 0)$, which follows a curved surface in the three-dimensional space. The maximal extension of this homoclinic orbit in the u-direction has decreased with respect to that of the standard KdV case (cf. Fig. 1), indicating that the higher-order terms have a diminishing effect on the height of the solitary wave. Likewise, the homoclinic orbit also extends much less into the u'-direction, meaning that the sides of the solitary wave are less steep than for the standard KdV soliton.

In the interior region of the homoclinic orbit (following the curved surface in the three-dimensional phase space) there exist periodic solutions of cnoidal type, which we intend to discuss in a forthcoming publication.

Concentrating on the height of the solitary wave, we see that it is drastically reduced as a result of the higher-order terms of order $O(\alpha^2, \alpha\beta, \alpha^3, \alpha^2\beta)$. Figure 3 illustrates this point. The KdV soliton itself (lower curve) shows a linear increase of the height, $h(c) \propto (c-1)$, as can also be read from the analytic solution (7). The third-order solitary wave (lower curve) shows a much slower, approximately logarithmic increase $h(c) \propto \ln c$. On the basis of this we conjecture that the inclusion of corrections of arbitrarily high order in α and β might even stop the growth altogether [25]. If this is indeed the case, the height of the solitary waves at high speed c saturates at a definite, finite value, as observed in practice.

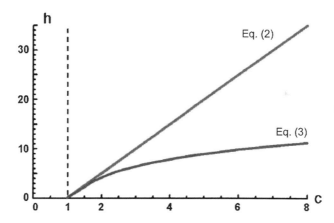

Fig. 3 Height of the solitary wave as function of its speed c for the standard KdV (2) and its third-order generalization (3). In the upper curve the height $h(c)$ grows linearly without bound, following the analytic solution (7). The lower curve shows the restraining influence of the higher-order terms in α and β. Note: The wave amplitudes achievable in water experiments are restricted to the narrow zone $h \leq 1.38$ (or according to theory, $h \leq 1.95$) and corresponding wave velocities $c \leq 1.3$ (or $c \leq 1.4$), where the two curves have hardly separated yet [15, 31–33]

3 Back to the PDE: Stability of the Solitary Waves

In this section we investigate the stability of the solitary waves. In order to do so, we insert the previously obtained profiles as initial condition into the KdV equations (2) and (3), respectively, and follow their time evolution.

We begin with the standard KdV equation. Inserting into (2) the profile of the corresponding soliton with $c = 2.5$ (a bit faster, and hence higher, than the one depicted in Fig. 1), we observe that the wave travels towards the right without any loss of form or speed, see Fig. 4 left column. After $\tau = 40$ time units the wave has propagated over a distance of precisely $\Delta \xi = c\tau = 100$ units. We have intentionally given some slight perturbation to the initial profile [by using the numerically obtained profile, not the analytic formula of (7)], but this is quickly smoothed out and the wave attains the familiar $sech^2$ shape. Evidently, this solitary wave is stable. In fact we could have taken a much higher wave, with arbitrarily high speed c, and it would still have been stable.

In Fig. 4 (right column) we repeat the same procedure for the third-order KdV equation, again for $c = 2.5$, inserting the profile of an orbit close to the numerically obtained homoclinic one into (3). Also in this case, we have included some initial perturbation, as shown in the close-up at $\tau = 0$: Note that the right slope does not join smoothly onto zero, which represents a distinct deviation from the normal solitary wave form. In the course of time, however, this small perturbation is being repaired by the system itself. The profile sheds off a train of small wavelets (which can be seen trailing behind the main wave, reminiscent of the "violent agitation" left behind by the solitary wave observed by John Scott Russell [9]) and after a while it attains the form of a true solitary wave. It has lost some of its height in the process, with its initial speed decreased to a value slightly below $c = 2$ (as can be read from the snapshots at $\tau = 20$ and $\tau = 40$), but travels on without any further loss of form or speed. Its profile at $\tau = 40$ (see close-up) is smooth and symmetric, with a height of approximately 4.10, in agreement with the height-velocity relation of the lower curve in Fig. 3. Thus, this solitary wave is still stable.

Let us now demonstrate that, unlike the standard KdV equation, the third-order equation (3) does *not admit* stable solitary waves of *arbitrary* height and speed. This is illustrated in Fig. 5, where we start from a (perturbed) initial profile corresponding to $c = 2.8$. In this case, the system is *not* capable of repairing its perturbation, and breaks down after the comparatively short time span of $\Delta \tau = 0.18$. The right plot in Fig. 5 (just before the breakdown) shows that the wave has shed off a few ripples, but is not able to detach itself from the trailing wavelets. Instead, a deep trough develops at its rear end, frustrating its progress to the right. Simultaneously, small disturbances appear on the main wave itself, as can be seen in the close-up. So the solitary wave has ceased to be stable. The details of this destabilization process will be published elsewhere.[1]

---

[1] We note that a similar loss of stability was observed in [30], in a study of the generalized KdV equation (3) with different values for the coefficients ρ_1 to σ_4 (not representing water waves) and $\alpha = 0.6$, $\beta = 0.05$. The solitary waves in that case were found to become unstable when the propagation speed exceeded the value $c = 1.1$.

Model Reduction of a Higher-Order KdV Equation

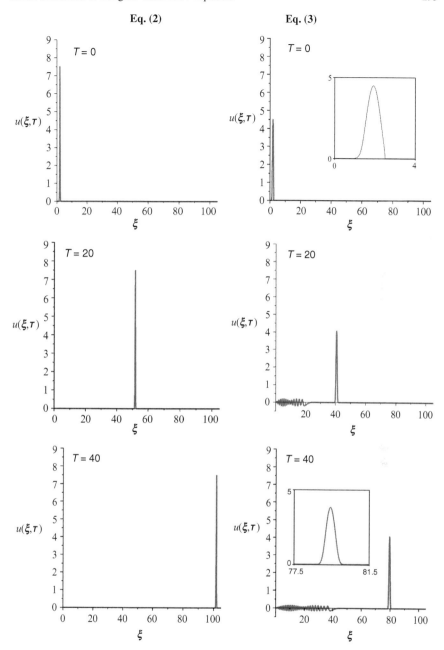

Fig. 4 Time evolution of two initial conditions close to a solitary wave profile. *Left column*: For the standard KdV equation (2) the wave is stable, proceeding without any noticeable loss of form, and keeping its initial speed $c = 2.5$. *Right column*: For the third-order KdV equation (3), the wave starts out with a sizeable perturbation (see close-up at $\tau = 0$) but becomes a true solitary wave later on, shedding off a train of small waves that quickly fall behind. In the process, the wave loses some of its height and speed. The final, stable solitary wave has a height of 4.1 (against an initial height of 4.5) and speed $c = 1.9$ (initially 2.5)

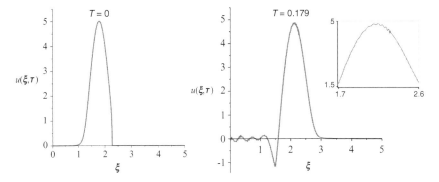

Fig. 5 Time evolution of an initial condition close to a solitary wave profile for (3) with initial velocity $c = 2.8$. In this case, the wave does not succeed in detaching itself from the train of wavelets. A deepening trough at its rear end keeps it back. The development of ripples around its maximum (see close-up) is an additional mark of the unstable character of the wave

For higher values of the speed (beyond the instability threshold) the amplitude of the wavelets trailing the leading disturbance quickly grows and the identity of the solitary wave profile is completely lost. It is important to remark that this is not due to numerical inaccuracies, since the prediction of the instability threshold remains practically the same for different choices of space and time discretization in the numerical algorithm.

4 Conclusion

In this paper we have examined the effect of including higher-order terms in the small parameters α and β in the KdV equation. These terms preserve the familiar shape of the soliton solutions of the KdV equation, but make them flatter and wider, while radically altering the relation between the wave's amplitude and its speed of propagation (see Fig. 3). The higher-order terms also cause the loss of the pure "soliton" property: When two of them meet, they do not emerge completely unscathed from the interaction [20, 26, 27].

The widening of the profile is physically quite relevant, especially at high propagation speeds. For the standard KdV equation the width of the wave decreases with growing speed as width$(c) \propto (c-1)^{-1/2}$ [this can be inferred from (7)], implying that solitary waves in the limit of high velocities become infinitesimally thin [25]. This unrealistic feature is corrected by the higher-order terms in α and β. We conjecture that the width, like the height, saturates at a finite nonzero value in the limit for large c when terms of arbitrarily high order in α and β are taken into account.

The most important change, however, is the occurrence of a stability threshold. While the soliton solutions of the standard KdV (2) are *always* stable, irrespective of their height (in contrast to the experimental observations in real water channels), the higher-order terms in α and β cause a breakdown of stability beyond a certain critical wave height.

For the third-order equation studied in the present paper (3) the threshold value for the height is found to be around $h_{crit} \approx 4$. This is larger than what is observed in practice, where solitary water waves cannot grow (in our normalized units) beyond a height $h = 1.38$ [15], and also larger than the theoretical upper limit 1.95 [31–33]. This may be due to the particular values of α and β used in our study. In this light, it would be interesting to investigate whether the actual values of α and β associated with water waves can be found by using them as fitting parameters to make the numerically predicted instability threshold coincide with the experimental one.

We conclude therefore that the higher-order terms in (3) introduce several new, realistic features that are absent in the standard KdV equation. This opens a promising field for further study.

Acknowledgements TB, KvdW and GK acknowledge support from the Carathéodory Programme of the University of Patras, grant number C167. KA thanks the Hellenic Scholarships Foundation (IKY) for partial financial support of this work.

References

1. Korteweg, D.J., de Vries, G.: On the change of form of long waves advancing in a rectangular canal, and on a new type of long stationary waves. Phil. Mag. **39** (1895) 422–443
2. Whitham, G.B.: Linear and Nonlinear Waves. Wiley, New York (1974)
3. Ablowitz, M.J., Segur, H.: Solitons and the Inverse Scattering Transform. SIAM, Philadelphia (1981)
4. Novikov, S., Manakov, S.V., Pitaevskii, L.P., Zakharov, V.E.: Theory of Solitons: The inverse Scattering Method. Consultants Bureau, New York (1984)
5. Drazin, P.J., Johnson, R.S.: Solitons: an Introduction. Cambridge University Press, Cambridge (1989)
6. Ablowitz, M.J., Clarkson, P.A.: Solitons, Nonlinear Evolution Equations and Inverse Scattering. Cambridge University Press, Cambridge (1991)
7. Remoissenet, M.: Waves Called Solitons: Concepts and Experiments, 3rd Ed. Springer, Berlin (1999)
8. Billingham, J., King, A.C.: Wave Motion. Cambridge University Press, Cambridge (2000)
9. Russell, J.S.: Report on Waves. In: Rep. 14th Meet. British Assoc. Adv. Sci. York. John Murray, London (1845) 11–390
10. Hammack, J.L., Segur, H.: The Korteweg-de Vries equation and water waves, Part 2: Comparison with Experiments. J. Fluid Mech. **65** (1974) 289–314
11. Zabusky, N.J., Kruskal, M: Interaction of solitons in a collisionless plasma and the recurrence of initial states. Phys. Rev. Lett. **15** (1965) 240–243
12. Amick, C.J., Toland, J.F.: On solitary water-waves of finite amplitude. Arch. Rat. Mech. Anal. **76** (1981) 9–95
13. Johnson, R.S.: A Modern Introduction to the Mathematical Theory of Water Waves. Cambridge University Press, Cambridge (1997)
14. Constantin, A., Escher, J.: Particle trajectories in solitary water waves. Bull. Am. Math. Soc. **44** (2007) 423-431
15. Nelson, R.C.: Depth limited design wave heights in very flat regions. Coastal Engng. **23** (1994) 43–59

16. Alvarez-Samaniego, B., Lannes, D.: Large time existence for 3D water-waves and asymptotics. Invent. Math. **171** (2008) 485-541
17. Constantin, A., Johnson, R.S.: On the non-dimensionalisation, scaling and resulting interpretation of the classical governing equations for water waves. J. Nonl. Math. Phys. **15** (2008) 58–73
18. Kodama, Y.: On integrable systems with higher order corrections. Phys. Lett. A **107** (1985) 245–249
19. Marchant, T.R., Smyth, N.F.: The extended Korteweg-de Vries equation and the resonant flow over topography. J. Fluid Mech. **221** (1990) 263–288
20. Kichenassamy, S., Olver, P.J.: Existence and nonexistence of solitary wave solutions to higher-order model evolution equations. SIAM J. Math. Anal. **23** (1992) 1141–1161
21. Marchant, T.R., Smyth, N.F.: Soliton solutions for the extended Korteweg-de Vries equation. IMA J. Appl. Math. **56** (1996) 157–176
22. Fokas, A.S., Liu, Q.M.: Asymptotic integrability of water waves. Phys. Rev. Lett. **77** (1996) 2347–2351
23. Kraenkel, R.A.: First-order perturbed Korteweg-de Vries solitons. Phys. Rev. E **57** (1998) 4775–4777
24. Marchant, T.R.: Asymptotic solitons of the extended Korteweg-de Vries equation. Phys. Rev. E **59** (1999) 3745–3748
25. Andriopoulos, K., Bountis, T., van der Weele, K., Tsigaridi, L.: The shape of soliton-like solutions of a higher-order KdV equation describing water waves. J. Nonlin. Math. Phys. **16, s-1** (2009) 1–12
26. Marchant, T.R.: High-order interaction of solitary waves on shallow water. Studies in Applied Math. **109** (2002) 1–17
27. Marchant, T.R.: Asymptotic solitons for a third-order Korteweg-de Vries equation. Chaos, Solitons and Fractals **22** (2004) 261–270
28. Fokas, A.S.: On a class of physically important integrable equations. Physica D **87** (1995) 145–150
29. Tzirtzilakis, E., Xenos, M., Marinakis, V., Bountis, T.C.: Interactions and stability of solitary waves in shallow water. Chaos, Solitons and Fractals **14** (2002) 87–95
30. Tzirtzilakis, E., Marinakis, V., Apokis, C., Bountis, T.C.: Soliton-like solutions of higher order wave equations of the Korteweg-de Vries type. J. Math. Phys. **43** (2002) 6151–6165
31. McCowan, J.: On the highest wave of permanent type, Phil. Mag. **38** (1894) 351–358
32. Massel, S.R.: On the largest wave height in water of constant depth. Ocean Engng. **23** (1996) 553–573
33. Abohadima, S., Isobe, M.: Limiting criteria of permanent progressive waves. Coastal Engng. **44** (2002) 231–237
34. Toland, J.F.: Stokes waves. Topol. Methods Nonlinear Anal. **7** (1996) 1–48
35. Constantin, A.: The trajectories of particles in Stokes waves. Invent. Math. **166** (2006) 523–535

Coarse Collective Dynamics of Animal Groups

Thomas A. Frewen, Iain D. Couzin, Allison Kolpas, Jeff Moehlis, Ronald Coifman, and Ioannis G. Kevrekidis

Abstract The coarse-grained, computer-assisted analysis of models of collective dynamics in animal groups involves (a) identifying appropriate observables that best describe the state of these complex systems and (b) characterizing the dynamics of such observables. We devise "equation-free" simulation protocols for the analysis of a prototypical individual-based model of collective group dynamics. Our approach allows the extraction of information at the macroscopic level via parsimonious usage of the detailed, "microscopic" computational model. Identification of meaningful coarse observables ("reduction coordinates") is critical to the success of such an approach, and we use a recently-developed dimensionality-reduction approach (diffusion maps) to detect good observables based on data generated by

---

I.G. Kevrekidis (✉)
Department of Chemical Engineering, Princeton University, Princeton, NJ 08544, USA
e-mail: yannis@princeton.edu
and
PACM and Mathematics, Princeton University, Princeton, NJ 08544, USA

T.A. Frewen
Department of Chemical Engineering, Princeton University, Princeton, NJ 08544, USA
e-mail: tfrewen@gmail.com

I.D. Couzin
Department of Ecology and Evolutionary Biology, Princeton University, Princeton,
NJ 08544, USA
e-mail: icouzin@princeton.edu

A. Kolpas
Department of Mathematical Sciences, University of Delaware, Newark, DE 19716, USA
e-mail: akolpas@math.udel.edu

J. Moehlis
Department of Mechanical Engineering, University of California, Santa Barbara,
CA 93106, USA
e-mail: moehlis@engineering.ucsb.edu

R. Coifman
Department of Mathematics, Yale University, New Haven, CT 06520, USA
e-mail: coifman@fmah.com

A.N. Gorban and D. Roose (eds.), *Coping with Complexity: Model Reduction and Data Analysis*, Lecture Notes in Computational Science and Engineering 75, DOI 10.1007/978-3-642-14941-2_16, © Springer-Verlag Berlin Heidelberg 2011

local model simulation bursts. This approach can be more generally applicable to the study of coherent behavior in a broad class of collective systems (e.g., collective cell migration).

1 Introduction

Many animal groups such as fish schools and bird flocks display remarkable collective behavior such as coherent group motion and transitions between different group configurations [1–4]. A small number of *informed* individuals in such animal groups, namely those with a preferred direction of motion, have been shown to be capable of influencing the group by facilitating the transfer of information, such as the location of a migration route, resources, or predators, to uninformed group members [5]. Understanding the mechanisms of information transfer in these and other biological systems is a problem of fundamental interest.

We consider a 1-dimensional model of animal group motion based on the 2-dimensional individual-based model in [6]. The direction of travel for each individual in the group is computed based on the occupancy of surrounding "zones of information." A single informed individual in the group has a preferred direction of travel. Model simulations indicate "stick-slip"–type dynamic behavior: the group appears to be "stuck in place" some of the time, with no net motion of its centroid; at other times the group clearly travels in the preferred direction of the informed individual.

We analyze the dynamics of this model problem through the computation of an effective free energy surface, obtained in terms of a suitable *reaction coordinate* that characterizes the state of the system. Such a reaction coordinate is first proposed by trial and error, after extended computational exploration of the dynamics. We then use *diffusion maps* [7] to extract the appropriate observable (the reaction coordinate) in an automated fashion; this involves the computation of the leading eigenvectors of the weighted Laplacian on a graph constructed from direct simulation data. Such data-mining procedures can directly link with, and facilitate, coarse-graining algorithms; in particular, there is a natural connection between these manifold-learning techniques and the equation-free framework [8] for complex/multiscale system modeling. The coarse variables identified through diffusion maps can be used to design, initialize, and process the results of short bursts of detailed simulation, whose purpose is to extract coarse-grained, macroscopic information from the fine scale, microscopic solver. We will also briefly discuss procedures for translating between physical system variables and these data-based coarse variables (observables).

2 Model Description

We consider a 1-dimensional model of collective group motion where each individual i in a group of size N is characterized by its position $c_i(t)$, speed $s_i(t) > 0$, and direction of travel $v_i(t)$. These quantities are updated for each individual at

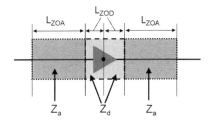

Fig. 1 Zones of information surrounding an individual in our spatially one-dimensional model. Position along the spatial axis is indicated by a *black dot*; the large *triangle* indicates direction of travel. The *zone of deflection* Z_d extends a distance L_{zod} to the *left* and *right* of the individual. The zone of attraction Z_a extends a distance L_{zoa} beyond the edges of Z_d. In our simulations we use $L_{zod} = 0.1$, $L_{zoa} = 4$, $N = 10$ individuals, and an agent response time $\Delta t = 0.1$

each time step (the time step size Δt may be interpreted as the response time of an individual). The individual update rules are based on the distribution of neighbors in spatial "zones" of information surrounding each individual as indicated in Fig. 1 (model parameter values are indicated in caption).

In an inner *zone of deflection* Z_d, collisions with near neighbors are avoided. When this zone is occupied for individual i its desired direction of travel at the next time step $d_i(t + \Delta t)$ may be computed using

$$d_i(t + \Delta t) = -\frac{\sum_{\substack{j \neq i \\ j \in Z_d}} \frac{c_j(t) - c_i(t)}{|c_j(t) - c_i(t)|}}{\left| \sum_{\substack{j \neq i \\ j \in Z_d}} \frac{c_j(t) - c_i(t)}{|c_j(t) - c_i(t)|} \right|}. \tag{1}$$

This equation reduces to

$$d_i(t + \Delta t) = \frac{N_{L,Z_d}(i) - N_{R,Z_d}(i)}{|N_{L,Z_d}(i) - N_{R,Z_d}(i)|} \tag{2}$$

where $N_{L,Z_d}(i)$ ($N_{R,Z_d}(i)$) is the number of individuals to the left (right) of individual i within its zone of deflection. Individuals with an occupied zone of deflection prioritize avoidance by ignoring the positions and directions of travel of others outside of Z_d. Individual i only interacts with those in an outer *zone of attraction* Z_a if its zone of deflection is empty. Individuals align with and are attracted towards neighbors in this outer zone. The desired direction of travel for individual i (with no neighbors in Z_d) is calculated from

$$d_i^0(t + \Delta t) = \frac{N_{R,Z_a}(i) - N_{L,Z_a}(i)}{|N_{R,Z_a}(i) - N_{L,Z_a}(i)|} + \sum_{j \in Z_a} \frac{v_j(t)}{|v_j(t)|} \tag{3a}$$

$$d_i(t + \Delta t) = \frac{d_i^0(t + \Delta t)}{|d_i^0(t + \Delta t)|} \tag{3b}$$

where $N_{L,Z_a}(i)$ ($N_{R,Z_a}(i)$) is the number of individuals to the left (right) of individual i within its zone of attraction, and the second term on the right hand side in (3a) is a contribution based on the direction of travel of neighbor j ($v_j(t)$) at the current time step; the summation includes individual i. The desired directions of travel for each individual in the group are computed using (2) and (3).

Individuals within the group having a preferred direction of travel (here, to the right) are called *informed*; all other group members are *naive*. We consider a single informed group individual here; the "level" of information of this individual is set by the parameter $p_{inf} = 0.2$ which is the probability with which its desired direction of travel is set to its preferred direction at each time step (irrespective of the dynamics indicated by the zone rules). This parameter determines the balance between interactions with neighbors and a preferred travel direction for the informed individual.

As in [9], information uncertainty is incorporated in the model by flipping the desired travel direction of each individual with a small probability $p_{flip} = 0.001$ using a random number drawn from the uniform distribution to give the actual direction of travel for each individual at the next time step $v_i(t + \Delta t)$. The positions of the individuals are updated using

$$c_i(t + \Delta t) = c_i(t) + s_i(t + \Delta t)v_i(t + \Delta t)\Delta t. \tag{4}$$

The initial positions and directions of travel for each individual are set randomly. The speed of each individual is reset at each time step by drawing from the normal distribution $N(\mu_s = 0.85, \sigma_s^2 = 6.25 \times 10^{-4})$.

3 Stick-Slip Phenomena

The variation of the group centroid with the number of time steps, along with the position of the informed individual, simulated with the model described in the previous section, is shown in Fig. 2 for three different values of p_{inf}. The simulation results clearly visually suggest stick-slip-type behavior, with instances where the group is stuck in place (individuals "vibrate" in place) and also periods where the swarm flows in the preferred direction of the informed individual (here, to the right). In Fig. 2 flat (sloped) regions of the centroid profile correspond to *stick* (*slip*) regimes. At higher values of p_{inf} the group travels further with less stick; we analyze the $p_{inf} = 0.2$ case in greater detail since it shows a balance between instances of stick and slip. The position of the informed individual at each step (black jagged line for $p_{inf} = 0.2$) in Fig. 2 suggests that the group sticks (slips) when the informed individual is distant from (close to) the group centroid. Extensive observation of transient simulations characterized by such behavior suggests,

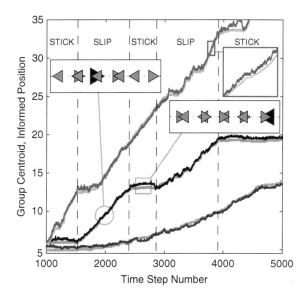

Fig. 2 Variation of group centroid and informed position as a function of simulation time step number for different values of p_{inf}, indicating stick-slip group motion. Jagged *light grey lines* (*top, middle,* and *bottom*) indicate centroid position, jagged *dark grey/black lines* (*top, middle,* and *bottom*) indicate position of the informed individual (*bottom*: $p_{inf} = 0.1$, *middle*: $p_{inf} = 0.2$, *top*: $p_{inf} = 0.4$) Stick-slip regimes for $p_{inf} = 0.2$ case are indicated in the text at the top of the figure, and separated by *dashed vertical lines*. *Grey open circle* (*square*) shows sample group configuration in slip (stick) regime with informed individual *shaded black*; *square* region at *top right* provides a close-up view of the group centroid and informed position for the $p_{inf} = 0.4$ case

as a possible *reaction coordinate* that characterizes the state of the system, $R(t)$ given by

$$R(t) = c_{inf}(t) - \overline{c}(t) \tag{5}$$

where $c_{inf}(t)$ is the position of the informed individual, and $\overline{c}(t) = \frac{1}{N}\sum_{i=1}^{N} c_i(t)$ is the group centroid. The evolution of this human experience-based coordinate (which we will call the "MAN" reaction coordinate) $R(t)$ with the simulation time step number (along with the corresponding centroid evolution) is plotted in Fig. 3. When the value of $R(t)$ is less than about 0.3 (middle panel) the group tends to slip – the informed individual is either at the rear or close to the centroid of the group. For values of $R(t)$ above approximately 0.3 the group is "stuck" – and the informed individual lies at the leading edge of the group.

We selected the "MAN" reaction coordinate $R(t)$ by inspecting the results of many simulations; an alternative, data-based computational procedure that automates the detection of such a coordinate (diffusion maps) will be described in Sect. 5 below.

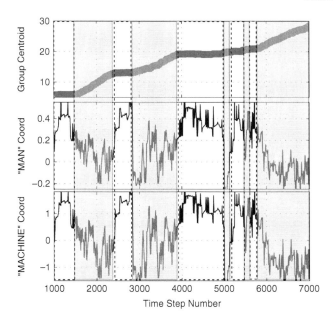

Fig. 3 Evolution of the group centroid ($p_{inf} = 0.2$) as a function of simulation time step number (*top panel*); corresponding evolution of the "MAN" coordinate $R(t)$ defined in (5) (*middle panel*); and corresponding evolution of the "MACHINE" coordinate (*bottom panel*). *Dashed boxes* mark "stick" regimes, *shaded boxes* correspond to "slip" regimes

4 Coarse Variable Analysis

We assume that the evolution of the system can be described in terms of a single, slowly evolving random variable q; the remaining system variables (or their statistics) are assumed to quickly become slaved to q. The evolution of the system can then be approximately described in terms of a time-dependent probability density function $f(q,t)$ for the slow variable q that evolves according to the following *effective* Fokker–Planck equation [10]:

$$\frac{\partial f}{\partial t} = \frac{\partial}{\partial q}\left(\frac{\partial}{\partial q}[D(q)f(q,t)] - V(q)f(q,t)\right). \tag{6}$$

If the effective drift $V(q)$ and the effective diffusion coefficient $D(q)$ can be explicitly written down as functions of q, then (6) can be used to compute interesting long-time properties of the system (e.g., the equilibrium distribution). Assuming that (6) provides a good approximation, we use $V(q) = \lim_{\Delta t \to 0} \frac{<Q(t+\Delta t)-q \mid Q(t)=q>}{\Delta t}$, and $D(q) = \lim_{\Delta t \to 0} \frac{<[Q(t+\Delta t)-q]^2 \mid Q(t)=q>}{\Delta t}$ to estimate the average drift, V, and diffusion coefficient D from short-time bursts of appropriately initialized detailed ("microscopic") simulations [11–14]. The steady solution of (6) is proportional to

$\exp[-\beta\Phi(q)]$, where the effective free energy $\Phi(q)$ is defined as

$$\beta\Phi(q) = -\int_0^q \frac{V(q')}{D(q')}dq' + \ln D(q) + \text{constant}. \tag{7}$$

Consequently, computing the effective free energy and the equilibrium probability distribution could be accomplished without the need for long-time detailed simulations. Processing the results of multiple, short simulation bursts allows computational estimation of the effective drift $V(q)$ and the effective diffusion coefficient $D(q)$ (using the above definitions averaged over many replicas). $\Phi(q)$ is then computed by numerical evaluation of the formula (7). The estimation of drift and diffusion coefficients for an effective Fokker–Planck equation assumes the knowledge of suitable slow reaction coordinates that describes the system dynamics.

The effective free energy profile shown in Fig. 4 (top panel) is computed by binning long time simulation data in the (empirically determined, "MAN") coordinate $R(t)$ and using the relationship (at equilibrium) $P(R) \propto \exp(-\beta\Phi)$ where $P(R)$ is the probability distribution of R. This effective free energy profile is characterized by a number of shallow local minima, roughly equally separated by a value of $L_{zod} = 0.1$ in the reaction coordinate. When the group is "stuck", the rate of change of $\overline{c}(t)$ (the second quantity in (5)) is close to zero, and changes in $R(t)$ (for this state) largely consist of changes in the informed position ($c_{inf}(t)$) by approximately $\pm\mu_s\Delta t$. In the slip state both terms in (5) vary; $\overline{c}(t)$ changes at each time step by approximately $\frac{1}{N}(N_R - N_L)\mu_s\Delta t$, where N_R (N_L) is the total number of individuals traveling to the right (left). The speed of the group, estimated from the difference in the group centroid location over time windows of 250 and of 500 steps, is plotted as a function of the reaction coordinate $R(t)$ in the inset in (Fig. 4) (top panel); there is clearly a change in group speeds between the stick and slip regimes.

The bistable effective free energy profile shown in Fig. 4 (bottom panel) bins long time simulation data according to values of the centroid speed, and uses the equilibrium relationship between the effective free energy and the stationary distribution of centroid speed.

5 Diffusion Maps and Data-Based Analysis

We now describe a *diffusion map* approach that automates the detection of suitable coarse coordinates by data-mining detailed simulation results.

We first define a pairwise similarity matrix $\mathbf{K}$ between simulation datapoints where each "datapoint" is a vector $\mathbf{x}$ with components consisting of the (sorted) distances of all naive individuals to the informed individual in the group

$$K_{i,j} = K\left(\mathbf{x}_i, \mathbf{x}_j\right) = \exp\left[-\left(\frac{\|\mathbf{x}_i - \mathbf{x}_j\|}{\sigma}\right)^2\right]; \tag{8}$$

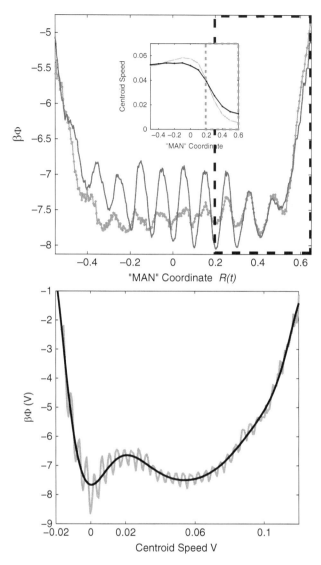

Fig. 4 *Top*: Effective free energy profile computed from long time simulations (*thick black line*) and from coarse estimation (*thin grey line*) in terms of the "MAN" coordinate $R(t)$ (5). The local minima in the effective free energy profile are separated by a value of approximately $L_{ZOD} = 0.1$. Inset: plot of centroid speed vs. the "MAN" coordinate $R(t)$; line shade distinguishes number of time steps over which centroid speed is estimated (*grey line*: 250 steps, *black line*: 500 steps). *Dashed boxes* in this figure suggest the "stick" regimes. *Bottom*: Effective free energy $\beta \Phi(V)$ (*grey line*) in terms of the centroid speed (computed from binning centroid speeds from a long time simulation and using $P(V) \propto \exp(-\beta \Phi)$). The simulation database consists of 200,000 time steps, with the centroid speed estimated over intervals of 200 steps. The *black line* is a polynomial fit to the effective potential

σ is a parameter that determines the "neighborhood size" within which the Euclidean distance between two data vectors provides a meaningful measure of their similarity. We use a small representative sample of the data in the construction of $\mathbf{K}$. Defining the diagonal normalization matrix $D_{i,i} = \sum_j K_{i,j}$, we construct the Markovian matrix $\mathbf{M} = \mathbf{D}^{-1}\mathbf{K}$. We show below that the components of the simulation data in a few of the leading eigenvectors of the matrix $\mathbf{M}$ may be used as a low dimensional representation of the data.

The matrix $\mathbf{M}$ is adjoint to the symmetric matrix $\mathbf{M_s}$

$$\mathbf{M_s} = \mathbf{D}^{1/2}\mathbf{M}\mathbf{D}^{-1/2} = \mathbf{D}^{-1/2}\mathbf{K}\mathbf{D}^{-1/2} \tag{9}$$

which shares its eigenvalues; eigenvectors of $\mathbf{M}$, denoted $\boldsymbol{\phi}_j$, are related to those of $\mathbf{M}_s$, denoted $\boldsymbol{\psi}_j$ as follows

$$\boldsymbol{\phi}_j = \mathbf{D}^{-1/2}\boldsymbol{\psi}_j \tag{10}$$

The *diffusion map* curve shown in Fig. 5 plots components of the top two, non-trivial, eigenvectors of $\mathbf{M}$ obtained by eigendecomposition of $\mathbf{M_s}$ with the use of (10). It is clear that points in this diffusion map are ordered along a curve according to the value of the diffusion map coordinate $\Phi(2)$ associated with each simulation datapoint. The diffusion map calculation has thereby provided an alternative (we will call it the "MACHINE") reaction coordinate through eigenprocessing of simulation data. We find a remarkable one-to-one correspondence between this "MACHINE" coordinate $\Phi(2)$ and the "MAN" coordinate $R(t)$, shown in the top inset in Fig. 5. Information on the coordinate $R(t)$ is not passed to the diffusion map calculation routine, yet this data-mining approach locates a coordinate that clearly and strongly correlates with $R(t)$. We also note that the Nyström formula [15] allows us to approximate the diffusion map coordinates of new simulation points, outside the original sample (used to construct $\mathbf{M_s}$).

6 Conclusions

Coarse-graining of the collective dynamics of animal group models may be enhanced through the detection of good *coarse variables* (observables). These can be proposed based on experience, obtained through extensive observation of system simulations; alternatively, they can be obtained using manifold learning techniques, such as the diffusion maps demonstrated here. In our illustrative example the essential dynamics exhibited bistability between a "slip" and a "stick" state; we note, however, that the observed bistability was especially sensitive to the "agent response time" parameter Δt. The bistable dynamics could be effectively characterized in terms of a single reaction coordinate; an effective free energy surface was obtained that succinctly summarizes the bistable dynamics. What was particularly gratifying here was the strong correlation between an empirical coordinate, based on extensive observations of the problem, and the coordinate discovered by the computer-assisted

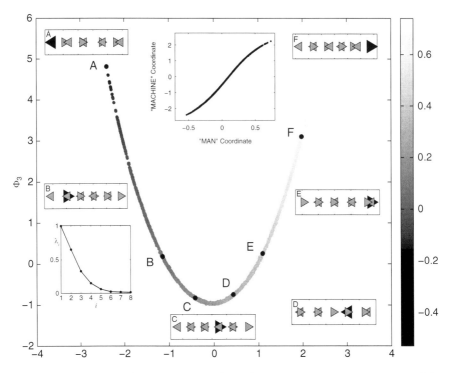

Fig. 5 Diffusion map results. The main plot shows the top two eigenvector components of the simulation data, with points shaded according to value of "MAN" coordinate $R(t)$ (scale shown in shadebar). The *bottom left* inset shows the leading eigenvalues of the corresponding Markov matrix. Individual configurations at several representative points (A,B,C,D,E,F) along the diffusion map curve are included. The instantaneous directions of the naive individuals (*grey triangles*) and of the informed individual (*black triangle*) are indicated by the triangle orientation. Top inset: plot of "MAN" coordinate $R(t)$ vs. the "MACHINE" coordinate $\Phi(2)$ for all simulation data points

Diffusion Map approach. In [16], this framework was shown to be equally effective in coarse-graining the dynamics of a 1-dimensional individual-based model of animal group formation with signaling constraints. This model has no informed individuals but for certain parameters exhibits "stick-slip" dynamics. This suggests that data-mining techniques can be naturally and efficiently linked with computational multiscale algorithms, in a way that can help accelerate the extraction of coarse-grained, system-level information from direct, individual-based models.

Acknowledgements This work was partially supported by the National Science Foundation and by the US AFOSR.

References

1. Couzin, I. D., Krause, J.: Self-organization and collective behavior in vertebrates. Adv. Study Behav. **32** (2003) 1–75
2. Deneubourg, J. L., Goss, S.: Collective patterns and decision making. Ethol. Ecol. Evol. **1** (1989) 295–311
3. Partridge, B. L.: The structure and function of fish schools. Sci. Am. **246** (1982) 114–123
4. Parrish, J. K., Edelstein-Keshet, L.: Complexity, pattern, and evolutionary trade-offs in animal aggregation. Science **284** (1999) 99–101
5. Couzin, I.: Collective Minds. Nature **445** (2007) 715
6. Couzin, I. D., Krause, J., Franks, N. R., Levin, S.: Effective leadership and decision making in animal groups on the move. Nature **433** (2005) 513–516
7. Coifman, R., Lafon, S., Lee, A., Maggioni, M., Nadler, B., Warner, F., Zucker, S. Geometric Diffusions as a tool for Harmonic Analysis and structure definition of data. Part I: Diffusion maps. Proc. Natl. Acad. Sci. USA **102** (2005) 7426–7431
8. Kevrekidis, I. G., Gear, C. W., Hummer, G.: Equation-free: The computer-aided analysis of complex multiscale systems. AIChE J. **50** (2004) 1346–1355
9. Kolpas, A., Moehlis, J., Kevrekidis, I. G.: Coarse-grained analysis of stochasticity-induced switching between collective motion states. Proc. Natl. Acad. Sci. USA **104** (2007) 5931–5935
10. Risken, H.: The Fokker-Planck Equation, Methods of solution and applications. Springer, Berlin (1989)
11. Haataja, M., Srolovitz, D., Kevrekidis, I. G.: Apparent hysteresis in a driven system with self-organized drag. Phys. Rev. Lett. **92** (2004) 160603
12. Hummer, G., Kevrekidis, I. G.: Coarse Molecular Dynamics of a Peptide Fragment: Free Energy, Kinetics and Long Time Dynamics Computations. J. Chem. Phys. **118** (2003) 10762–10773
13. Kopelevich, D., Panagiotopoulos, A. Z., Kevrekidis, I. G.: Coarse-Grained Kinetic Computations for Rare Events: Application to Micelle Formation. J. Chem. Phys. **122** (2005) 044908
14. Sriraman, S., Kevrekidis, I. G., Hummer, G.: Coarse Nonlinear Dynamics and Metastability of Filling-Emptying Transitions: Water in Carbon Nanotubes. Phys. Rev. Lett. **95** (2005) 130603
15. Baker, C.: The Numerical Treatment of Integral Equations. Clarendon Press, Oxford (1977)
16. Kolpas, A., Frewen, T. A., Moehlis, J., Kevrekidis, I. G.: Coarse analysis of collective motion with different communication mechanisms. Math. Biosci. **214** (2008) 49–57

Self-Simplification in Darwin's Systems

Alexander N. Gorban

Abstract We prove that a non-linear kinetic system with *conservation of supports* for distributions has generically limit distributions with final support only. The conservation of support has a biological interpretation: *inheritance*. We call systems with inheritance "Darwin's systems". Such systems are apparent in many areas of biology, physics (the theory of parametric wave interaction), chemistry and economics. The finite dimension of limit distributions demonstrates effects of *natural selection*. Estimations of the asymptotic dimension are presented. After some initial time, solution of a kinetic equation with conservation of support becomes a finite set of narrow peaks that become increasingly narrow over time and move increasingly slowly. It is possible that these peaks do not tend to fixed positions, and the path covered tends to infinity as $t \to \infty$. The *drift equations* for peak motion are obtained. They describe the asymptotic layer near the *omega*-limit distributions with finite support.

1 Introduction: Unusual Conservation Law

How can we prove that all the attractors of an infinite-dimensional system belong to a finite-dimensional manifold? How can we estimate the dimensions of attractor? There exist two methods to obtain such estimations.

First, if we find that k-*dimensional volumes are contracted* due to dynamics, then (after some additional technical steps) we can claim that the *Hausdorff dimension* of the maximal attractor is less than k. This idea is in the essence of *inertial manifold* theory [11]. The standard way to prove this k-dimensional volumes contraction is to check that the symmetrized Jakobian operator has discrete spectrum and the sum of any k its eigenvalues has negative real part (under some additional conditions of uniform boundness of solutions).

A.N. Gorban
Department of Mathematics, University of Leicester, LE1 7RH Leicester, UK
e-mail: ag153@le.ac.uk

A.N. Gorban and D. Roose (eds.), *Coping with Complexity: Model Reduction and Data Analysis*, Lecture Notes in Computational Science and Engineering 75, DOI 10.1007/978-3-642-14941-2_17, © Springer-Verlag Berlin Heidelberg 2011

Second, if we find a representation of our system as a nonlinear kinetic system with *conservation of supports* of distributions, then (again, after some additional technical steps) we can state that the asymptotics is finite-dimensional: the distribution evolves in a sum of several narrow peaks of density. This conservation of support has a *quasi-biological interpretation, inheritance* (if a gene was not presented initially in an isolated population without mutations, then it cannot appear at later time). The finite-dimensional asymptotic demonstrates effects of *"natural" selection*. It is very natural to call them *Darwin's systems*.

In the 1970s to the 1980s, theoretical work developed another "common" field simultaneously applicable to physics, biology and mathematics. For physics it is (so far) part of the theory of a special kind of approximation, demonstrating, in particular, interesting mechanisms of discreteness in the course of the evolution of distributions with initially smooth densities. However, what for physics is merely a convenient approximation is a fundamental law in biology: inheritance. The consequences of inheritance (collected in the selection theory [3, 13, 17, 21, 25, 27, 28, 41, 47–49]) give one of the most important tools for biological reasoning.

Consider a community of animals. Let it be biologically isolated. Mutations can be neglected in the first approximation. In this case, new genes do not emerge. Support of the distribution of genes does not increase.

An example from physics is also possible and leads to a very seminal approach to nonlinear wave theory. Let waves with wave vectors k be excited in some system. Denote K a set of wave vectors k of excited waves. Let the wave interaction does not lead to the generation of waves with new $k \notin K$. Such an approximation is applicable to a variety of situations, and has been described in detail for wave turbulence in [53, 54].

What is common in these examples is the evolution of a distribution with a support that does not increase over time.

What does not increase must, as a rule, decrease, if the decrease is not prohibited. This naive thesis can be converted into rigorous theorems for the case under consideration [21]. It is proved that the support decreases in the limit $t \to \infty$ if it was sufficiently large initially. (At finite times the distribution supports are conserved and decrease only in the limit $t \to \infty$.) Conservation of the support usually results in the following effect: dynamics of an initially infinite-dimensional system at $t \to \infty$ can be described by finite-dimensional systems and distribution degenerates into a sum of finite number of narrow peaks, the peaks' width tends to zero and the ω-limit distributions are finite sums of δ-functions.

This is description of the final, ω-limit distribution. More precisely, it is given by the selection theorem, and the dimension of the limiting systems can be evaluated by the properties of the reproduction coefficient functions.

In this paper we are focused on another problem: how a Darwin system approaches this finite-dimensional asymptotic? The first naive expectation is as follows: dynamics of infinite system tends to a finite-dimensional dynamics, which is predefined by the support of initial distribution and the adaptation landscape. This is, on some sense, another example of finite-dimensional inertial manifold.

Such asymptotic behavior of Darwin systems is possible, but surprisingly it was demonstrated [20, 21, 46] that even in rather simple examples Darwin's systems do not obligatory tend to one finite-dimensional asymptotic dynamical system, but can wander near infinitely many such systems. This wandering becomes slower in time, the density peaks become narrower, but the way of the wandering may tend to infinity because the velocity does not tends to zero in logarithmic time, $\log t$. We call this wandering the "drift effect". Below we describe the asymptotic behavior of systems with inheritance approaching their ω-limit distributions.

The simplest model for "reproduction + small mutations" where mutations are presented as diffusion is also studied. The limit of zero mutations is singular, and in the systems with small mutations the zero limit of the drift velocity of their drift at $t \to \infty$ substitutes by a small finite one in the presence of drift effect. Moreover, there exists a *scale invariance*, and dynamics for large t does not depend on nonzero mutation intensity, if the last is sufficiently small: to change this intensity, we need just to rescale time.

The structure of the paper is as follows. In Sect. 2 Darwin's systems are formally described and selection theorems are presented. In Sect. 2.2 the optimality principles for supports of ω-limit distributions are developed. These principles have a "weak" form; the set of possible supports is estimated from above and it is not obvious that this estimation is effective. A theorem of selection efficiency is presented in Sect. 2.5. The sense of this theorem is as follows: for almost every system the support of all ω-limit distributions is small (in an appropriate strong sense). Its geometrical interpretation suggested by M. Gromov is explained in Sect. 2.6.

Minimax estimations of the number of points in the support of ω-limit distributions are given in Sect. 2.3. The idea is to study systems under a ε-small perturbation, to estimate the maximal number of points for each realization of the perturbed system, and then to estimate the minimum of these maxima among various realizations. These minimax estimates can be constructive and do not use integration of the system. The set of reproduction coefficients $\{k(\mu) \,|\, \mu \in M\}$ is compact in $C(X)$. Therefore, this set can be approximated by a finite–dimensional linear space L_ε with any given accuracy ε.

The number of coexisting inherited units ("quasi-species") is estimated from above as $\dim L_\varepsilon$. This estimate is true both for stationary and non-stationary coexistence. In its general form this estimate was proved in 1980 [20,21], but the reasoning of this type has a long history. Perhaps, Gause [19] was the first to suggest the direct connection between the number of species and the number of resources. One can call this number "dimension of the environment". He proposed the famous concurrent exclusion principle.

More details about early history of the concurrent exclusion principle are presented in the review paper of Hardin [29].

MacArthur and Levins [39] suggested that the number of coexisting species is limited by the number of ecological resources. Later [40], they studied the continuous resource distribution (niche space) where the number of species is limited by the fact that the niches must not overlap too much. In 1999, Meszena and Metz [43] developed further the idea of environmental feedback dimensionality (perhaps, independently of [20, 21]).

In 2006 the idea of robustness in concurrent exclusion was approached again, as a "unified theory" of "competitive exclusion and limiting similarity" [42]. All these achievements are related to estimation of dimension of the set $\{k(\mu) \mid \mu \in M\}$ or of some its subsets. This dimension plays the role of "robust dimension of population regulation".

Drift equations without mutations are derived in Sect. 3.1 and asymptotic equations For Darwin's systems with small mutations with scaling invariance theorem are presented in Sect. 3.2.

The special structure of Darwin's systems requires to introduce three types of stability of the limit behavior (Sect. 4): internal stability (stability in the limiting finite-dimensional systems), external stability or uninvadability (stability with respect to strongly small perturbations that extend the support), and stable realizability (stability with respect to small shifts and extensions of the density peaks). Internal stability (Sect. 4.1) is just a particular case of Lyapunov stability applied to finite-dimensional asymptotic. External stability (uninvadability, Sect. 4.2) was first introduced and studied by Haldane [27], then, after papers [48, 49] it was intensively used in biology [4–9, 12, 32, 50, 51], and general evolutionary games theory [15, 31, 44, 45]. In physics, this notion was introduced in an entirely independent series of works on the S-approximation in the spin wave theory and on wave turbulence [38, 53, 54], which studied wave configurations in the approximation of an "inherited" wave vector.

The stable realizability (stability with respect to small shifts and extensions of the density peaks, Sect. 4.3) was first introduced and studied in full generality in early 1980s [21]. Later, the important particular cases were independently introduced and studied in series of papers [15, 44, 45]. In these papers the idea of drift equations appeared for the gaussian peaks in the dynamics of continuous symmetric evolutionary games. The authors [44, 45] called this property of "stable realizability" by "evolutionary robustness" and claimed the necessity of this additional type of stability very energetically: "Furthermore, we provide new conditions for the stability of rest points and show that even strict equilibria may be unstable".

In Sect. 5 a rich family of examples is described. Those examples are the generalized Lottka–Volterra–Gause infinite dimensional systems with distributed coefficients. The main benefit from this special structure is the generalized Volterra averaging principle [52]. This principle allows to substitute the time averages of linear functionals by their values at steady states. For the distributed Lottka–Volterra–Gause systems the drift equations are written explicitly.

In Sect. 7 a brief description of the main results is presented.

2 Inheritance and Selection Theorems

2.1 Asymptotically Stable States

The simplest and most common class of equations in applications for which the distribution support does not grow over time is constructed as follows. To each

Self-Simplification in Darwin's Systems

distribution μ is assigned a function k_μ by which distributions can be multiplied. Let us write down the equation:

$$\frac{d\mu}{dt} = k_\mu \times \mu. \tag{1}$$

The multiplier k_μ is called a *reproduction coefficient*. It depends on μ, and this dependence can be rather general and non-linear.

Two remarks can be important:

1. The apparently simple form of (1) does not mean that this system is linear or even close to linear. The operator $\mu \mapsto k_\mu$ is a general non-linear operator, and the only restriction is its continuity in an appropriate sense (see below).
2. On a finite set $X = \{x_1, \ldots, x_n\}$, non-negative measures μ are simply non-negative vectors $\mu_i \geq 0$ ($i = 1, \ldots, n$), and (1) appears to be a system of equations of the following type:

$$\frac{d\mu_i}{dt} = k_i(\mu_1, \ldots, \mu_n) \times \mu_i, \tag{2}$$

and the only difference from a general dynamic system is the special behavior of the right-hand side of (2) near zero values of μ_i.

The right-hand side of (1) is the product of the function k_μ and the distribution μ, and hence $d\mu/dt$ should be zero when μ is equal to zero; therefore the support of μ is conserved in time (over finite times).

X the *space of inherited units*. Most of the selection theorems are proved for compact metric space X with a metric $\rho(x, y)$. Further, for the drift and mutations equations we assume that X is a domain in finite-dimensional real space R^n (closed and bounded). As a particular case of compact space, a finite set X can be discussed.

μ is distribution on X. Each distribution on a compact space X is a continuous linear functional on the space of continuous real functions $C(X)$. We follow the Bourbaki approach [10]: a measure is a continuous functional, an integral. Book [10] contains all the necessary notions and theorems. Space $C(X)$ is a Banach space endowed with the maximum norm

$$\|f\| = \max_{x \in X} |f(x)|. \tag{3}$$

If $\mu \in C^*(X)$ and $f \in C(X)$, then $[\mu, f]$ is the value of μ at a function f. If X is a bounded closed subset of a finite-dimensional space R^n, then we represent this functional as the integral

$$[\mu, f] = \int \mu(x) f(x) \, dx, \tag{4}$$

which is the standard notation for distribution (or generalized function) theory. The "density" $\mu(x)$ is not assumed to be an absolute continuous function with respect to the Lebesgue measure dx (or even a function), and the notation in (4) has the same sense as $[\mu, f]$. If the measure is defined as a function on a σ-algebra of sets, then the following notation is used:

$$[\mu, f] = \int f(x)\,\mu(\mathrm{d}x).$$

We use the notation $[\mu, f]$ for general spaces X and the representation (4) on domains in R^n. The product $k \times \mu$ is defined for any $k \in C(X)$, $\mu \in C^*(X)$ by the equality: $[k\mu, f] = [\mu, kf]$.

The support of μ, supp μ, is the smallest closed subset of X with the following property: if $f(x) = 0$ on supp μ, then $[\mu, f] = 0$, i.e., $\mu(x) = 0$ outside supp μ.

In the space of measures we use *weak* convergence*, i.e., the convergence of averages:

$$\mu_i \to \mu^* \text{ if and only if } [\mu_i, \varphi] \to [\mu^*, \varphi] \tag{5}$$

for all continuous functions $\varphi \in C(X)$. This weak* convergence of measures generates *weak* topology* on the space of measures (sometimes called weak topology of conjugated space, or wide topology).

Strong topology on the space of measures $C^*(X)$ is defined by the norm $\|\mu\| = sup_{\|f\|=1}[\mu, f]$.

The properties of the mapping $\mu \mapsto k_\mu$ should be specified, and the existence and uniqueness of solutions of (1) under given initial conditions should be identified. In specific situations the answers to these questions are not difficult.

The sequence of continuous functions $k_i(x)$ is considered to be convergent if it converges uniformly. The sequence of measures μ_i is called convergent if for any continuous function φ the integrals $[\mu_i, \varphi]$ converge [weak* convergence (5)]. The mapping $\mu \mapsto k_\mu$ assigning the reproduction coefficient k_μ to the measure μ is assumed to be continuous with respect to these convergencies.

The space of measures is assumed to have a bounded set M that is positively invariant relative to system (1): if $\mu(0) \in M$, then $\mu(t) \in M$ (we also assume that M is non-trivial, i.e., it is neither empty nor a one-point set i.e., it is neither empty nor a one-point set but includes at least one point with its vicinity). This M serves as the phase space of system (1). (Let us remind that the set of measures M is bounded if the set of integrals $\{[\mu, f] \mid \mu \in M, \|f\| \le 1\}$ is bounded, where $\|f\|$ is the norm (3).) We study dynamic of system (1) in bounded positively invariant set M.

Most of the results for systems with inheritance use a **theorem on weak* compactness**: *The bounded set of measures is precompact with respect to weak* convergence (i.e., its closure is compact).* Therefore the set of corresponding reproduction coefficients $k_M = \{k_\mu \mid \mu \in M\}$ is precompact.

The simplest example of an emerging discrete distribution from a continuous initial distribution gives us the following equation:

$$\frac{\partial \mu(x,t)}{\partial t} = \left[f_0(x) - \int_a^b f_1(x)\mu(x,t)\,\mathrm{d}x \right] \mu(x,t), \tag{6}$$

Self-Simplification in Darwin's Systems

where the functions $f_0(x)$ and $f_1(x)$ are positive and continuous on the closed segment $[a, b]$. Let the function $f_0(x)$ reach the global maximum on the segment $[a, b]$ at a single point x_0. If $x_0 \in \operatorname{supp} \mu(x, 0)$, then:

$$\mu(x, t) \to \frac{f_0(x_0)}{f_1(x_0)} \delta(x - x_0), \qquad \text{when } t \to \infty, \tag{7}$$

where $\delta(x - x_0)$ is the δ-function.

If $f_0(x)$ has several global maxima, then the right-hand side of (7) can be the sum of a finite number of δ-functions. Here a natural question arises: is it worth considering such a possibility? Indeed, such a case seems to be very unlikely to occur. More details on this are given below.

The limit behavior of a typical system with inheritance (1) can be much more complicated than (7). Here we can mention that any finite-dimensional system with a compact phase space can be embedded in a system with inheritance (2). An additional possibility for the limit behavior is, for example, the drift effect (Sect. 3.1).

The first step in the routine investigation of a dynamical system is a question about fixed points and their stability. The first observation concerning the system (1) is that it can only be asymptotically stable for steady-state distributions, the support of which is discrete (i.e., the sums of δ-functions). This can be proved for all consistent formalizations. Thus, we have the first theorem.

Theorem 1. *The support of asymptotically stable distributions for the system (1) is always discrete.* $\square$

For the proof of this theorem and other selection theorems we refer to [20,21,23, 24]

The perturbation discussed is small not only in the weak* topology, but also in the strong sense, and thus it is sufficient to consider strongly small perturbations to prove that the asymptotically stable distribution should be discrete. Hence, this statement is true if the operator $\mu \mapsto k_\mu$ is continuous for strong topology on the space of measures. This is a significantly weaker requirement than being continuous in weak* topology.

This simple observation has many strong generalizations to general ω-limit points, to equations for vector measures, etc.

2.2 Optimality Principle for Limit Diversity

Description of the limit behavior of a dynamical system can be much more complicated than enumerating stable fixed points and limit cycles. The leading rival to adequately formalize the limit behavior is the concept of the "ω-limit set". It was discussed in detail in the classical monograph [2]. The fundamental textbook on dynamical systems [30] and the introductory review [34] are also available.

Let $f(t)$ be the dependence of the position of point in the phase space on time t (i.e., the *motion* of the dynamical system). A point y is a ω-limit point of the motion $f(t)$, if there exists such a sequence of times $t_i \to \infty$, that $f(t_i) \to y$.

The set of all ω-limit points for the given motion $f(t)$ is called the ω-limit set. If, for example, $f(t)$ tends to the equilibrium point y^* then the corresponding ω-limit set consists of this equilibrium point. If $f(t)$ is winding onto a closed trajectory (the limit cycle), then the corresponding ω-limit set consists of the points of the cycle and so on.

General ω-limit sets are not encountered oft in specific situations. This is because of the lack of efficient methods to find them in a general situation. Systems with inheritance is a case, where there are efficient methods to estimate the limit sets from above. This is done by the optimality principle.

Let $\mu(t)$ be a solution of (1). Note that

$$\mu(t) = \mu(0) \exp \int_0^t k_{\mu(\tau)} \, d\tau. \tag{8}$$

Here and below we do not display the dependence of distributions μ and of the reproduction coefficients k on x when it is not necessary. Fix the notation for the average value of $k_{\mu(\tau)}$ on the segment $[0, t]$

$$\langle k_{\mu(t)} \rangle_t = \frac{1}{t} \int_0^t k_{\mu(\tau)} \, d\tau. \tag{9}$$

Then the expression (8) can be rewritten as

$$\mu(t) = \mu(0) \exp(t \langle k_{\mu(t)} \rangle_t).$$

If μ^* is the ω-limit point of the solution $\mu(t)$, then there exists such a sequence of times $t_i \to \infty$, that $\mu(t_i) \to \mu^*$. Let it be possible to chose a convergent subsequence of the sequence of the average reproduction coefficients $\langle k_{\mu(t)} \rangle_t$, which corresponds to times t_i. We denote as k^* the limit of this subsequence. Then, the following statement is valid: on the support of μ^* the function k^* vanishes and on the support of $\mu(0)$ it is non-positive:

$$k^*(x) = 0 \text{ if } x \in \text{supp} \, \mu^*,$$
$$k^*(x) \leq 0 \text{ if } x \in \text{supp} \, \mu(0). \tag{10}$$

Taking into account the fact that $\text{supp} \, \mu^* \subseteq \text{supp} \, \mu(0)$, we come to the formulation of *the optimality principle* (10): *The support of limit distribution consists of points of the global maximum of the average reproduction coefficient on the initial distribution support. The corresponding maximum value is zero.*

We should also note that not necessarily all points of maximum of k^* on $\text{supp} \, \mu(0)$ belong to $\text{supp} \, \mu^*$, but all points of $\text{supp} \, \mu^*$ are the points of maximum of k^* on $\text{supp} \, \mu(0)$.

Self-Simplification in Darwin's Systems

If $\mu(t)$ tends to the fixed point μ^*, then $\langle k_{\mu(t)} \rangle_t \to k_{\mu^*}$ as $t \to \infty$, and supp μ^* consists of the points of the global maximum of the corresponding reproduction coefficient k_{μ^*} on the support of μ^*. The corresponding maximum value is zero.

If $\mu(t)$ tends to the limit cycle $\mu^*(t)$ ($\mu^*(t + T) = \mu^*(t)$), then all the distributions $\mu^*(t)$ have the same support. The points of this support are the points of maximum (global, zero) of the averaged over the cycle reproduction coefficient

$$k^* = \langle k_{\mu^*(t)} \rangle_T = \frac{1}{T} \int_0^T k_{\mu^*(\tau)} \, d\tau,$$

on the support of $\mu(0)$.

The supports of the ω-limit distributions are specified by the functions k^*. It is obvious where to get these functions from for the cases of fixed points and limit cycles. There are at least two questions: what ensures the existence of average reproduction coefficients at $t \to \infty$, and how to use the described extremal principle (and how efficient is it). The latter question is the subject to be considered in the following sections. In the situation to follow the answers to these questions have the validity of theorems.

Due to the theorem about weak* compactness, the set of reproduction coefficients $k_M = \{k_\mu \,|\, \mu \in M\}$ is precompact, hence, the set of averages (9) is precompact too, because it is the subset of the closed convex hull $\overline{\mathrm{conv}}(k_M)$ of the compact set. This compactness allows us to claim the existence of the *average reproduction coefficient* k^* for the description of the ω-limit distribution μ^* with the optimality principle (10).

2.3 How Many Points Does the Limit Distribution Support Hold?

The limit distribution is concentrated in the points of (zero) global maximum of the average reproduction coefficient. The average is taken along the solution, but the solution is not known beforehand. With the convergence towards a fixed point or to a limit cycle this difficulty can be circumvented. In the general case the extremal principle can be used without knowing the solution, in the following way [21]. Considered is a set of all dependencies $\mu(t)$ where μ belongs to the phase space, the bounded set M. The set of all averages over t is $\{\langle k_{\mu(t)} \rangle_t\}$. Further, taken are all limits of sequences formed by these averages – the set of averages is closed. The result is the closed convex hull $\overline{\mathrm{conv}}(k_M)$ of the compact set k_M. This set involves all possible averages (9) and all their limits. In order to construct it, the true solution $\mu(t)$ is not needed.

The weak optimality principle is expressed as follows. Let $\mu(t)$ be a solution of (1) in M, μ^* is any of its ω-limit distributions. Then in the set $\overline{\mathrm{conv}}(k_M)$ there is such a function k^* that its maximum value on the support supp μ_0 of the initial distribution μ_0 equals to zero, and supp μ^* consists of the points of the global maximum of k^* on supp μ_0 only (10).

Of course, in the set $\overline{\text{conv}}(k_M)$ there are usually many functions that are irrelevant to the time average reproduction coefficients for the given motion $\mu(t)$. Therefore, the weak extremal principle is really weak – it gives too many possible supports of μ^*. However, even such a principle can help to obtain useful estimates of the number of points in the supports of ω-limit distributions.

It is not difficult to suggest systems of the form (1), in which any set can be the limit distribution support. The simplest example: $k_\mu \equiv 0$. Here ω-limit (fixed) is any distribution. However, almost any arbitrary small perturbation of the system destroys this pathological property.

In the realistic systems, especially in biology, the coefficients fluctuate and are never known exactly. Moreover, the models are in advance known to have a finite error which cannot be exterminated by the choice of the parameters values. This gives rise to an idea to consider not individual systems (1), but ensembles of similar systems [21].

Let us estimate the maximum for each individual system from the ensemble (in its ω-limit distributions), and then, estimate the minimum of these maxima over the whole ensemble – (*the minimax estimation*). The latter is motivated by the fact, that if the inherited unit has gone extinct under some conditions, it will not appear even under the change of conditions.

Let us consider an ensemble that is simply the ε-neighborhood of the given system (1). The minimax estimates of the number of points in the support of ω-limit distribution are constructed by approximating the dependencies k_μ by finite sums

$$k_\mu = \varphi_0(x) + \sum_{i=1}^{n} \varphi_i(x)\psi_i(\mu). \tag{11}$$

Here φ_i depend on x only, and ψ_i depend on μ only. Let $\varepsilon_n > 0$ be the distance from k_μ to the nearest sum (11) (the "distance" is understood in the suitable rigorous sense, which depends on the specific problem). So, we reduced the problem to the estimation of the diameters $\varepsilon_n > 0$ of the set $\overline{\text{conv}}(k_M)$.

The minimax estimation of the number of points in the limit distribution support gives the answer to the question, "How many points does the limit distribution support hold": *If $\varepsilon > \varepsilon_n$ then, in the ε-vicinity of k_μ, the minimum of the maxima of the number of points in the ω-limit distribution support does not exceed n.*

In order to understand this estimate it is sufficient to consider system (1) with k_μ of the form (11). In this case for any dependence $\mu(t)$ the averages (9) have the form

$$\langle k_{\mu(t)} \rangle_t = \frac{1}{t} \int_0^t k_{\mu(\tau)} \, d\tau = \varphi_0(x) + \sum_{i=1}^{n} \varphi_i(x)a_i. \tag{12}$$

where a_i are some numbers. The ensemble of the functions (12) for various a_i forms a n-dimensional linear manifold. How many points of the global maximum (equal to zero) could a function of this family have?

Generally speaking, it can have any number of maxima. However, it seems obvious, that "usually" one function has only one point of global maximum, while it is "improbable" that the maximum value is zero. At least, with an arbitrary small perturbation of the given function, we can achieve for the point of the global maximum to be unique and the maximum value be non-zero.

In a one-parametric family of functions there may occur zero value of the global maximum, which cannot be eliminated by a small perturbation, and individual functions of the family may have two global maxima.

In the general case we can state, that "usually" each function of the n-parametric family (12) can have not more than n points of the zero global maximum (of course, there may be less, and the global maximum is, as a rule, not equal to zero at all for the majority of functions of the family). What "usually" means here requires a special explanation given in the next section.

In application k_μ is often represented by an integral operator, linear or nonlinear. In this case the form (11) corresponds to the kernels of integral operators, represented in a form of the sums of functions' products. For example, the reproduction coefficient of the following form

$$k_\mu = \varphi_0(x) + \int K(x, y)\mu(y)\, \mathrm{d}y \,,$$

$$\text{where } K(x, y) = \sum_{i=1}^{n} \varphi_i(x)g_i(y), \tag{13}$$

has also the form (11) with $\psi_i(\mu) = \int g_i(y)\mu(y)\, \mathrm{d}y$.

The linear reproduction coefficients occur in applications rather frequently. For them the problem of the minimax estimation of the number of points in the ω-limit distribution support is reduced to the question of the accuracy of approximation of the linear integral operator by the sums of kernels-products (13).

2.4 Almost Finite Sets and "Almost Always"

The supports of the ω-limit distributions for the systems with inheritance were characterized by the optimality principle. These supports consist of points of global maximum of the average reproduction coefficient. We can a priori (without studying the solutions in details) characterize the compact set that includes all possible average reproduction coefficients. Hence, we get a problem: how to describe the set of global maximum for all functions from generic compact set of functions. First of all, any closed subset $M \subset X$ is a set of global maximum of a continuous function, for example, of the function $f(x) = -\rho(x, M)$, where $\rho(x, M)$ is the distance between a set and a point: $\rho(x, M) = \inf_{y \in M} \rho(x, y)$, and $\rho(x, y)$ is the distance between points. Nevertheless, we can expect that one generic function has one point of global maximum, in a generic one-parametric family might exist functions with two points of global maximum, etc. How these expectations meet the exact results? What does the notion "generic" mean? What can we say about sets of global maximum of functions from a generic compact family? In this section we answer these questions.

Here are some examples of correct but useless statements about "generic" properties of function: Almost every continuous function is not differentiable; Almost every C^1-function is not convex. Their meaning for applications is most probably this: the genericity used above for continuous functions or for C^1-function is irrelevant to the subject.

Most frequently the motivation for definitions of genericity is found in such a situation: given n equations with m unknowns, what can we say about the solutions? The answer is: in a typical situation, if there are more equations, than the unknowns ($n > m$), there are no solutions at all, but if $n \le m$ (n is less or equal to m), then, either there is a $(m - n)$-parametric family of solutions, or there are no solutions.

The best known example of using this reasoning is the *Gibbs phase rule* in classical chemical thermodynamics. It limits the number of co-existing phases. There exists a well-known example of such reasoning in mathematical biophysics too. Let us consider a medium where n species coexist. The medium is assumed to be described by m parameters s_j. Dynamics of these parameters depends on the organisms. In the simplest case, the medium is a well-mixed solution of m substances. Let the organisms interact through the medium, changing its parameters – concentrations of m substances. It can be formalized by a system of equation:

$$
\frac{d\mu_i}{dt} = k_i(s_1, \ldots, s_m) \times \mu_i \ (i = 1, \ldots n);
$$
$$
\frac{ds_j}{dt} = q_j(s_1, \ldots, s_m, \mu_1, \ldots, \mu_n) \ (j = 1, \ldots m), \tag{14}
$$

In a steady state, for each of the coexisting species we have an equation with respect to the state of the medium: the corresponding reproduction coefficient k_i is zero. So, the number of such species cannot exceed the number of parameters of the medium. In a typical situation, in the m-parametric medium in a steady state there can exist not more than m species. This is the concurrent exclusion principle in the Gause form [19]. Here, the main hypothesis about interaction of organisms with the media is that the number of essential components of the media is bounded from above by m and increase of the number of species does not extend the list of components further. Dynamics of parameters depends on the organisms, but their nomenclature is fixed.

This concurrent exclusion principle allows numerous generalizations [14, 37, 39, 40, 42]. Theorem of the natural selection efficiency may be also considered as its generalization.

Analogous assertion for a non-steady state coexistence of species in the case of (14) is not true. It is not difficult to give an example of stable coexistence under oscillating conditions of n species in the m-parametric medium at $n > m$.

Self-Simplification in Darwin's Systems

But, if $k_i(s_1, \ldots, s_m)$ are linear functions of $s_1, \ldots, s_m$, then for non-stable conditions we have the concurrent exclusion principle, too. In that case, the average in time of the reproduction coefficient is the reproduction coefficient for the average state of the medium:

$$\langle k_i(s_1(t), \ldots, s_m(t)) \rangle = k_i(\langle s_1 \rangle, \ldots, \langle s_m \rangle)$$

because of linearity. If $\langle x_i \rangle \neq 0$ then $k_i(\langle s_1 \rangle, \ldots, \langle s_m \rangle) = 0$, and we obtain the non-stationary concurrent exclusion principle "in average". And again, it is valid "almost always".

The non-stationary concurrent exclusion principle "in average" is valid for linear reproduction coefficients. This is a combination of the Volterra [52] averaging principle and the Gause principle,

It is worth to mention that, for our basic system (1), if k_μ are linear functions of μ, then the average in time of the reproduction coefficient $k_{\mu(t)}$ is the reproduction coefficient for the average $\mu(t)$ because of linearity. Therefore, the optimality principle (10) for the average reproduction coefficient k^*, transforms into the following optimality principle for the reproduction coefficient $k_{\langle \mu \rangle}$ of the average distribution $\langle \mu \rangle$

$$k_{\langle \mu \rangle}(x) = 0 \text{ if } x \in \operatorname{supp} \mu^*,$$
$$k_{\langle \mu \rangle}(x) \leq 0 \text{ if } x \in \operatorname{supp} \mu(0). f$$

(the generalized *Volterra averaging principle* [52]).

Formally, various definitions of genericity are constructed as follows. All systems (or cases, or situations and so on) under consideration are somehow parameterized – by sets of vectors, functions, matrices etc. Thus, the "space of systems" Q can be described. Then the *thin sets* are introduced into Q, i.e., the sets, which we shall later neglect. The union of a finite or countable number of thin sets, as well as the intersection of any number of them should be thin again, while the whole Q is not thin. There are two traditional ways to determine thinness.

1. A set is considered thin when it has *measure zero*. This is reasonable for a finite-dimensional case, when there is the standard Lebesgue measure – the length, the area, the volume.
2. But most frequently we deal with the functional parameters. In that case it is common to restore to the second definition, according to which the sets of Baire first category are negligible. The construction begins with nowhere dense sets. The set Y is nowhere dense in Q, if in any nonempty open set $V \subset Q$ (for example, in a ball) there exists a nonempty open subset $W \subset V$ (for example, a ball), which does not intersect with Y: $W \cap Y = \emptyset$. Roughly speaking, Y is "full of holes" – in any neighborhood of any point of the set Y there is an open hole. Countable union of nowhere dense sets is called the set of first category. The second usual way is to define thin sets as the *sets of first category*. A *residual set* (a "thick" set) is the complement of a set of the first category.

For the second approach, the Baire category theorem is important: In a non-empty complete metric space, any countable intersection of dense, open subsets is non-empty.

But even the real line R can be divided into two sets, one of which has zero measure, the other is of first category. The genericity in the sense of measure and the genericity in the sense of category considerably differ in the applications where both of these concepts can be used. The conflict between the two main views on genericity stimulated efforts to invent new and stronger approaches.

Systems (1) were parameterized by continuous maps $\mu \mapsto k_\mu$. Denote by Q the space of these maps $M \to C(X)$ with the topology of uniform convergence on M. It is a Banach space. Therefore, we shall consider below thin sets in a Banach space Q. First of all, let us consider n-dimensional affine compact subsets of Q as a Banach space of affine maps $\Psi : [0, 1]^n \to Q$ ($\Psi(\alpha_1, \ldots \alpha_n) = \sum_i \alpha_i f_i + \varphi$, $\alpha_i \in [0, 1]$, $f_i, \varphi \in Q$) in the maximum norm. For the image of a map Ψ we use the standard notation imΨ.

Definition 1. A set $Y \subset Q$ is n-thin, if the set of affine maps $\Psi : [0, 1]^n \to Q$ with non-empty intersection im$\Psi \cap Y \neq \emptyset$ is the set of first category.

All compact sets in infinite-dimensional spaces and closed linear subspaces with codimension greater then n are n-thin. If dim $Q \leq n$, then only empty set is n-thin in Q. The union of a finite or countable number of n-thin sets, as well as the intersection of any number of them is n-thin, while the whole Q is not n-thin.

Let us consider compact subsets in Q parametrized by points of a compact space K. It can be presented as a Banach space $C(K, Q)$ of continuous maps $K \to Q$ in the maximum norm.

Definition 2. A set $Y \subset Q$ is completely thin, if for any compact K the set of continuous maps $\Psi : K \to Q$ with non-empty intersection im$\Psi \cap Y \neq \emptyset$ is the set of first category.

A set Y in the Banach space Q is completely thin, if for any compact set K in Q and arbitrary positive $\varepsilon > 0$ there exists a vector $q \in Q$, such that $\|q\| < \varepsilon$ and $K+q$ does not intersect Y: $(K + q) \cap Y = \emptyset$. All compact sets in infinite-dimensional spaces and closed linear subspaces with infinite codimension are completely thin. Only empty set is completely thin in a finite-dimensional space. The union of a finite or countable number of completely thin sets, as well as the intersection of any number of them is completely thin, while the whole Q is not completely thin.

Proposition 1. *If a set Y in the Banach space Q is completely thin, then for any compact metric space K the set of continuous maps $\Psi : K \to Q$ with non-empty intersection im$\Psi \cap Y \neq \emptyset$ is completely thin in the Banach space $C(K, Q)$.* □

Below the wording "almost always" means: the set of exclusions is completely thin. The main result presented in this section sounds as follows: almost always the sets of global maxima of functions from a compact set are uniformly almost finite.

Self-Simplification in Darwin's Systems

Proposition 2. *Let X have no isolated points. Then almost always a function $f \in C(X)$ has nowhere dense set of zeros $\{x \in X \mid f(x) = 0\}$ (the set of exclusions is completely thin in $C(X)$).* $\square$

After combination Proposition 2 with Proposition 1 we get the following

Proposition 3. *Let X have no isolated points. Then for any compact space K and almost every continuous map $\Psi : K \to C(X)$ all functions $f \in \mathrm{im}\Psi$ have nowhere dense sets of zeros (the set of exclusions is completely thin in $C(K, C(X))$).* $\square$

In other words, in almost every compact family of continuous functions all the functions have nowhere dense sets of zeros.

Let us consider a space of closed subsets of the compact metric space X endowed by the Hausdorff distance. The Hausdorff distance between closed subsets of X is

$$\mathrm{dist}(A, B) = \max\{\sup_{x \in A} \inf_{x \in B} \rho(x, y), \ \sup_{x \in B} \inf_{x \in A} \rho(x, y)\}.$$

The almost finite sets were introduced in [21] for description of the typical sets of maxima for continuous functions from a compact set. This definition depends on an arbitrary sequence $\varepsilon_n > 0$, $\varepsilon_n \to 0$. For any such sequence we construct a class of subsets $Y \subset X$ that can be approximated by finite set faster than $\varepsilon_n \to 0$, and for families of sets we introduce a notion of *uniform* approximation by finite sets faster than $\varepsilon_n \to 0$:

Definition 3. Let $\varepsilon_n > 0$, $\varepsilon_n \to 0$. The set $Y \subset X$ can be approximated by finite sets faster than $\varepsilon_n \to 0$ ($\varepsilon_n > 0$), if for any $\delta > 0$ there exists a finite set S_N such that $\mathrm{dist}(S_N, Y) < \delta\varepsilon_N$. The sets of family $\mathbb{Y}$ can be uniformly approximated by finite sets faster than $\varepsilon_n \to 0$, if for any $\delta > 0$ there exists such a number N that for any $Y \in \mathbb{Y}$ there exists a finite set S_N such that $\mathrm{dist}(S_N, Y) < \delta\varepsilon_N$.

The simplest example of almost finite set on the real line for a given $\varepsilon_n \to 0$ ($\varepsilon_n > 0$) is the sequence ε_n/n. If $\varepsilon_n < \mathrm{const}/n$, then the set Y on the real line which can be approximated by finite sets faster than $\varepsilon_n \to 0$ have zero Lebesgue measure. At the same time, it is nowhere dense, because it can be covered by a finite number of intervals with an arbitrary small sum of lengths (hence, in any interval we can find a subinterval free of points of Y).

Let us study the sets of global maxima $\mathrm{argmax} f$ for continuous functions $f \in C(X)$. For each $f \in C(X)$ and any $\epsilon > 0$ there exists $\phi \in C(X)$ such that $\|f - \phi\| \le \epsilon$ and $\mathrm{argmax}\phi$ consists of one point. Such a function ϕ can be chosen in the form

$$\phi(x) = f(x) + \frac{\epsilon}{1 + \rho(x, x_0)^2},$$

where x_0 is an arbitrary element of $\mathrm{argmax} f$. In this case $\mathrm{argmax}\phi = \{x_0\}$.

Hence, the set $\mathrm{argmax} f$ can be reduced to one point by an arbitrary small perturbations of the function f. On the other hand, it is impossible to extend significantly the set $\mathrm{argmax} f$ by a sufficiently small perturbation, the dependence of this set on f is semicontinuous in the following sense.

Proposition 4. *For given $f \in C(X)$ and any $\varepsilon > 0$ there exists $\delta > 0$ such that, whenever $\| f - \phi \| < \delta$, then*

$$\max_{x \in \operatorname{argmax} \phi} \ \min_{y \in \operatorname{argmax} f} \rho(x, y) < \varepsilon. \quad \square \tag{15}$$

These constructions can be generalized onto n-parametric affine compact families of continuous functions. Let us consider affine maps of the cube $[0, 1]^k$ into $C(X)$, $\Phi : [0, 1]^k \to C(X)$. The space of all such maps is a Banach space endowed with the maximum norm.

Proposition 5. *For any affine map $\Phi : [0, 1]^k \to C(X)$ and an arbitrary $\epsilon > 0$ there exists such a continuous function $\psi \in C(X)$, that $\| \psi \| < \epsilon$ and the set $\operatorname{argmax}(f + \psi)$ includes not more than $k + 1$ points for all $f \in \operatorname{im}\Phi$. $\square$*

To prove this Proposition we used the following Lemma which is of general interest.

Lemma 1. *Let $Q \subset C(X)$ be a compact set of functions, $\varepsilon > 0$. Then there are a finite set $Y \subset X$ and a function $\phi \in C(X)$ such that $\| \phi \| < \varepsilon$, and any function $f \in Q + \phi$ achieves its maximum only on Y: $\operatorname{argmax} f \subset Y$. $\square$*

Note, that Proposition 5 and Lemma 1 demonstrate us different sources of discreteness: in Lemma 1 it is the approximation of a compact set by a finite net, and in Proposition 5 it is the connection between the number of parameters and the possible number of global maximums in a k-parametric family of functions. There is no direct connection between N and k values, and it might be that $N \gg k$. For smooth functions in finite-dimensional real space polynomial approximations can be used instead of Lemma 1 in order to prove the analogue of Proposition 5.

The rest of this Sect. 2.4 is devoted to application of Proposition 5 to evaluation of maximizers for functions from a compact sets of functions. For any compact K the space of continuous maps $C(K, C(X))$ is isomorphic to the space of continuous functions $C(K \times X)$. Each continuous map $F : K \to C(X)$ can be approximated with an arbitrary accuracy $\varepsilon > 0$ by finite sums of the following form ($k \geq 0$):

$$F(y)(x) = \sum_{i=1}^{k} \alpha_i(y) f_i(x) + \varphi(x) + o,$$

$$y \in K, \ x \in X, \ 0 \leq \alpha_i \leq 1, \ f_i, \varphi \in C(X), \ |o| < \varepsilon. \tag{16}$$

Each set $f_i, \varphi \in C(X)$ generates a map $\Phi : [0, 1]^k \to C(X)$. A dense subset in the space of these maps satisfy the statement of Proposition 5: each function from $\operatorname{im}\Phi$ has not more than $k + 1$ points of global maximum. Let us use for this set of maps Φ notation $\mathbf{P}_k$, for the correspondent set of the maps $F : K \to C(X)$, which have the form of finite sums (16), notation $\mathbf{P}_k^K$, and $\mathbf{P}^K = \cup_k \mathbf{P}_k^K$.

For each $\Phi \in \mathbf{P}_k^K$ and any $\varepsilon > 0$ there is $\delta = \delta_\Phi(\varepsilon) > 0$ such that, whenever $\| \Psi - \Phi \| < \delta_\Phi(\varepsilon)$, the set $\operatorname{argmax} f$ belongs to a union of $k + 1$ balls of radius ε for any $f \in \operatorname{im}\Psi$ (Proposition 4).

Let us introduce some notations: for $k \geq 0$ and $\varepsilon > 0$

$$\mathbf{U}_{k,\varepsilon}^K = \{\Psi \in C(K, C(X)) \,|\, \|\Psi - \Phi\| < \delta_\Phi(\varepsilon) \text{ for some } \Phi \in \mathbf{P}_k^K\};$$

for $\varepsilon_i > 0$, $\varepsilon_i \to 0$

$$\mathbf{V}_{\{\varepsilon_i\}}^K = \bigcup_{k=0}^{\infty} \mathbf{U}_{k,\varepsilon_k}^K;$$

and, finally,

$$\mathbf{W}_{\{\varepsilon_i\}}^K = \bigcap_{s=1}^{\infty} \mathbf{V}_{\{\frac{1}{2^s}\varepsilon_i\}}^K.$$

The set $\mathbf{P}^K$ is dense in $C(K, C(X))$. Any $F \in \mathbf{P}^K$ has the form of finite sum (16), and any $f \in \operatorname{im} F$ has not more than $k+1$ point of global maximum, where k is the number of summands in presentation (16). The sets $\mathbf{V}_{\{\varepsilon_i\}}^K$ are open and dense in the Banach space $C(K, C(X))$ for any sequence $\varepsilon_i > 0$, $\varepsilon_i \to 0$. The set $\mathbf{W}_{\{\varepsilon_i\}}^K$ is intersection of countable number of open dense sets. For any $F \in \mathbf{W}_{\{\varepsilon_i\}}^K$ the sets of the family $\{\operatorname{argmax} f \,|\, f \in \operatorname{im} F\}$ can be uniformly approximated by finite sets faster than $\varepsilon_n \to 0$. It is proven that this property is typical in the Banach space $C(K, C(X))$ in the sense of category.

In order to prove that the set of exclusions is completely thin in $C(K, C(X))$ it is sufficient to use the approach of Proposition 1. Note that for arbitrary compact space Q the set of continuous maps $Q \to C(K, C(X))$ in the maximum norm is isomorphic to the spaces $C(Q \times K, C(X))$ and $C(Q \times K \times X)$. The space $Q \times K$ is compact. We can apply the previous construction to the space $C(Q \times K, C(X))$ for arbitrary compact Q and get the result: the set of exclusion is completely thin in $C(K, C(X))$.

In the definition of $\mathbf{W}_{\{\varepsilon_i\}}^K$ we use only one sequence $\varepsilon_i > 0$, $\varepsilon_i \to 0$. Of course, for any finite or countable set of sequences the intersection of correspondent sets $\mathbf{W}_{\{\varepsilon_i\}}^K$ is also a residual set, and we can claim that almost always the sets of $\{\operatorname{argmax} f \,|\, f \in \operatorname{im} F\}$ can be uniformly approximated by finite sets faster than $\varepsilon_n \to 0$ for all given sequences.

2.5 Selection Efficiency

The first application of the extremal principle for the ω-limit sets is the theorem of the selection efficiency. The dynamics of a system with inheritance leads indeed to a selection in the limit $t \to \infty$. In the typical situation, a diversity in the limit $t \to \infty$ becomes less than the initial diversity. There is an efficient selection for the "best". The basic effects of selection are formulated below. Let X be compact metric space without isolated points.

Theorem 2. (Theorem of selection efficiency.)

1. *For almost every system (1) the support of any ω-limit distribution is nowhere dense in X (and it has the Lebesgue measure zero for Euclidean space).*
2. *Let $\varepsilon_n > 0$, $\varepsilon_n \to 0$ be an arbitrary chosen sequence. The following statement is true for almost every system (1). Let the support of the initial distribution be the whole X. Then the support of any ω-limit distribution can be approximated by finite sets uniformly faster than $\varepsilon_n \to 0$.*

The set of exclusive systems that do not satisfy the statement 1 or 2 is completely thin.

Remark. These properties hold for the continuous reproduction coefficients. It is well-known, that it is dangerous to rely on the genericity among continuous functions. For example, almost all continuous functions are nowhere differentiable. But the properties 1, 2 hold also for the smooth reproduction coefficients on the manifolds and sometimes allow to replace the "almost finiteness" by simply finiteness.

Scheme of Proof. To prove the first statement, it is sufficient to refer to Proposition 2.3. In order to clarify the second part of this theorem, note that:

1. Support of an arbitrary ω-limit distribution μ^* consist of points of global maximum of the average reproduction coefficient on a support of the initial distribution. The corresponding maximum value is zero.
2. Almost always a function has only one point of global maximum, and corresponding maximum value is not 0.
3. In a one-parametric family of functions almost always there may occur zero values of the global maximum (at one point), which cannot be eliminated by a small perturbation, and individual functions of the family may stably have two global maximum points.
4. For a generic n-parameter family of functions, there may exist stably a function with n points of global maximum and with zero value of this maximum.
5. Our phase space M is compact. The set of corresponding reproduction coefficients k_M in $C(X)$ for the given map $\mu \to k_\mu$ is compact too. The average reproduction coefficients belong to the closed convex hull of this set $\overline{\mathrm{conv}}(k_M)$. And it is compact too.
6. A compact set in a Banach space can be approximated by compacts from finite-dimensional linear manifolds. Generically, in a space of continuous functions, a function, which belongs such a n-dimensional compact, can have not more than n points of global maximum with zero maximal value.

The rest of the of proof of the second statement is purely technical. Some technical details are presented in the previous section. The easiest demonstration of the "natural" character of these properties is the demonstration of instability of exclusions: If, for example, a function has several points of global maxima, then with an arbitrary small perturbation (for all usually used norms) it can be transformed into a function with the unique point of global maximum. However "stable" does not

Self-Simplification in Darwin's Systems

always mean "dense". The discussed properties of the system (1) are valid in a very strong sense: the set of exclusion is completely thin. $\qquad\square$

2.6 Gromov's Interpretation of Selection Theorems

In his talk [26], M. Gromov offered a geometric interpretation of the selection theorems. Let us consider dynamical systems in the standard m-simplex σ_m in $m + 1$-dimensional space R^{m+1}:

$$\sigma_m = \{x \in R^{m+1} \mid x_i \geq 0, \sum_{i=1}^{m+1} x_i = 1\}.$$

We assume that simplex σ_m is positively invariant with respect to these dynamical systems: if the motion starts in σ_m at some time t_0, then it remains in σ_m for $t > t_0$. Let us consider the motions that start in the simplex σ_m at $t = 0$ and are defined for $t > 0$.

For large m, almost all volume of the simplex σ_m is concentrated in a small neighborhood of the center of σ_m, near the point $c = \left(\frac{1}{m}, \frac{1}{m}, \ldots, \frac{1}{m}\right)$. Hence, one can expect that a typical motion of a general dynamical system in σ_m for sufficiently large m spends almost all the time in a small neighborhood of c.

Indeed, the m-dimensional volume of σ_m is $V_m = \frac{1}{m!}$. The part of σ_m, where $x_i \geq \varepsilon$, has the volume $(1 - \varepsilon)^m V_m$. Hence, the part of σ_m, where $x_i < \varepsilon$ for all $i = 1, \ldots, m + 1$, has the volume $V_\varepsilon > (1 - (m + 1)(1 - \varepsilon)^m) V_m$. Note, that $(m + 1)(1 - \varepsilon)^m \sim m \exp(-\varepsilon m) \to 0$, if $m \to \infty$ ($1 > \varepsilon > 0$). Therefore, for $m \to \infty$, $V_\varepsilon = (1 - o(1)) V_m$. The volume W_ρ of the part of σ_m with Euclidean distance to the center c less than $\rho > 0$ can be estimated as follows: $W_\rho > V_\varepsilon$ for $\varepsilon \sqrt{m + 1} = \rho$, hence $W_\rho > (1 - (m + 1)(1 - \rho/\sqrt{m + 1})^m) V_m$. Finally, $(m + 1)(1 - \rho/\sqrt{m + 1})^m \sim m \exp(-\rho \sqrt{m})$, and $W_\rho = (1 - o(1)) V_m$ for $m \to \infty$. Let us mention here the opposite concentration effect for a m-dimensional ball B_m: for $m \to \infty$ the most part of its volume is concentrated in an arbitrary small vicinity of its boundary, the sphere. This effect is the essence of the famous equivalence of microcanonical and canonical ensembles in statistical physics (for detailed discussion see [22]).

Let us consider dynamical systems with an additional property ("inheritance"): all the faces of the simplex σ_m are also positively invariant with respect to the systems with inheritance. It means that if some $x_i = 0$ initially at the time $t = 0$, then $x_i = 0$ for $t > 0$ for all motions in σ_m. The essence of selection theorems is as follows: a typical motion of a typical dynamical system with inheritance spends almost all the time in a small neighborhood of low-dimensional faces, even if it starts near the center of the simplex.

Let us denote by $\partial_r \sigma_m$ the union of all r-dimensional faces of σ_m. Due to the selection theorems, a typical motion of a typical dynamical system with inheritance spends almost all time in a small neighborhood of $\partial_r \sigma_m$ with $r \ll m$. It should not obligatory reside near just one face from $\partial_r \sigma_m$, but can travel in neighborhood of different faces from $\partial_r \sigma_m$ (the drift effect). The minimax estimation of the number of points in ω-limit distributions through the diameters $\varepsilon_n > 0$ of the set $\overline{\mathrm{conv}}(k_M)$ is the estimation of r.

3 Drift and Mutations

3.1 Drift Equations

So far, we talked about the support of an individual ω-limit distribution. For almost all systems it is small. But this does not mean, that the union of these supports is small even for one solution $\mu(t)$. It is possible that a solution is a finite set of narrow peaks getting in time more and more narrow, moving slower and slower, but not tending to fixed positions, rather continuing to move along its trajectory, and the path covered tends to infinity as $t \to \infty$.

This effect was not discovered for a long time because the slowing down of the peaks was thought as their tendency to fixed positions For the best of our knowledge, the first detailed publication of the drift equations and corresponded types of stability appeared in book [21], first examples of coevolution drift on a line were published in the series of papers [46].

There are other difficulties related to the typical properties of continuous functions, which are not typical for the smooth ones. Let us illustrate them for the distributions over a straight line segment. Add to the reproduction coefficients k_μ the sum of small and narrow peaks located on a straight line distant from each other much more than the peak width (although it is ε-small). However small is chosen the peak's height, one can choose their width and frequency on the straight line in such a way that from any initial distribution μ_0 whose support is the whole segment, at $t \to \infty$ we obtain ω-limit distributions, concentrated at the points of maximum of the added peaks.

Such a model perturbation is small in the space of continuous functions. Therefore, it can be put as follows: *by small continuous perturbation the limit behavior of system (1) can be reduced onto a ε-net for sufficiently small ε*. But this can not be done with the small smooth perturbations (with small values of the first and the second derivatives) in the general case. The discreteness of the net, onto which the limit behavior is reduced by small continuous perturbations, differs from the discreteness of the support of the individual ω-limit distribution. For an individual distribution the number of points is estimated, roughly speaking, by the number of essential parameters (11), while for the conjunction of limit supports – by the number of stages in approximation of k_μ by piece-wise constant functions.

Thus, in a typical case the dynamics of systems (1) with smooth reproduction coefficients transforms a smooth initial distributions into the ensemble of narrow peaks. The peaks become more narrow, their motion slows down, but not always they tend to fixed positions.

The equations of motion for these peaks can be obtained in the following way [21]. Let X be a domain in the n-dimensional real space, and the initial distributions

μ_0 be assumed to have smooth density. Then, after sufficiently large time t, the position of distribution peaks are the points of the average reproduction coefficient maximum $\langle k_\mu \rangle_t$ (9) to any accuracy set in advance. Let these points of maximum be x^α, and

$$q_{ij}^\alpha = -t \frac{\partial^2 \langle k_\mu \rangle_t}{\partial x_i \partial x_j}\bigg|_{x=x^\alpha}.$$

It is easy to derive the following differential relations just by differentiation in time of the extremum conditions: at points $x^\alpha(t)$ gradient of the average reproduction coefficient $\langle k_\mu \rangle_t$ vanishes: $\partial \langle k_\mu \rangle_t(x)/\partial x_i \big|_{x=x^\alpha(t)} = 0$.

$$\sum_j q_{ij}^\alpha \frac{\mathrm{d}x_j^\alpha}{\mathrm{d}t} = \frac{\partial k_{\mu(t)}}{\partial x_i}\bigg|_{x=x^\alpha} ;$$
$$\frac{\mathrm{d}q_{ij}^\alpha}{\mathrm{d}t} = -\frac{\partial^2 k_{\mu(t)}}{\partial x_i \partial x_j}\bigg|_{x=x^\alpha}. \qquad (17)$$

The exponent coefficients q_{ij}^α remain time dependent even when the distribution tends to a δ-function. It means (in this case) that peaks became infinitely narrow. Nevertheless, it is possible to change variables and represent the weak* tendency to stationary discrete distribution as usual tendency to a fixed points, see (21) below.

These relations (17) do not form a closed system of equations, because the right-hand parts are not functions of x_i^α and q_{ij}^α. For sufficiently narrow peaks there should be separation of the relaxation times between the dynamics *on* the support and the dynamics *of* the support: the relaxation of peak amplitudes (it can be approximated by the relaxation of the distribution with the finite support, $\{x^\alpha\}$) should be significantly faster than the motion of the locations of the peaks, the dynamics of $\{x^\alpha\}$. Let us write the first term of the corresponding asymptotics [21].

For the finite support $\{x^\alpha\}$ the distribution is $\mu = \sum_\alpha N_\alpha \delta(x - x^\alpha)$. Dynamics of the finite number of variables, N_α obeys the system of ordinary differential equations

$$\frac{\mathrm{d}N_\alpha}{\mathrm{d}t} = k_\alpha(N)N_\alpha \qquad (18)$$

where N is vector with components N_α, $k_\alpha(N)$ is the value of the reproduction coefficient k_μ at the point x^α:

$$k_\alpha(N) = k_\mu(x^\alpha) \text{ for } \mu = \sum_\alpha N_\alpha \delta(x - x^\alpha).$$

For finite-dimensional dynamics (18) we have to find the relevant SBR (Sinai–Bowen–Ruelle) invariant measure (or "physical measure") [30, 34] for averaging and substitute the average time along the solutions of (18)

$$\frac{1}{t} \int_0^t k_{\mu^*(N)(\tau)} \, \mathrm{d}\tau \text{ where } \mu^*(N) = \sum_\alpha N_\alpha \delta(x - x^\alpha)$$

by the average with respect to the SBR measure on space of vectors N. For this average, we use notation $k^*(\{x^\alpha\}) = \langle k_{\mu*}\rangle$.

In the simplest case the finite-dimensional attractor is just one stable fixed point and the average $k^*(\{x^\alpha\})$ is a value at this point. Let the dynamics of the system (18) for a given set of initial conditions be simple: the motion $N(t)$ goes to the stable fixed point $N = N^*(\{x^\alpha\})$. Then we can take $k^*(\{x^\alpha\}) = k_{\mu*}$ where $\mu^* = \sum_\alpha N_\alpha^* \delta(x - x^\alpha)$.

One can use in the right hand side of (17) the following approximation for $k^*(\{x^\alpha\})$ instead of $k_{\mu(t)}$.

$$\mu(t) = \mu^*(\{x^\alpha(t)\}) = \sum_\alpha N_\alpha^* \delta(x - x^\alpha(t)). \tag{19}$$

This is a standard averaging hypothesis. We can use it because density peaks are sufficiently narrow, hence, (i) the difference between true $k_{\mu(t)}$ and the reproduction coefficient for the measure with finite support $k(\sum_\alpha N_\alpha(t)\delta(x - x^\alpha$ is negligible and (ii) dynamics of peak motion is much slower than relaxation of the finite-dimensional system (18) to its attractor. The relations (17) transform into the ordinary differential equations

$$\sum_j q_{ij}^\alpha \frac{dx_j^\alpha}{dt} = \left.\frac{\partial k^*(\{x^\beta\})(x)}{\partial x_i}\right|_{x=x^\alpha} ;$$
$$\frac{dq_{ij}^\alpha}{dt} = -\left.\frac{\partial^2 k^*(\{x^\beta\})(x)}{\partial x_i \partial x_j}\right|_{x=x^\alpha} . \tag{20}$$

The matrix variables q_{ij}^α are usually not bounded. For example, near a nondegenerated fixed point $\{x^\alpha\}$ they go to infinity linearly in time. On the other hand, relaxation of $\{x^\alpha\}$ to their stationary positions, for example, is not exponential due to (20). To return to the standard situation with compact phase space and exponential relaxation it is useful to switch to the logarithmic time $\tau = \ln t$ and to new variables

$$b_{ij}^\alpha = \frac{1}{t} q_{ij}^\alpha = -\left.\frac{\partial^2 \langle k(\mu)\rangle_t}{\partial x_i \partial x_j}\right|_{x=x^\alpha} .$$

For large t we obtain from (20)

$$\sum_j b_{ij}^\alpha \frac{dx_j^\alpha}{d\tau} = \left.\frac{\partial k^*(\{x^\beta\})(x)}{\partial x_i}\right|_{x=x^\alpha} ;$$
$$\frac{db_{ij}^\alpha}{d\tau} = -\left.\frac{\partial^2 k^*(\{x^\alpha\})(x)}{\partial x_i \partial x_j}\right|_{x=x^\beta} - b_{ij}^\alpha . \tag{21}$$

In these equations it becomes obvious that dynamics of matrix b_{ij}^α is the differential pursuit of Hessian $\partial^2 k^*(\{x^\alpha\})(x)/\partial x_i \partial x_j \mid_{x=x^\beta}$.

Self-Simplification in Darwin's Systems 333

Equations for drift in logarithmic time (21) are the main equations in the theory of the asymptotic layer For Darwin's systems near their limit behavior.

The way of constructing the drift equations (20,21) for a specific system (1) is as follows:

1. For finite sets $\{x^\alpha\}$ one studies systems (18) and finds the equilibrium solutions $*(\{x^\alpha\})$ or the relevant SBR measure;
2. For given measures $\{x^\alpha\}$ (19) one calculates the reproduction coefficients $k^*(\{x^\alpha\})(x)$ together with the first and second first derivatives of these functions in x at points x^α. That is all, the drift equations (21) are set up.

The drift equations (20,21) describe the dynamics of the peaks positions x^α and of the coefficients q_{ij}^α. For given x^α, q_{ij}^α and N_α^* the distribution density μ can be approximated as the sum of narrow Gaussian peaks:

$$\mu = \sum_\alpha N_\alpha^* \sqrt{\frac{\det Q^\alpha}{(2\pi)^n}} \exp\left(-\frac{1}{2}\sum_{ij} q_{ij}^\alpha (x_i - x_i^\alpha)(x_j - x_j^\alpha)\right), \qquad (22)$$

where Q^α is the inverse covariance matrix (q_{ij}^α).

If the limit dynamics of the system (18) for finite supports at $t \to \infty$ can be described by a more complicated attractor, then instead of reproduction coefficient $k^*(\{x^\alpha\})(x) = k_{\mu^*}$ for the stationary measures μ^* (19) one can use the average reproduction coefficient with respect to the corresponding Sinai–Ruelle–Bowen measure. If finite systems (18) have several attractors for given $\{x^\alpha\}$, then the dependence $k^*(\{x^\alpha\})$ is multi-valued, and there may be bifurcations and hysteresis with the function $k^*(\{x^\alpha\})$ transition from one sheet to another. There are many interesting effects concerning peaks' birth, disintegration, divergence, and death, and the drift equations (20, 21) describe the motion in a non-critical domain, between these critical effects.

Inheritance (conservation of support) is never absolutely exact. Small variations, mutations, immigration in biological systems are very important. Excitation of new degrees of freedom, modes diffusion, noise are present in physical systems. How does small perturbation in the inheritance affect the effects of selection? The answer is usually as follows: there is such a value of perturbation of the right-hand side of (1), at which they would change nearly nothing, just the limit δ-shaped peaks transform into sufficiently narrow peaks, and zero limit of the velocity of their drift at $t \to \infty$ substitutes by a small finite one.

3.2 Drift in Presence of Mutations and Scaling Invariance

The simplest model for "inheritance + small variability" is given by a perturbation of (1) with diffusion term

$$\frac{\partial \mu(x,t)}{\partial t} = k_{\mu(x,t)} \times \mu(x,t) + \varepsilon \sum_{ij} d_{ij}(x) \frac{\partial^2 \mu(x,t)}{\partial x_i \partial x_j}. \qquad (23)$$

where $\varepsilon > 0$ and the matrix of diffusion coefficients d_{ij} is symmetric and positively definite.

There are almost always no qualitative changes in the asymptotic behavior, if ε is sufficiently small. With this the asymptotics is again described by the drift equations (20, 21), modified by taking into account the diffusion as follows:

$$\sum_j q_{ij}^\alpha \frac{dx_j^\alpha}{dt} = \left. \frac{\partial k^*(\{x^\beta\})(x)}{\partial x_i} \right|_{x=x^\alpha} ;$$

$$\frac{dq_{ij}^\alpha}{dt} = - \left. \frac{\partial^2 k^*(\{x^\beta\})(x)}{\partial x_i \partial x_j} \right|_{x=x^\alpha} - 2\varepsilon \sum_{kl} q_{ik}^\alpha d_{kl}(x^\alpha) q_{lj}^\alpha . \tag{24}$$

Now, as distinct from (20), the eigenvalues of the matrices $Q^\alpha = (q_{ij}^\alpha)$ cannot grow infinitely. This is prevented by the quadratic terms in the right-hand side of the second equation (24).

Dynamics of (24) does not depend on the value $\varepsilon > 0$ qualitatively, because of the obvious scaling property. If ε is multiplied by a positive number ν, then, upon rescaling $t' = \nu^{-1/2}t$ and $q_{ij}^{\alpha}{}' = \nu^{-1/2}q_{ij}^\alpha$, we have the same system again. Multiplying $\varepsilon > 0$ by $\nu > 0$ changes only peak's velocity values by a factor $\nu^{1/2}$, and their width by a factor $\nu^{1/4}$. The paths of peaks' motion do not change at this for the drift approximation (24) (but the applicability of this approximation may, of course, change).

4 Three Main Types of Stability

4.1 *Internal Stability*

Stable steady-state solutions of equations of the form (1) may be only the sums of δ-functions – this was already mentioned. There is a set of specific conditions of stability, determined by the form of equations.

Consider a stationary distribution for (1) with a finite support

$$\mu^*(x) = \sum_\alpha N_\alpha^* \delta(x - x^{*\alpha}).$$

Steady state of μ^* means, that

$$k_{\mu^*}(x^{*\alpha}) = 0 \text{ for all } \alpha. \tag{25}$$

The *internal stability* means, that this distribution is stable with respect to perturbations not increasing the support of μ^*. That is, the vector N_α^* is the stable fixed point for the dynamical system (18). Here, as usual, it is possible to distinguish

Self-Simplification in Darwin's Systems   335

between the Lyapunov stability, the asymptotic stability and the first approximation stability (negativeness of real parts for the eigenvalues of the matrix $\partial \dot{N}_\alpha^* / \partial N_\alpha^*$ at the stationary points).

4.2 External Stability: Uninvadability

The *external stability* (*uninvadability*) means stability to an expansion of the support, i.e., to adding to μ^* of a small distribution whose support contains points not belonging to supp μ^*. It makes sense to speak about the external stability only if there is internal stability. In this case it is sufficient to restrict ourselves with δ-functional perturbations. The external stability has a very transparent physical and biological sense. It is stability with respect to *introduction* into the systems of a new inherited unit (gene, variety, specie ...) in a small amount.

The *necessary condition for the external stability* is: the points $\{x^{*\alpha}\}$ are points of the global maximum of the reproduction coefficient $k_{\mu^*}(x)$. It can be formulated as the optimality principle

$$k_{\mu^*}(x) \leq 0 \text{ for all } x; \ k_{\mu^*}(x^{*\alpha}) = 0. \tag{26}$$

The *sufficient condition for the external stability* is: the points $\{x^{*\alpha}\}$ and only these points are points of the global maximum of the reproduction coefficient $k_{\mu^*}(x^{*\alpha})$. At the same time it is the condition of the external stability in the first approximation and the optimality principle

$$k_{\mu^*}(x) < 0 \text{ for } x \notin \{x^{*\alpha}\}; \ k_{\mu^*}(x^{*\alpha}) = 0. \tag{27}$$

The only difference from (26) is the change of the inequality sign from $k_{\mu^*}(x) \leq 0$ to $k_{\mu^*}(x) < 0$ for $x \notin \{x^{*\alpha}\}$. The necessary condition (26) means, that the small δ-functional addition will not grow in the first approximation. According to the sufficient condition (27) such a small addition will exponentially decrease.

If X is a finite set, then the combination of the external and the internal stability is equivalent to the standard stability for a system of ordinary differential equations.

4.3 Stable Realizability: Evolutionary Robustness

External stability of a internally stable limit distribution is insufficient for its stability with respect to the drift: It does not imply convergence to x^* when starting from a distribution of small deviations from x^*, regardless of how small these deviations are. The standard idea of asymptotic stability is: "after small deviation the system returns to the initial regime, and do not deviate to much on the way of returning". The crucial question for the measure dynamics is: in which topology the deviation is small? The small shift of the narrow peak of distribution in the continuous space of strategies can be considered as a small deviation in the weak* topology, but it is definitely large deviation in the strong topology, for example, if the shift is not small in comparison with the peak with.

For the continuous X there is one more kind of stability important from the applications viewpoint. Substitute δ-shaped peaks at the points $\{x^{*\alpha}\}$ by narrow Gaussians and shift slightly the positions of their maxima away from the points $x^{*\alpha}$. How will the distribution from such initial conditions evolve? If it tends to μ without getting too distant from this steady state distribution, then we can say that the third type of stability – *stable realizability* – takes place. It is worth mentioning that the perturbation of this type is only weakly* small, in contrast to perturbations considered in the theory of internal and external stability. Those perturbations are small by their norms. Let us remind that the norm of the measure μ is $\|\mu\| = \sup_{|f|\leq 1}[\mu, f]$. If one shifts the δ-measure of unite mass by any nonzero distance ε, then the norm of the perturbation is 2. Nevertheless, this perturbation weakly* tends to 0 with $\varepsilon \to 0$.

In order to formalize the condition of stable realizability it is convenient to use the drift equations in the form (21). Let the distribution μ^* be internally and externally stable in the first approximations. Let the points $x^{*\alpha}$ of global maxima of $k_{\mu^*}(x)$ be non-degenerate in the second approximation. This means that the matrices

$$b_{ij}^{*\alpha} = -\left(\frac{\partial^2 k_{\mu^*}(x)}{\partial x_i \partial x_j}\right)_{x=x^{*\alpha}} \tag{28}$$

are strictly positively definite for all α.

Under these conditions of stability and non-degeneracy the coefficients of (21) can be easily calculated using Taylor series expansion in powers of $(x^\alpha - x^{*\alpha})$. The stable realizability of μ^* in the first approximation means that the fixed point of the drift equations (21) with the coordinates

$$x^\alpha = x^{*\alpha}, \ b_{ij}^\alpha = b_{ij}^{*\alpha} \tag{29}$$

is stable in the first approximation. It is the usual stability for the system (21) of ordinary differential equations, and these conditions with the notion of the stable realizability became clear from the logarithmic time drift equations (21) directly.

The specific structure of (21) allows us to simplify stability analysis for steady states. Let the steady state be externally stable steady states in the first approximations and let matrices $b^{*\alpha}$ be strictly positive definite. Equation (21) have the structure

$$\begin{aligned} \mathcal{B}\dot{\mathcal{X}} &= F(\mathcal{X}); \\ \dot{\mathcal{B}} &= \Phi(\mathcal{X}) - \mathcal{B}, \end{aligned} \tag{30}$$

where $\mathcal{X}$ is a vector composed from vectors x^α, and $\mathcal{B}$ is a block-diagonal matrix composed from matrices b^α. The steady values are $\mathcal{X}^*$ and $\mathcal{B}^*$: $F(\mathcal{X}^*) = 0, \mathcal{B}^* =$

Self-Simplification in Darwin's Systems   337

$\Phi(\mathcal{X}^*)$. Direct calculations gives for Jacobian J:

$$J = \begin{pmatrix} \mathcal{B}^{-1} \frac{DF(\mathcal{X})}{D\mathcal{X}}\Big|_{\mathcal{X}^*} & 0 \\ \frac{D\Phi(\mathcal{X})}{D\mathcal{X}}\Big|_{\mathcal{X}^*} & -1 \end{pmatrix}. \tag{31}$$

This form of Jacobian immedially implies the following proposition.

Proposition 6. *Stability of (21) (in the first approximation) near a steady state could be defined by the spectrum of matrix* $\mathcal{B}^{-1} \frac{DF(\mathcal{X})}{D\mathcal{X}}\Big|_{\mathcal{X}^*}$: *If the real parts of its eigenvalues are negative then the system is stable. If some of them are positive then the system is unstable.* $\square$

To explain the sense of the stable realizability we used in the book [25] the idea of the "Gardens of Eden" from Conway "Game of Life" [18]. That are Game of Life patterns which have no father patterns and therefore can occur only at generation 0, from the very beginning. It is not known if a pattern which has a father pattern, but no grandfather pattern exists. It is the same situation, as for internal and external stable (uninvadable) state which is not stable realizable: it cannot be destroyed by mutants invasion and by the small variation of conditions, but, at the same time, it is not attractive for drift, and, hence, can not be realized in this asymptotic motion. It can be only created.

5 Explicit Drift Equations for Distributed Lottka–Volterra–Gause Systems

Construction of the drift system (21) goes through several operations. The most complicated of them is averaging: for a system with finite support $\{x^\alpha\}$ (18) we have to find the relevant finite-dimensional average $k^*(\{x^\alpha\}) = \langle k(\sum_\alpha N_\alpha \delta(x - x^\alpha)) \rangle$.

The most difficult operation in the construction of drift equations is the qualitative study of the finite-dimensional system and its SBR measures. Even in the simplest case of unique and globally attractive stable steady state in (18) the study of global stability may be difficult and even solution of equations for steady states may be computationally expensive.

There is a lucky exclusion: if the reproduction coefficient is a value of a linear integral operator then the steady-state can be found from linear equations and average values of k coincide with its values at steady-states. Replicator systems with linear reproduction coefficients include all classical Lottka–Volterra–Gause systems and can have an arbitrary complex dynamics. Nevertheless, equilibria of these systems satisfy linear equations and average reproduction coefficient is equal to its steady-state value.

Let us write down these equations:

$$\frac{d\mu(x)}{dt} = \mu(x)\left[r(x) + \int_X K(x,\xi)\mu(\xi)\,d\xi\right]. \tag{32}$$

Here we assume that $q(x)$ and $K(x,\xi)$ are continuous functions. The space of measures is assumed to have a bounded set of positive measures M $\mu(x \geq 0$ that is positively invariant relative to system (32): if $\mu(0) \in M$, then $\mu(t) \in M$ (we also assume that M is non-trivial, i.e., it is neither empty nor a one-point set but includes at least one point with its vicinity). This M serves as the phase space of system (32).

A steady state μ with support $\mathrm{supp}\mu$ satisfies a linear equation:

$$r(x) + \int_X K(x,\xi)\mu(\xi)\,d\xi = 0 \ \text{ for } \ x \in \mathrm{supp}\,\mu. \tag{33}$$

For a finite set $\{x^\alpha\}$ we introduce a matrix $K(\{x^\alpha\}) = [K_{\alpha\beta}(\{x^\alpha\})] = [K(x^\alpha, x^\beta)]$ and a vector $r = r_\alpha = r(x^\alpha)$. Equation (18) for (32) has a form:

$$\frac{dN_\alpha}{dt} = \left(r_\alpha + \sum_\beta K_{\alpha\beta} N_\beta\right) N_\alpha. \tag{34}$$

The positive stationary solution (if it exists) is given by

$$N^*(\{x^\alpha\}) = -K^{-1}(\{x^\alpha\})r(\{x^\alpha\}). \tag{35}$$

For strictly positive bounded solutions of (34) $N(t)$ with $N(t) > \varepsilon > 0$ the time average of $N(t)$ coincides with N^* and the time average of any linear functional $l(N(t))$ is $l(N^*)$. Hence, for this type of finite-dimensional dynamics we can use in (21)

$$k^*(\{x^\alpha\})(x) = r(x) + \sum_\alpha K(x,x^\beta)N_\beta^*(\{x^\alpha\}). \tag{36}$$

Here functions $N_\beta^*(\{x^\alpha\})$ are explicitly derived from the coefficients (35), hence. the drift equations (21) could be also found in explicit form, by differentiation. The coefficients of those equations includes nothing more than functions $r(x)$, $K(x,x^\beta)$, their rational combinations and derivatives.

Just for simplicity let us demonstrate this for system of two quasispecies. Let X be a disjoint union of two intervals on real line. The replicator system (32) in this case is

$$\frac{d\mu_1(x)}{dt} = \mu_1(x)\left[r_1(x) + \int K_{11}(x,\xi)\mu_1(\xi)\,d\xi + \int K_{12}(x,v)\mu_2(v)\,dv\right];$$

$$\frac{d\mu_2(y)}{dt} = \mu_2(y)\left[r_2(x) + \int K_{21}(y,\xi)\mu_1(\xi)\,d\xi + \int K_{22}(y,v)\mu_2(v)\,dv\right]. \tag{37}$$

Self-Simplification in Darwin's Systems 339

Two quasispecies are two peaks, one for μ_1 with coordinate $x = x^1$ and variance $\mathrm{var}_1 \approx 1/(tb^1)$ and another for μ_2 with coordinate $y = x^2$ and variance $\mathrm{var}_2 \approx 1/(tb^2)$ for large t.

The finite-dimensional system (34) transforms in

$$\dot{N}_1 = [r^1(x^1) + K_{11}(x^1, x^1)N_1 + K_{12}(x^1, x^2)N_2]N_1;$$
$$\dot{N}_2 = [r^2(x^2) + K_{21}(x^2, x^1)N_1 + K_{22}(x^2, x^2)N_2]N_2. \tag{38}$$

The steady state solution is

$$N_1^*(x^1, x^2) = \frac{K_{22}r_1 - K_{12}r_2}{K_{12}K_{21} - K_{11}K_{22}};$$
$$N_2^*(x^1, x^2) = \frac{K_{11}r_2 - K_{21}r_1}{K_{12}K_{21} - K_{11}K_{22}}, \tag{39}$$

where coefficients on the right hand side are calculated for $x = x^1$, $y = x^2$. The inequalities $N_{1,2}^* > 0$ should hold (this is a condition on the coefficients values). We omit here stability and positivity analysis for 2D system (38).

Function $k^*(\{x^\alpha\})(x)$ is represented by two functions $k_1^*(x^1, x^2)(x)$, $k_2^*(x^1, x^2)(y)$ (because X is a disjoint union of two intervals on real line):

$$k_1^*(x^1, x^2)(x) = r_1(x) + K_{11}(x, x^1)N_1^*(x^1, x^2) + K_{12}(x, x^2)N_2^*(x^1, x^2);$$
$$k_1^*(x^1, x^2)(y) = r_2(y) + K_{21}(y, x^1)N_1^*(x^1, x^2) + K_{22}(y, x^2)N_2^*(x^1, x^2). \tag{40}$$

Now, we can write down the drift equation in logarithmic time (21):

$$\dot{x}^1 = \frac{1}{b^1}(\partial_x k_1^*(x^1, x^2)(x))_{x=x^1}, \quad \dot{x}^2 = \frac{1}{b^2}(\partial_y k_2^*(x^1, x^2)(y))_{y=x^2};$$
$$\dot{b}^1 = -(\partial_x^2 k_1^*(x^1, x^2)(x))_{x=x^1} - b^1, \quad \dot{b}^2 = -(\partial_y^2 k_2^*(x^1, x^2)(y))_{y=x^2} - b^2. \tag{41}$$

Dynamics of $b^{1,2}$ is a differential pursuit of the (minus) second derivatives of functions $k_{1,2}^*$ at peak positions. The velocity of peaks drift is proportional to the first derivatives of these functions with the coefficients $1/b^{1,2}$. Therefore, velocity is proportional to the peak variance or more precise, to $\frac{\mathrm{var}}{t}$. Already such simple systems as (41) demonstrate various regimes of coevolution [46].

6 Simple Example of Arbitrary Complex Dynamics of Drift

Let X be a closed domain in R^n with nonempty interior. For a smooth vector field $v(x)$ in X we would like to construct such a Darwin's system (1) that drift (in logarithmic time $\tau = \ln t$) approximates dynamics defined by differential equation

$\dot{x} = v(x)$. In order to consider this dynamics in X, some additional assumptions are needed. To guarantee positive invariance of X we can assume that there exists such $\varepsilon > 0$ that if $x \in X$ and $0 < \delta \leq \varepsilon$ then $x + \delta v(x) \in X$. To consider function $v(x)$ in a vicinity of X we will use an arbitrary smooth continuation of $v(x)$ on R^n.

For any measure μ on X we use notations:

$$M_0(\mu) = \int_X \mu \, dx, \quad M_1(\mu) = \int_X x \mu \, dx, \quad M_2(\mu) = \int_X x^2 \mu \, dx.$$

Let us select the reproduction coefficient in the following form:

$$K(\mu)(x) = -(x - v(M_1(\mu)))^2 M_0(\mu) + C(\mu), \tag{42}$$

where functional $C(\mu)$ is selected in such a way that $M_0(\mu)$ satisfies exactly equation $\dot{M}_0 = (1 - M_0)M_0$ if $\dot{\mu} = K(\mu)\mu$. Darwin's equation with the reproduction coefficient (42) gives for time derivative of M_0

$$\dot{M}_0 = \int_X \dot{\mu}(x) \, dx = -M_2 M_0 + 2(x, v(M_1))M_0 - v^2(M_1)M_0 + C M_0.$$

It is straightforward to check that for functional

$$C(\mu) = 1 - M_0 + M_2 - 2(M_1, v(M_1)) + M_0 v^2(M_1)$$

dynamics of $M_0(\mu)$ satisfies the simple equation $\dot{M}_0 = (1 - M_0)M_0$. Therefore, for positive initial condition after sufficiently long time the value of M_0 is arbitrarily closed to one. After some rearranging of coefficients, we get for time averages of $K(\mu)$:

$$\langle K(\mu)(x) \rangle_t = \frac{1}{t} \int_0^t K(\mu(\tau))(x) \, d\tau$$

$$= -\left(x - \frac{\langle M_0 v(M_1) \rangle_t}{\langle M_0 \rangle_t}\right)^2 \langle M_0 \rangle_t - \langle M_0 v^2(M_1) \rangle_t + \frac{\langle M_0 v(M_1) \rangle_t^2}{\langle M_0 \rangle_t} + \langle C \rangle_t. \tag{43}$$

This average reproduction coefficient achieves its maximum at point

$$x^* = \frac{\langle M_0 v(M_1) \rangle_t}{\langle M_0 \rangle_t}.$$

After sufficiently long time $x^* \approx \langle v(M_1) \rangle_t$, hence, for analysis of drift dynamics we have to study motion of the point $\langle v(M_1) \rangle_t$. By definition of time average, the velocity of this point in logarithmic time is

Self-Simplification in Darwin's Systems

$$\frac{\mathrm{d}\langle v(M_1)\rangle_t}{\mathrm{d}\ln t} = v(M_1) - \frac{\langle v(M_1)\rangle_t}{t}.$$

For large t the second term tends to zero and we found that the time derivative of the reproduction coefficient maximizer x^* is $v(M_1)$: $\dot{x}^* = v(M_1)$ with arbitrarily chosen accuracy.

This is not yet an equation for peak motion. We need additional asymptotic identity $x^* \approx M_1$. It is not always true because it is possible that $\operatorname{supp}\mu \neq X$. Nevertheless, if at the initial moment $\operatorname{supp}\mu = X$ then for sufficiently large t $x^* \approx M_1$ because $\mu(t) = \mu(0)\exp(t\langle K(\mu)(x)\rangle_t)$ and almost all measure $\mu(t)$ is concentrated in an arbitrarily small vicinity of x^*. Therefore, $x^* \approx M_1$.

Finally, for drift dynamics we obtain equation: with an arbitrarily chosen accuracy in logarithmic time

$$\dot{x}^* = v(x^*).$$

This simple example demonstrates that the drift of density peaks for Darwin's equations may be arbitrarily complex.

7 Conclusion

Darwin's equation demonstrate a mechanism of self-simplification of complex system. This mechanism, under the name "natural selection" was extracted from analysis of biological evolution by Charles Robert Darwin and Alfred Russel Wallace and published in 1859. Selection mechanism is based on specific separation of time: the support of distributions changes very slowly (small mutations) or cannot increase at al (inheritance).

Such a separation of time scales implies typical asymptotic behavior: supports of the ω-limit distributions are discrete. The asymptotic layer near the ω-limit distributions is drift of finite number of narrow density peaks. This drift becomes slower in time, but its dynamics in logarithmic time could be arbitrarily complex.

The equations for peak dynamics, the drift equations, (20, 21, 24) describe dynamics of the shapes of the peaks and their positions. For systems with small variability ("mutations") the drift equations (24) has the scaling property: the change of the intensity of mutations is equivalent to the change of the time scale.

Some further exact results of the mathematical selection theory can be found in [23, 35, 36]. Karev [33] recently developed an entropic description of limit behaviour of replicator systems.

There exists an important class of generalization of all selection theorems for distributions with vector space of values. In biological language this means that non-inherited properties are taken into account: distribution in size, age, space of birth and so on. The results are, essentially, the same: the Perron–Frobenius theorem and it generalizations allow to reduce the vector Darwin's systems back to the scalar optimality principle. The key role in this reduction plays the Birkhoff contraction theorem [1, 16].

Many examples of Darwin's systems outside the theory of biological evolution in physics and other applications, such as weak turbulence or wave turbulence theory [38, 53, 54] or ecological applications [47], are already known. Nevertheless, this mechanism is, perhaps, still underestimated and we will meet them in many other areas.

References

1. Birkhoff, G.: Extensions of Jentzsch's theorem. Trans. Amer. Math. Soc. **85** (1957) 219–227
2. Birkhoff, G.D.: Dynamical systems. AMS Colloquium Publications, Providence (1927)
3. Bishop, D.T., Cannings, C.: A generalized war of attrition. J. Theoret. Biol. **70(1)** (1978) 85–124
4. Bomze, I.M.: Dynamical aspects of evolutionary stability. Monatshefte für Mathematik **110** (1990) 189–206
5. Bomze, I.M.: Detecting all evolutionarily stable strategies. J. Optim. Theor. Appl. **75(2)** (1992) 313–329.
6. Bomze, I.M.: Uniform barriers and evolutionarily stable sets. In: Game theory, experience, rationality. Foundations of social sciences, economics and ethics. In honor of John C. Harsanyi. Edited by Werner Leinfellner and Eckehart Köhler. Vienna Circle Institute Yearbook, 5. Kluwer, Dordrecht (1998) 225–243
7. Bomze, I.M.: Regularity vs. degeneracy in dynamics, games, and optimization: a unified approach to different aspects. SIAM Review **44** (2002) 394–414
8. Bomze, I.M., Bürger, R.: Stability by mutation in evolutionary games, Evolutionary game theory in biology and economics. Games Econom. Behav. **11(2)** (1995) 146–172
9. Bomze, I.M., Pötscher, B.M.: Game Theoretical Foundations of Evolutionary Stability. Lecture Notes in Economics and Math. Systems **324**, Springer, Berlin (1989)
10. Bourbaki, N.: Elements of mathematics – Integration I. Springer, Berlin (2003)
11. Constantin, P., Foias, C., Nicolaenko, B., Temam, R.: Integral manifolds and inertial manifolds for dissipative partial differential equations. Applied Math. Sci. **70** Springer, New York (1988)
12. Cressma, R.: The Stability Concept of Evolutionary Game Theory. Lecture Notes in Biomathematics **94**. Springer, Berlin (1992)
13. Darwin, Ch.: On the origin of species by means of natural selection, or preservation of favoured races in the struggle for life: A Facsimile of the First Edition. Harvard, Cambridge (1964)
14. Diekmann, O., Gyllenberg, M., Metz, J.A.J.: Steady state analysis of structured population models. Theor. Popul. Biol. **63** (2003) 309–338
15. Eshel. I., Sansone, E.: Evolutionary and dynamic stability in continuous population games. J. Math. Biol. **46(5)** (2003) 445–459
16. Eveson, S.P., Nussbaum, R.D.: Applications of the Birkhoff–Hopf theorem to the spectral theory of positive linear operators. Math. Proc. Cambridge Philos. Soc. **117** (1995) 491–512
17. Ewens, W.J.: Mathematical Population Genetics. Springer, Berlin (1979)
18. Gardner, M.: On cellular automata, self-reproduction, the garden of Eden and the game of life. Sci. Am. **224(2)** (1971) 112–115
19. Gause, G.F.: The struggle for existence. Williams and Wilkins, Baltimore (1934) Online: http://www.ggause.com/Contgau.htm.
20. Gorban, A.N.: Dynamical systems with inheritance. In: Some problems of community dynamics, Khlebopros, R.G. (ed.). Institute of Physics RAS, Siberian Branch, Krasnoyarsk (1980) (in Russian)
21. Gorban, A.N.: Equilibrium encircling. Equations of chemical kinetics and their thermodynamic analysis. Nauka, Novosibirsk (1984) (in Russian)
22. Gorban, A.N.: Order–disorder separation: Geometric revision. Physica A **374** (2007) 85–102

23. Gorban, A.N.: Selection Theorem for Systems with Inheritance. Math. Model. Nat. Phenom., **2 (4)** (2007) 1–45. E-print: http://arxiv.org/abs/cond-mat/0405451
24. Gorban, A.N., Karlin, I.V.: Invariant Manifolds for Physical and Chemical Kinetics. Lect. Notes Phys. **660** Springer, Berlin (2005)
25. Gorban, A.N., Khlebopros, R.G.: Demon of Darwin: Idea of optimality and natural selection. Nauka (FizMatGiz), Moscow (1988) (in Russian)
26. Gromov, M.: A dynamical model for synchronisation and for inheritance in micro-evolution: a survey of papers of A.Gorban. The talk given in the IHES seminar, Initiation to functional genomics: biological, mathematical and algorithmical aspects, Institut Henri Poincaré (Nov. 16, 2000)
27. Haldane, J.B.S.: The Causes of Evolution. Princeton Science Library, Princeton University Press, NJ (1990)
28. Hammerstein, P.: Darwinian adaptation, population genetics and the streetcar theory of evolution. J. Math. Biol. **34** (1996) 511–532
29. Hardin, G.: The Competitive Exclusion Principle. Science (29 Apr. 1960) 1292–1297
30. Hasselblatt, B., Katok, A. (Eds.): Handbook of Dynamical Systems. **1A** Elsevier, Amsterdam (2002)
31. Hofbauer, J., Sigmund, K.: Evolutionary game dynamics. Bull. (New Series) American Math. Soc. **40(4)** (2003) 479–519
32. Hofbauer, J., Schuster, P., Sigmund, K.: A note on evolutionary stable strategies and game dynamics. J. Theor. Biol. **81** (1979) 609–612
33. Karev G.P.: On mathematical theory of selection: continuous time population dynamics. J. Math. Biol. **60(1)** (2010) 107–129
34. Katok, A., Hasselblat, B.: Introduction to the Modern Theory of Dynamical Systems. Encyclopedia of Math. and its Applications **54**, Cambridge University Press, Cambridge (1995)
35. Kuzenkov, O.A.: A dynamical system on the set of Radon probability measures. Diff. Equat. **31(4)** (1995) 549–554
36. Kuzenkov, O.A.: Weak solutions of the Cauchy problem in the set of Radon probability measures. Diff. Equat. **36(11)** (2000) 1676–1684
37. Levin, S.M.: Community equilibria and stability, and an extension of the competitive exclusion principle. Am. Nat. **104(939)** (1970) 413–423
38. L'vov, V.S.: Wave turbulence under parametric excitation applications to magnets. Springer, Berlin (1994)
39. MacArthur, R., Levins, R.: Competition, habitat selection, and character displacement in a patchy environment. Proc. Nat. Acad. Sci. **51** (1964) 1207–1210
40. MacArthur, R., Levins, R.: The limiting similarity, convergence, and divergence of coexisting species. Am. Nat. **101(921)** (1967) 377–385
41. Mayr, E.: Animal Species and Evolution. Harvard University Press, Cambridge, MA (1963)
42. Meszena, G., Gyllenberg, M., Pasztor, L., Metz, J.A.J.: Competitive exclusion and limiting similarity: A unified theory. Theoretical Population Biology **69** (2006) 68–87
43. Meszena, G., Metz, J.A.J.: Species diversity and population regulation: the importance of environmental feedback dimensionality. International Institute for Applied SystemsAnalysis, Interim Report IR-99-045 (1999)
44. Oechssler, J., Riedel, F.: Evolutionary dynamics on infinite strategy spaces. Economic Theory **17(1)** (2001) 141–162
45. Oechssler, J., Riedel, F.: On the Dynamic Foundation of Evolutionary Stability in Continuous Models. Journal of Economic Theory **107** (2002) 223–252
46. Rozonoer, L.I., Sedyh, E.I.: On the mechanisms of evolution of self-reproduction systems. 1, 2, 3. Automation and Remote Control **40** (1979) 243–251; 419–429; 741–749
47. Semevsky, F.N., Semenov, S.M.: Mathematical modeling of ecological processes. Gidrometeoizdat, Leningrad (1982) (in Russian)
48. Smith, J.M.: Evolution and the theory of games. Cambridge University Press, Cambridge (1982)
49. Smith, J.M., Price, G.R.: The Logic of Animal Conflict. Nature **246(5427)** (1973) 15–18

50. Taylor, P.: Evolutionary stable strategies and game dynamics. J. Appl. Probability **16** (1979) 76–83
51. Vickers, G.T., Cannings, C.: Patterns of ESSs. I, II. J. Theoret. Biol. **132(4)** (1988) 387–408; 409–420
52. Volterra, V.: Lecons sur la théorie mathematique de la lutte pour la vie. Gauthier-Villars, Paris (1931)
53. Zakharov, V.E., L'vov, V.S., Starobinets, S.S.: Turbulence of spin-waves beyond threshold of their parametric-excitation. Uspekhi Fizicheskikh Nauk **114(4)** (1974) 609–654; English translation: Sov. Phys. – Usp. **17(6)** (1975) 896–919
54. Zakharov, V.E., L'vov, V.S., Falkovich, G.E.: Kolmogorov spectra of turbulence. Vol. 1. Wave Turbulence. Springer, Berlin (1992)

Author Index

Alexander N. Gorban, 168, 310
Alison S. Tomlin, 10
Allison Kolpas, 299
Anatoly Ya. Rodionov, 63
Andrei N. Sobolevskii, 63

Cecilia Clementi, 112
Christophe Vandekerckhove, 150
Constantinos Theodoropoulos, 37

Dávid Csercsik, 253
David J. Packwood, 168
Dirk Lebiedz, 241
Dirk Roose, 268

Eliodoro Chiavazzo, 230

Francisco J. Uribe, 206

Gábor Szederkényi, 253
Giorgos Kanellopoulos, 287
Giovanni Samaey, 150
Grigory L. Litvinov, 63

Iain D. Couzin, 299
Ilya Karlin, 230
Ioannis G. Kevrekidis, 112, 299

Jeff Moehlis, 299
Jeremy Levesley, 168
Jochen Siehr, 241
Jonas Unger, 241

Katalin M. Hangos, 253
Koichi Yamashita, 132
Kostis Andriopoulos, 287
Ko van der Weele, 287

Marshall Slemrod, 1

Payel Das, 112

Ronald Coifman, 299

Sergei Manzhos, 132

Tassos Bountis, 287
Thomas A. Frewen, 112, 299
Tilo Ziehn, 10
Tucker Carrington, 132

Ulrich Maas, 90

Viatcheslav Bykov, 90
Victor P. Maslov, 63
Vladimir Gol'dshtein, 90
Volkmar Reinhardt, 241

Wim Vanroose, 150

Yves Frederix, 268

Subject Index

Air pollution models, 11
Analysis of variance, 12
Auto-ignition, 238
Averaging, 1

Balanced truncation, 39
Bellman equations, 65
BGK collision, 179
Boltzmann equation, 2, 170
Brenner's theory, 215

Chapman–Enskog analysis, 169
Chapman–Enskog expansion, 154
Chapman–Enskog method, 210
Chemical reaction networks, 253
Coarse-grained information, 114
Coarse grid correction, 152
Coarse reverse integration, 115
Combustion models, 11
Compactness, 3
Complexity reduction, 15

Darwin's systems, 312
Data mining, 113
Data structure family, 64
Density functional theory, 141
Diffusion estimation, 273
Diffusion maps, 125, 305
Dilute gases, 207
Dimensionality reduction, 133, 140
Dominant invariant subspace, 44
Drift equations, 330
Drift-filtered diffusion, 275

Electronic Schrödinger equation, 134
Equation-free, 42, 152
Euler-Maruyama scheme, 274

Fast system, 3
Fokker-Planck equation, 114, 304

Gauss–Jacobi formula, 67
Generalized KdV equation, 288
Geodesic distance, 125

Hankel norm, 40
Hardware design, 82
Hausdorff dimension, 311
High dimensional model, 9, 136

Idempotent dequantization, 78
Idempotent semirings, 65
Interval analysis, 85
Intrinsic low-dimensional manifolds, 102
Invariance equation, 176
Invariant manifold, 93, 210, 232
Invariants, 6

KdV equation, 5, 287
Krylov projection, 41

Lattice Boltzmann models, 151, 169
Legendre transform, 80
Linear irreversible thermodynamics, 210
Linearly decomposed vector field, 105
Local equilibrium, 154
Lotka-Volterra, 256

Macroscopic, 2
Macroscopic stability, 196
Mass action reaction networks, 255
Maximum-likelihood estimation, 277

348 Subject Index

Maxwell distribution, 170
Mixed integer linear programming, 254
Model predictive control, 55
Model reduction, 37
Model robustness, 15
Model uncertainty, 9
Monte Carlo, 14
Multilevel algorithm, 160
Multimode expansion, 135
Multiple scales, 93
Multiscale diffusion, 269
Multiscale homogenization, 272
Multivariate functions, 133

Navier-Stokes equations, 170, 207, 211
Neural networks, 133, 137
Newton–Cotes formula, 67
Numerical semirings, 68

Observables, 307
Observer, 2
Optimal-order polynomials, 16
Oscillations, 4
Output variance, 12

Parameter tuning, 15, 22
Parameterized algorithm, 64
Physical reduction, 38
Plane shock waves, 207
Potential energy surface, 134
Principal componenet analysis, 144
Projection, 39
Projective integration, 4
Proper orthogonal decomposition, 41

Quasi-random sampling, 9
Quasi-steady state, 22, 105

Random sampling, 13
Random trajectories, 18
Reaction coordinate, 300
Reaction kinetics, 233

Reaction trajectories, 242
Recursive projection method, 42, 43
Reduction coordinates, 113
Relaxation redistribution method, 234
Reversible reaction, 257
Richardson iteration, 156
Riemann integral, 79

Scalable Isomap, 124
Scaling invariance, 106
Schrödinger equation, 65
Selection efficiency, 327
Semenov's explosion model, 100
Sensitivity analysis, 10, 12
Simulated annealing Monte Carlo, 120
Singularly perturbed system, 95
Slow drift dynamics, 118
Slow manifolds, 39, 169
Smoothing, 152
Software design, 84
Solitary wave, 288
Sparse data, 133
Species reconstruction, 243
Stability of slow manifolds, 95
Steady states, 152
Stick-Slip, 302
Stiffness, 94
Stochastic homogenization, 269
Subsmapling, 273

Thin sets, 323
Trajectory optimization, 242
Troe reactions, 249
Tubular reactor, 37
Turbulence, 29

Wall termination, 26
Water wave models, 287
Weak optimality, 319
Weighted direct graphs, 71

Young measure, 3

Editorial Policy

1. Volumes in the following three categories will be published in LNCSE:

i) Research monographs
ii) Tutorials
iii) Conference proceedings

Those considering a book which might be suitable for the series are strongly advised to contact the publisher or the series editors at an early stage.

2. Categories i) and ii). Tutorials are lecture notes typically arising via summer schools or similar events, which are used to teach graduate students. These categories will be emphasized by Lecture Notes in Computational Science and Engineering. **Submissions by interdisciplinary teams of authors are encouraged.** The goal is to report new developments – quickly, informally, and in a way that will make them accessible to non-specialists. In the evaluation of submissions timeliness of the work is an important criterion. Texts should be well-rounded, well-written and reasonably self-contained. In most cases the work will contain results of others as well as those of the author(s). In each case the author(s) should provide sufficient motivation, examples, and applications. In this respect, Ph.D. theses will usually be deemed unsuitable for the Lecture Notes series. Proposals for volumes in these categories should be submitted either to one of the series editors or to Springer-Verlag, Heidelberg, and will be refereed. A provisional judgement on the acceptability of a project can be based on partial information about the work: a detailed outline describing the contents of each chapter, the estimated length, a bibliography, and one or two sample chapters – or a first draft. A final decision whether to accept will rest on an evaluation of the completed work which should include

– at least 100 pages of text;
– a table of contents;
– an informative introduction perhaps with some historical remarks which should be accessible to readers unfamiliar with the topic treated;
– a subject index.

3. Category iii). Conference proceedings will be considered for publication provided that they are both of exceptional interest and devoted to a single topic. One (or more) expert participants will act as the scientific editor(s) of the volume. They select the papers which are suitable for inclusion and have them individually refereed as for a journal. Papers not closely related to the central topic are to be excluded. Organizers should contact the Editor for CSE at Springer at the planning stage, see *Addresses* below.

In exceptional cases some other multi-author-volumes may be considered in this category.

4. Only works in English will be considered. For evaluation purposes, manuscripts may be submitted in print or electronic form, in the latter case, preferably as pdf- or zipped ps-files. Authors are requested to use the LaTeX style files available from Springer at http://www. springer.com/authors/book+authors?SGWID=0-154102-12-417900-0.

For categories ii) and iii) we strongly recommend that all contributions in a volume be written in the same LaTeX version, preferably LaTeX2e. Electronic material can be included if appropriate. Please contact the publisher.

Careful preparation of the manuscripts will help keep production time short besides ensuring satisfactory appearance of the finished book in print and online.

5. The following terms and conditions hold. Categories i), ii) and iii):

Authors receive 50 free copies of their book. No royalty is paid.
Volume editors receive a total of 50 free copies of their volume to be shared with authors, but no royalties.

Authors and volume editors are entitled to a discount of 33.3 % on the price of Springer books purchased for their personal use, if ordering directly from Springer.

6. Commitment to publish is made by letter of intent rather than by signing a formal contract. Springer-Verlag secures the copyright for each volume.

Addresses:

Timothy J. Barth
NASA Ames Research Center
NAS Division
Moffett Field, CA 94035, USA
barth@nas.nasa.gov

Michael Griebel
Institut für Numerische Simulation
der Universität Bonn
Wegelerstr. 6
53115 Bonn, Germany
griebel@ins.uni-bonn.de

David E. Keyes
Mathematical and Computer Sciences
and Engineering
King Abdullah University of Science
and Technology
P.O. Box 55455
Jeddah 21534, Saudi Arabia
david.keyes@kaust.edu.sa

and

Department of Applied Physics
and Applied Mathematics
Columbia University
500 W. 120 th Street
New York, NY 10027, USA
kd2112@columbia.edu

Risto M. Nieminen
Department of Applied Physics
Aalto University School of Science
and Technology
00076 Aalto, Finland
risto.nieminen@tkk.fi

Dirk Roose
Department of Computer Science
Katholieke Universiteit Leuven
Celestijnenlaan 200A
3001 Leuven-Heverlee, Belgium
dirk.roose@cs.kuleuven.be

Tamar Schlick
Department of Chemistry
and Courant Institute
of Mathematical Sciences
New York University
251 Mercer Street
New York, NY 10012, USA
schlick@nyu.edu

Editor for Computational Science
and Engineering at Springer:
Martin Peters
Springer-Verlag
Mathematics Editorial IV
Tiergartenstrasse 17
69121 Heidelberg, Germany
martin.peters@springer.com

Lecture Notes in Computational Science and Engineering

1. D. Funaro, *Spectral Elements for Transport-Dominated Equations.*

2. H.P. Langtangen, *Computational Partial Differential Equations.* Numerical Methods and Diffpack Programming.

3. W. Hackbusch, G. Wittum (eds.), *Multigrid Methods V.*

4. P. Deuflhard, J. Hermans, B. Leimkuhler, A.E. Mark, S. Reich, R.D. Skeel (eds.), *Computational Molecular Dynamics: Challenges, Methods, Ideas.*

5. D. Kröner, M. Ohlberger, C. Rohde (eds.), *An Introduction to Recent Developments in Theory and Numerics for Conservation Laws.*

6. S. Turek, *Efficient Solvers for Incompressible Flow Problems.* An Algorithmic and Computational Approach.

7. R. von Schwerin, *Multi Body System SIMulation.* Numerical Methods, Algorithms, and Software.

8. H.-J. Bungartz, F. Durst, C. Zenger (eds.), *High Performance Scientific and Engineering Computing.*

9. T.J. Barth, H. Deconinck (eds.), *High-Order Methods for Computational Physics.*

10. H.P. Langtangen, A.M. Bruaset, E. Quak (eds.), *Advances in Software Tools for Scientific Computing.*

11. B. Cockburn, G.E. Karniadakis, C.-W. Shu (eds.), *Discontinuous Galerkin Methods.* Theory, Computation and Applications.

12. U. van Rienen, *Numerical Methods in Computational Electrodynamics.* Linear Systems in Practical Applications.

13. B. Engquist, L. Johnsson, M. Hammill, F. Short (eds.), *Simulation and Visualization on the Grid.*

14. E. Dick, K. Riemslagh, J. Vierendeels (eds.), *Multigrid Methods VI.*

15. A. Frommer, T. Lippert, B. Medeke, K. Schilling (eds.), *Numerical Challenges in Lattice Quantum Chromodynamics.*

16. J. Lang, *Adaptive Multilevel Solution of Nonlinear Parabolic PDE Systems.* Theory, Algorithm, and Applications.

17. B.I. Wohlmuth, *Discretization Methods and Iterative Solvers Based on Domain Decomposition.*

18. U. van Rienen, M. Günther, D. Hecht (eds.), *Scientific Computing in Electrical Engineering.*

19. I. Babuška, P.G. Ciarlet, T. Miyoshi (eds.), *Mathematical Modeling and Numerical Simulation in Continuum Mechanics.*

20. T.J. Barth, T. Chan, R. Haimes (eds.), *Multiscale and Multiresolution Methods.* Theory and Applications.

21. M. Breuer, F. Durst, C. Zenger (eds.), *High Performance Scientific and Engineering Computing.*

22. K. Urban, *Wavelets in Numerical Simulation.* Problem Adapted Construction and Applications.

23. L.F. Pavarino, A. Toselli (eds.), *Recent Developments in Domain Decomposition Methods.*

24. T. Schlick, H.H. Gan (eds.), *Computational Methods for Macromolecules: Challenges and Applications.*

25. T.J. Barth, H. Deconinck (eds.), *Error Estimation and Adaptive Discretization Methods in Computational Fluid Dynamics.*

26. M. Griebel, M.A. Schweitzer (eds.), *Meshfree Methods for Partial Differential Equations.*

27. S. Müller, *Adaptive Multiscale Schemes for Conservation Laws.*

28. C. Carstensen, S. Funken, W. Hackbusch, R.H.W. Hoppe, P. Monk (eds.), *Computational Electromagnetics.*

29. M.A. Schweitzer, *A Parallel Multilevel Partition of Unity Method for Elliptic Partial Differential Equations.*

30. T. Biegler, O. Ghattas, M. Heinkenschloss, B. van Bloemen Waanders (eds.), *Large-Scale PDE-Constrained Optimization.*

31. M. Ainsworth, P. Davies, D. Duncan, P. Martin, B. Rynne (eds.), *Topics in Computational Wave Propagation*. Direct and Inverse Problems.

32. H. Emmerich, B. Nestler, M. Schreckenberg (eds.), *Interface and Transport Dynamics*. Computational Modelling.

33. H.P. Langtangen, A. Tveito (eds.), *Advanced Topics in Computational Partial Differential Equations*. Numerical Methods and Diffpack Programming.

34. V. John, *Large Eddy Simulation of Turbulent Incompressible Flows*. Analytical and Numerical Results for a Class of LES Models.

35. E. Bänsch (ed.), *Challenges in Scientific Computing - CISC 2002.*

36. B.N. Khoromskij, G. Wittum, *Numerical Solution of Elliptic Differential Equations by Reduction to the Interface.*

37. A. Iske, *Multiresolution Methods in Scattered Data Modelling.*

38. S.-I. Niculescu, K. Gu (eds.), *Advances in Time-Delay Systems.*

39. S. Attinger, P. Koumoutsakos (eds.), *Multiscale Modelling and Simulation.*

40. R. Kornhuber, R. Hoppe, J. Périaux, O. Pironneau, O. Wildlund, J. Xu (eds.), *Domain Decomposition Methods in Science and Engineering.*

41. T. Plewa, T. Linde, V.G. Weirs (eds.), *Adaptive Mesh Refinement – Theory and Applications.*

42. A. Schmidt, K.G. Siebert, *Design of Adaptive Finite Element Software.* The Finite Element Toolbox ALBERTA.

43. M. Griebel, M.A. Schweitzer (eds.), *Meshfree Methods for Partial Differential Equations II.*

44. B. Engquist, P. Lötstedt, O. Runborg (eds.), *Multiscale Methods in Science and Engineering.*

45. P. Benner, V. Mehrmann, D.C. Sorensen (eds.), *Dimension Reduction of Large-Scale Systems.*

46. D. Kressner, *Numerical Methods for General and Structured Eigenvalue Problems.*

47. A. Boriçi, A. Frommer, B. Joó, A. Kennedy, B. Pendleton (eds.), *QCD and Numerical Analysis III.*

48. F. Graziani (ed.), *Computational Methods in Transport*.

49. B. Leimkuhler, C. Chipot, R. Elber, A. Laaksonen, A. Mark, T. Schlick, C. Schütte, R. Skeel (eds.), *New Algorithms for Macromolecular Simulation*.

50. M. Bücker, G. Corliss, P. Hovland, U. Naumann, B. Norris (eds.), *Automatic Differentiation: Applications, Theory, and Implementations*.

51. A.M. Bruaset, A. Tveito (eds.), *Numerical Solution of Partial Differential Equations on Parallel Computers*.

52. K.H. Hoffmann, A. Meyer (eds.), *Parallel Algorithms and Cluster Computing*.

53. H.-J. Bungartz, M. Schäfer (eds.), *Fluid-Structure Interaction*.

54. J. Behrens, *Adaptive Atmospheric Modeling*.

55. O. Widlund, D. Keyes (eds.), *Domain Decomposition Methods in Science and Engineering XVI*.

56. S. Kassinos, C. Langer, G. Iaccarino, P. Moin (eds.), *Complex Effects in Large Eddy Simulations*.

57. M. Griebel, M.A Schweitzer (eds.), *Meshfree Methods for Partial Differential Equations III*.

58. A.N. Gorban, B. Kégl, D.C. Wunsch, A. Zinovyev (eds.), *Principal Manifolds for Data Visualization and Dimension Reduction*.

59. H. Ammari (ed.), *Modeling and Computations in Electromagnetics: A Volume Dedicated to Jean-Claude Nédélec*.

60. U. Langer, M. Discacciati, D. Keyes, O. Widlund, W. Zulehner (eds.), *Domain Decomposition Methods in Science and Engineering XVII*.

61. T. Mathew, *Domain Decomposition Methods for the Numerical Solution of Partial Differential Equations*.

62. F. Graziani (ed.), *Computational Methods in Transport: Verification and Validation*.

63. M. Bebendorf, *Hierarchical Matrices. A Means to Efficiently Solve Elliptic Boundary Value Problems*.

64. C.H. Bischof, H.M. Bücker, P. Hovland, U. Naumann, J. Utke (eds.), *Advances in Automatic Differentiation*.

65. M. Griebel, M.A. Schweitzer (eds.), *Meshfree Methods for Partial Differential Equations IV*.

66. B. Engquist, P. Lötstedt, O. Runborg (eds.), *Multiscale Modeling and Simulation in Science*.

67. I.H. Tuncer, Ü. Gülcat, D.R. Emerson, K. Matsuno (eds.), *Parallel Computational Fluid Dynamics 2007*.

68. S. Yip, T. Diaz de la Rubia (eds.), *Scientific Modeling and Simulations*.

69. A. Hegarty, N. Kopteva, E. O'Riordan, M. Stynes (eds.), *BAIL 2008 – Boundary and Interior Layers*.

70. M. Bercovier, M.J. Gander, R. Kornhuber, O. Widlund (eds.), *Domain Decomposition Methods in Science and Engineering XVIII*.

71. B. Koren, C. Vuik (eds.), *Advanced Computational Methods in Science and Engineering*.

72. M. Peters (ed.), *Computational Fluid Dynamics for Sport Simulation*.

73. H.-J. Bungartz, M. Mehl, M. Schäfer (eds.), *Fluid Structure Interaction II - Modelling, Simulation, Optimization.*

74. D. Tromeur-Dervout, G. Brenner, D.R. Emerson, J. Erhel (eds.), *Parallel Computational Fluid Dynamics 2008.*

75. A.N. Gorban, D. Roose (eds.), *Coping with Complexity: Model Reduction and Data Analysis.*

For further information on these books please have a look at our mathematics catalogue at the following URL: www.springer.com/series/3527

Monographs in Computational Science and Engineering

1. J. Sundnes, G.T. Lines, X. Cai, B.F. Nielsen, K.-A. Mardal, A. Tveito, *Computing the Electrical Activity in the Heart.*

For further information on this book, please have a look at our mathematics catalogue at the following URL: www.springer.com/series/7417

Texts in Computational Science and Engineering

1. H. P. Langtangen, *Computational Partial Differential Equations.* Numerical Methods and Diffpack Programming. 2nd Edition

2. A. Quarteroni, F. Saleri, P. Gervasio, *Scientific Computing with MATLAB and Octave.* 3rd Edition

3. H. P. Langtangen, *Python Scripting for Computational Science.* 3rd Edition

4. H. Gardner, G. Manduchi, *Design Patterns for e-Science.*

5. M. Griebel, S. Knapek, G. Zumbusch, *Numerical Simulation in Molecular Dynamics.*

6. H. P. Langtangen, *A Primer on Scientific Programming with Python.*

7. A. Tveito, H. P. Langtangen, B. F. Nielsen, X. Cai, *Elements of Scientific Computing.*

For further information on these books please have a look at our mathematics catalogue at the following URL: www.springer.com/series/5151